Ukraine under Western Eyes

Ukrainian Research Institute
Harvard University

Harvard Series in Ukrainian Studies

Cambridge, Massachusetts

Ukraine under Western Eyes

The Bohdan and Neonila Krawciw Ucrainica Map Collection

Steven Seegel

Distributed by Harvard University Press
for the Harvard Ukrainian Research Institute

The Harvard Ukrainian Research Institute was established in 1973 as an integral part of Harvard University. It supports research associates and visiting scholars who are engaged in projects concerned with all aspects of Ukrainian studies. The Institute also works in close cooperation with the Committee on Ukrainian Studies, which supervises and coordinates the teaching of Ukrainian history, language, and literature at Harvard University.

Publication of this book has been made possible by the following endowed funds:
The Jaroslaw and Olha Duzey Publication Fund in Ukrainian Studies
The Ivan and Anastazia Jurkewych-Kocur and Wolodymyr and Osypa Rybchuk-Zazula Ukrainian Fund
The Lubomyra and Ihor Kocur Ukrainian Fund in memory of Dr. Lubomyra S. Kocur-Zazula by her husband and children
The Michael Maksymiw Bequest
The Dr. Evhen Omelsky Publication Fund in Ukrainian Studies

Printed in the U.S. on acid-free paper

Library of Congress Cataloging-in-Publication Data

Seegel, Steven.
Ukraine under western eyes : the Bohdan and Neonila Krawciw Ucrainica Map Collection / Steven Seegel.
p. cm.
Includes bibliographical references and index.
ISBN-13: 978-1-932650-04-4 (alk. paper)
ISBN-10: 1-932650-04-0 (alk. paper)
1. Cartography--Ukraine--History. 2. Bohdan and Neonila Krawciw Ucrainica Map Collection. 3. Kravtsiv, Bohdan, 1904---Map collections. 4. Krawciw, Neonila--Map collections. 5. Maps--Private collections--Ukraine. 6. Ukraine--Maps--Catalogs. I. Title.
GA1077.U47S44 2011
912.477--dc22
2011006650

Map on jacket front: Johann Baptist Homann, *Ukrania // quae et // Terra Cosaccorum // cum vicinis // Walachiae, Moldaviae, Minorisq. Tartariae Provinciis // exhibita // à Joh. Baptista Homanno // Noribergae*, 1730.

Jacket rear: Detail from Guillaume le Vasseur de Beauplan, *Typus Generalis Ukrainae sive Palatinatuum Podoliae, Kiovensis et Braczlaviensis terras nova delineatione exhibens*, 1680.

Contents

Preface

The Bohdan and Neonila Krawciw Map Collection is a monument in Ukrainian cultural preservation. The assembling of such a collection cannot be regarded as a simple case of nostalgia for a distant homeland, for the maps tell a story of Ukraine as a territory caught amid shifting borders, providing a record of different ways in which Ukrainian history might be conceptualized against the backdrop of its contested past. Taken as a whole, the maps show Ukraine's history in fragments and as artifacts. Ukraine in maps was imagined and created in many cases by outside forces, but produced by self-identifying ethnic and civic Ukrainians from the late nineteenth century on. The maps display Ukraine's many political trajectories and attendant mythologies, and trace the larger world's awareness of the borderland region and country the Krawciw family was forced to leave behind.

After Krawciw's death in 1975, his wife Neonila, their daughter Maria Dzwenyslawa, and her husband, Dr. Lubomir Jawny, graciously donated Bohdan Krawciw's personal archive and library of Ucrainica—twelve thousand volumes in all—to the Harvard University Libraries. These donations cover the fields of geography, cartography, history, anthropology, literature, art history, onomastics, and linguistics. Among the donated papers are newspaper clippings, correspondence, and photographs of artists and literary figures, in addition to materials on the Ukrainian press and community organizations since the late 1940s. Also included are Krawciw's notes for a bibliographical survey of major and minor figures in modern Ukrainian literature. To honor the initial gift in 1978 and to promote further scholarly research on Ukraine, the Harvard Ukrainian Research Institute established the biennial Krawciw Memorial Lecture.

Krawciw's extensive collection of maps of Ukraine was not included in this initial bequest. With firsthand knowledge of Galicia over *la longue durée*, Neonila Krawciw had grasped many of the nuances of early modern Ukraine's historical formation, and its modern political fragility.[1] Realizing the value of the map collection and the potential dangers involved in donating maps of Ukraine to a Soviet-controlled or otherwise institutionally insecure environment, she kept it private for nearly thirty years.

After Neonila's death in 2003 at the age of ninety-three, Maria Dzwenyslawa inherited the maps, as well as a collection of index cards that Neonila had prepared to organize and explain the maps using her husband's original notes. Maria Dzwenyslawa then prepared a preliminary catalogue, continuing to work on organizing the collection for nearly two years. In August 2005, in accordance with her parent's wishes that the collection neither be sold nor remain private, Maria Dzwenyslawa together with her husband made it available to students and scholars by donating the "Bohdan and Neonila Krawciw Antique Ucrainica Map Collection" to the Harvard Ukrainian Research Institute and Harvard's Pusey Library.[2] In November 2005, the map collection and numerous accompanying materials were brought to Harvard by Drs. Ksenya Kiebuzinski and Roman Procyk.

As an amateur enthusiast of cartography, Bohdan Krawciw intended to write a general history of Ukraine in maps, to be called *Monumenta Cartographica Ucrainae* (Cartographic Monuments of Ukraine). The work was to be a revision of the four-volume work produced in Kyiv by Veniamin Kordt, *Materialy po istorii russkoi kar-*

tografii (Materials on the History of Russian Cartography, 1899, 1906, 1910), and *Materialy do istorii kartografiï Ukraïny* (Materials for a History of the Cartography of Ukraine, 1931).[3] With the endeavor unfinished at the time of his death, he left behind thirteen personal notebooks, which divide into two groups. The first group includes a four-volume index of cartographers and publishers; a one-volume cartographical dictionary; a one-volume list of maps of Ukraine and neighboring countries, alphabetized by author; and a one-volume listing of maps featuring Ukraine and neighboring countries, alphabetized by title. The second group consists of a three-volume geographical index to maps of Ukraine and neighboring countries from the fifteenth to the eighteenth century, a one-volume index of place names for what he judged to be the most important Renaissance and early modern maps of Ukraine and its bordering areas, and a two-volume bibliography of secondary literature on the history of cartography, in reference to European history in general, and to the history of Ukraine in particular. Krawciw also penned a short essay on Beauplan, "Guillaume le Vasseur Sieur de Beauplan's 'Description of Ukraine' and His Military Maps of Ukraine," which appeared in the preface to a 1959 reprint of an English translation of Beauplan's original *Description de l'Ukranie*.[4] It is from these works, and with Bohdan Krawciw's intentions and those of his family in mind, that the catalogue of his collection is compiled.

In total, the collection itself includes nearly nine hundred maps, books, research files, and notebooks. The maps are in nine different languages, and many were researched and commented on in Krawciw's private notebooks. As the notebooks and cards indicate, both Bohdan and Neonila Krawciw defined the organization of the collection in a standardized manner, much like eighteenth-century *encyclopédistes*, in accordance with their own vision of a Ukrainian Enlightenment project. They divided the collection into two parts: part 1 covers maps produced chiefly before 1800, and is organized by cartographer (including publishers and engravers); part 2 deals primarily with nineteenth- and twentieth-century maps, and is arranged by geographical category. The appendix to this volume is organized according to this division.

In part 1 of the collection, the majority of the maps and atlases (in original or copy form) are in Latin, French, or English. They were engraved between approximately 1500 and 1800 both in major European mapmaking centers such as Amsterdam, London, Paris, Nuremberg, and Venice, and also in places as far-flung as Philadelphia, Stockholm, and St. Petersburg. This portion of the collection includes rare sixteenth-century maps of Ukrainian lands by Anthony Jenkinson, Abraham Ortelius, Wacław Grodecki, and Gerard Mercator.

Particularly prominent are the more than twenty original maps made by the French military engineer and cartographer Guillaume le Vasseur, Sieur de Beauplan, who worked for French monarchs and the elected Swedish Vasa kings of the Polish-Lithuanian Commonwealth. Beauplan's oft-reproduced maps of the Cossack Hetmanate and Zaporozhian Host (Sich) were the first major early modern military and scientific-technical maps of Ukraine.[5] For at least the following 150 years, information from Beauplan's maps provided a reference point for literary and aesthetic descriptions of Cossackdom and for other maps, forming a body of knowledge in Europe about Ukraine and its borderlands. The maps were used in the early modern era by such key European cartographers as Nicholas Sanson, Jodocus Hondius, Guillaume Delisle, Johann Baptist Homann, Pieter van der Aa, Matthäus Seutter, Tobias Lotter, Christoph Weigel, and Moses Pitt. Works by these cartographers also have a place in the collection.

Part 2 makes this collection unique in comparison to other antiquarian map collections, since Bohdan Krawciw included maps of Ukraine from the nineteenth and twentieth centuries. Many of the maps are excerpted from books in various languages, including atlases and encyclopedias reproduced and disseminated by members of the Ukrainian diaspora. This modern and far more political part of the collection is arranged by region rather than cartographer. It includes rare Russian and British battle plans of towns and harbors produced during the Crimean War; German, Austro-Hungarian, Russian, and Soviet ethnographic maps of Ukraine; Ukrainian national propaganda maps; and at least sixty administrative, geological, transportation, and geopolitical maps of Ukraine during the twentieth century.

Krawciw's collection focuses not only on independent Ukraine after 1918, but also on the eighteenth-century partitions of the Polish-Lithuanian Commonwealth and their consequences, the fabled lands of Tartary, the fate of Crimea before and after the Crimean War, the dissolution of the Ottoman Empire, and the repeated Great Power interventions in the regions of the Mediterranean, Black, and Caspian seas. In the process of collection and organization, the Krawciws concentrated principally on four states or empires: (1) the Russian Empire and Soviet Union; (2) Poland-Lithuania, later Poland; (3) the Ottoman Empire and Turkey; and (4) the Austrian Empire of the Habsburgs, which became Austria-Hungary after 1867. The broad scope of the Krawciws' attention is a testament to the multiple religious, linguistic, and political cultures of Ukraine, partitioned as it was among various states and empires, and to the many constructions of Ukrainian history and territory in a world of modern Great Power statecraft.

Significance of the Bohdan and Neonila Krawciw Map Collection

The Bohdan and Neonila Krawciw Map Collection is historically important in three ways. First, it is a visual representation of the Krawciw family's own history and reasoned mental mapping, closely intertwined by fact and imagination with that of Galicia, and, by extension, Ukraine. Second, it is a major contribution to the history of cartography in a shared East-Central European context, and to the international history of cartography, as a subject for future scholarly research. Third, it is a monument to, and reference for, Ukrainian studies and their critical promotion. Krawciw thought of his work as a project in preservation for future research, and, of course, a commemoration of Ukraine as nation and state. The Ukraine he visualized in maps naturally had a literary dimension as well, a point exemplified not only by his poetry, love of languages, encyclopedic and bibliographical works, and study of folklore and mythology, but also by his attention to geography and the technical and aesthetic aspects of the maps in his collection.

The third point of significance is the most complex, for it concerns the understandings of Ukraine as an ethnic nation that emerged particularly through modern nineteenth- and twentieth-century maps propagating the idea of an independent Ukraine. Through these maps, we trace not just Ukrainian, but European cultural and ethnolinguistic models as they developed through modern nation-state cartography. Insistence on romantic national history, "realist" demographic claims to territory, and ethnic categorization became an integral aspect of modern European nation-state mapping in the late nineteenth and twentieth centuries. A number of the modern maps in the collection offer Ukrainian demographic (reduced to language and territorial) claims to land, and forms of Ukrainian propaganda applied to lands and peoples in East-Central Europe and Russia.

The maps provide evidence for how Ukraine has been and can be studied and represented—or misrepresented. For instance, probably the most common Orientalist discourse about Ukraine, which persisted literally for centuries in European (including Russian) cultural imaginations, was the iconic representation of Ukraine as the land of the Cossacks. Certainly, the colonialist and imperialist geopolitics of Ukraine as a supposedly blank space was much affected by predominant European cartographic models, structural prepossessions, and religious and cultural tropes of the boundaries of Western civilization as applied to Ukraine as nation and state.[6] Given Ukraine's entangled history, the maps themselves, as well as the epistemology of collecting, therefore must be studied critically by individual researchers, and used cautiously.[7]

The Catalogue of the Collection

One of the stipulations of the Krawciw bequest to Harvard University was that a catalogue of the collection be produced. To this end, the Harvard Ukrainian Research Institute presents this volume. In the appendix, the book presents a detailed listing of all maps in the collection and the Krawciws' original index of place names found on the maps. George G. Grabowicz has provided a thought-provoking essay examining Krawciw's written works from a fresh perspective. Five chapters describe the history of the depiction of Ukraine in the maps of European cartographers from the Renaissance to the modern era, providing a broad historical context for the maps and tracing how the nation of Ukraine was perceived in various places and times. The book is illustrated by ninety-five maps from the Krawciw collection. More than a simple listing of the maps that Krawciw collected over his lifetime, this publication analyzes both the antecedents and scholarly implications of the collection, and provides detailed annotation and bibliographic references.

Acknowledgments

For their support of this commissioned project at the Harvard Ukrainian Research Institute, I would like to thank Michael Flier, George Grabowicz, Lubomyr Hajda, Tymish Holowinsky, Serhii Plokhy, and Roman Szporluk. I am grateful to Larry Wolff for his recommendations, to Patricia Herlihy and Ksenya Kiebuzinski for their counsel, and to Taras Koznarsky for his comments. I benefited greatly from the advice and technical assistance of David Cobb and Joseph Garver at Pusey Library, who offered their expertise in cartography and responded to my endless requests for maps and books. Heather Hendry and Louise Baptiste treated the maps for publication and exhibition. Matthew Herrington and Anna Gessen assisted with the data entry. Annika Frieberg and Lubomyr Hajda patiently read the entire manuscript. Thanks also to Marika Whaley, who arranged the book design and the layout of maps ultimately selected for the book. I offer a heartfelt note to the family of Eugene and Daymel Shklar for making possible my stay at Harvard University during the 2006–7 academic year. Finally, I would like to express special gratitude to General (ret.) Nicholas Krawciw, Maria Dzwenyslawa Jawny, and George Krawciw for their insights, and especially to Dr. Lubomir and Maria Dzwenyslawa Jawny for their generosity and hospitality. For all matters of fact and interpretation, I alone am responsible.

Steven Seegel
Greeley, Colo.

Notes

1. For an introduction to Galicia's multicultural heritage, see Chris Hann and Paul R. Magocsi, eds., Galicia: A Multicultured Land (Toronto: University of Toronto Press, 2005); and Andrei S. Markovits and Frank E. Sysyn, eds., *Nationbuilding and the Politics of Nationalism: Essays on Austrian Galicia* (Cambridge, Mass.: Harvard Ukrainian Research Institute, 1982).
2. The collection was renamed the "Bohdan and Neonila Krawciw Ucrainica Map Collection" in 2006, owing to the large number of maps published after 1800, which is the customary dividing line for classification as antique. Regarding the modern maps (after 1800), "these he collected by way of gathering any information available about Ukraine, both as a means of preserving Ukraine's identity and as an asset to any future research that he might undertake." Maria Dzwenyslawa Jawny, interview with Steven Seegel, 3 January 2008, Florham Park, New Jersey.
3. Kordt's important research in the history of Russian and Ukrainian cartography set a key precedent for recent scholarship on cartography in independent Ukraine. See the original works and compilations by the leading Ukrainian historians of cartography, Rostyslav I. Sossa, Iaroslav Dashkevych, and Mariia Vavrychyn. For recent Ukrainian scholarship, see Sossa, *Istoriia kartohrafuvannia terytoriï Ukraïny: pidruchnyk dlia studentiv vyshchykh navchal΄nykh zakladiv* (Kyiv: Lybid, 2007); Sossa, *Kartohrafuvannia terytoriï Ukraïny: istoriia, perspektyvy, naukovi osnovy* (Kyiv: Naukova dumka, 2005); and Sossa, *Istoriia kartohrafuvannia terytoriï Ukraïny: vid naidavnishykh chasiv do 1920 r.: korotkyi narys* (Kyiv: Naukova dumka, 2000). See also the album of maps by M. Vavrychyn, I. A. Dashkevych, and U. Kryshtalovych, eds., *Ukraïna na starodavnikh kartakh: kinets΄ XV – persha polovyna XVII st.* (Kyiv: DNVP "Kartohrafiia," 2004); and the work by Iaroslav Dashkevych et al., eds., *Istorychne kartoznavstvo Ukraïny: zbirnyk naukovykh prats΄* (Lviv and Kyiv: M. P. Kots, 2004); and M. Vavrychyn et al., eds., *Boplan i Ukraïna: zbirnyk naukovykh prats΄* (Lviv: Natsional΄na akademiia nauk Ukrainy, 1998).
4. Bohdan Krawciw, "Guillaume le Vasseur Sieur de Beauplan's 'Description of Ukraine' and His Military Maps of Ukraine," preface to *A Description of Ukraine*, by Guillaume le Vasseur, Sieur de Beauplan, ix–xiii (New York: Organization for the Defense of Four Freedoms of Ukraine, 1959).
5. The Harvard Ukrainian Research Institute published facsimiles of Beauplan's maps; see Guillaume le Vasseur, Sieur de Beauplan, *A Description of Ukraine*, introduction, translation, and notes by Andrew B. Pernal and Dennis F. Essar (Cambridge, Mass.: Harvard Ukrainian Research Institute, 1993).
6. The classic formulation is Oskar Halecki, *The Limits and Divisions of European History* (London; New York: Sheed & Ward, 1950); on the eighteenth-century geospatial imaginary, see Larry Wolff, *Inventing Eastern Europe: The Map of Civilization on the Mind of the Enlightenment* (Stanford: Stanford University Press, 1994). For a reification of the civilizational paradigm, see Samuel Huntington, "Ukraine: A Cleft Country," in *The Clash of Civilizations and the Remaking of the World Order*, 166 (New York: Simon & Schuster, 1996).
7. On the nineteenth-century epistemology of collecting, see Thomas Richards, *The Imperial Archive: Knowledge and the Fantasy of Empire* (London: Verso, 1993); John Elsner and Roger Cardinal, eds., *The Cultures of Collecting* (Cambridge: Harvard University Press, 1994); and Steven J. Seegel, "Cartography and the Collected Nation in Joachim Lelewel's Geographical Imagination: A Revised Approach to Intelligentsia," in *Words, Deeds and Values: The Intelligentsias in Russia and Poland during the Nineteenth and Twentieth Centuries*, ed. Fiona Björling and Alexander Pereswetoff-Morath, 23–31, Slavica Lundensia 22 (Lund: Lund University, 2005).

The Passions of Bohdan Krawciw

George G. Grabowicz

In the thirty-five years since his death, the figure of the poet and critic Bohdan Krawciw (1904–75) has been slowly coming into sharper focus, both in his native Ukraine, where he spent the first thirty-five years of his life, and in the West—"the emigration," as it was then known—where he spent the second half of his allotted life span. The seeming regularity of these periods (each just slightly more than a third of a century) only masks the interrupted and uneven pace of his entry into the canon. *Zibrani tvory*, the edition of his collected works initiated by fellow poet and critic Bohdan Boychuk, still remains incomplete: the first relatively complete volume of poetry (complete, but hardly flawless in its structure) appeared in 1978; the second volume of critical, primarily literary, studies, appeared soon thereafter in 1980; the third volume, containing Krawciw's journalistic writings, was published only in 1994, a few years after Ukraine's independence and at the height of its post-independence economic chaos and hyperinflation (to which the poor quality of the paper, typography, and binding of this volume still bear witness).[1] The projected fourth volume of translations, of which Krawciw's translations from Rilke would have been the centerpiece, has still not appeared—and this, as we shall see, is not just a lacuna in the present corpus, but an absence that continues to distort the poet's overall profile.[2] A 1993 Lviv edition of the poetry, based largely on volume 1 of *Zibrani tvory*, marks the first stage of the post-Soviet rediscovery of Krawciw in Ukraine.[3]

The configuration and assessment of Bohdan Krawciw's place in twentieth-century Ukrainian literature and culture that began with the reception of his first collections of poetry in the 1930s, then the provisional valuation of his (still incomplete) legacy in the influential anthology of twentieth-century Ukrainian émigré poetry, *Koordynaty* (1969), the first volumes of *Zibrani tvory*, the retrospective comments on the occasion of his death, and then his reemergence in post-independence Ukraine is thus, as always, an ongoing process.[4] It is the task of each generation, as T. S. Eliot observed, to engage in such reconsideration. The present edition facilitates this by documenting a dimension of Bohdan Krawciw's life and work that many of his readers and admirers did not even suspect existed: his role as map collector, specifically of cartographic Ucrainica. The dimension and importance of this activity, especially when combined with the enlightened decision to make it part of a universal cultural legacy by donating his collection to Harvard, the university with which he was affiliated in the last years of his life as research fellow in the newly established Ukrainian Research Institute (HURI), throws new light on some guiding principles of his life and brings some of its more known facets into sharper relief. In effect, it invites us to reconsider the whole—and its underlying values, motivations and strategies.

Within the context of Ukrainian interwar and émigré poetry, the pressures and turbulence of those years, and of Krawciw's no less turbulent biography, his poetic corpus, as conveyed, for example, by the first volume of *Zibrani tvory*, is not particularly large; nor, at first glance, is it particularly small. For lyrical poetry (he makes few forays into the narrative and none into the dramatic) it approaches what might be taken to be an average.[5] (Publishing practice, at any rate, tends to round off the product in this fashion: a volume per poet, with dimension determined by degree or principles of selection. But poetry, of course, is not to be judged by quantity or by averages, and the "product," by its very nature, varies greatly. And the principles of selection may also change profoundly.) Krawciw's poetry, in fact, was not written in a steady and predictable stream, but in spurts of activity, and is prone to circle back and eddy, and in its various stages differs considerably in terms of its intrinsic weight and the aesthetic and existential problems it poses. Most tellingly, with maturation Krawciw's understanding of poetry undergoes a palpable shift: from a poetry that in its early stages is often eclectic and echoes various thematic and rhetorical fashions of the day, to one that is highly focused, demanding, and inward-looking, created in a process of intense forging and long gestation; a poetry, in a word, of ever greater concentration. From the larger perspective, the nature of both his early and his late poetry, of what is really implied by this binary chronology, and of the deeper continuities between the two phases, also require rethinking—and for this one must start at the beginning.

Bohdan Krawciw was born on 5 May 1904 in western Ukraine into the family of a Uniate priest. Too young to serve in the soon-to-erupt war to end all wars, the defining experience of his young life, as he later recalled in a retrospective poem ("Simnadtsiatoho roku," [1917], 1936), was the news in the spring of 1917 that an independent Ukrainian Republic had come into being. Unable to participate in or witness the struggle, his passionate involvement in the reborn nation was displaced, as he recounts in the poem, to iconic mementos and images brought back from Ukraine by an uncle—the currency of the new nation, a postcard of Kyiv's Saint Sophia cathedral. The call of patriotism, of serving Ukraine and devoting his life to her, became a guiding principle for Krawciw in the interwar years. He completed the academic gymnasium in Lviv in 1923 and then first entered the secret Ukrainian University (1923–25) and in 1925 the Polish University of Jan Kazimierz in Lviv, studying literature and archeology (1925–28) and later law (1928–30). During this time he become involved in the Ukrainian student movement, in the Ukrainian scouting movement (Plast), and finally the Organization of Ukrainian Nationalists (OUN). In the late 1920s he was already highly placed in the latter organization.

In the aftermath of the armed Polish-Ukrainian conflict of 1918–19, the western Ukrainian National Republic was absorbed by Poland. Ukrainian political and cultural rights of self-determination, conceded by the Treaty of Versailles (and subscribed to by Poland), were subsequently foiled by Polish intransigence, and the resultant OUN militancy and terrorism were answered by massive Polish repressions and state terrorism. Given his position in the OUN, it was not surprising that Krawciw was soon arrested and sentenced to three years in prison (1930–33). Upon his release, he married in 1934 his sweetheart Nila Holovats´ka, but a few weeks after that was rearrested and sent to the notorious concentration camp of Bereza Kartuska. With his health impaired by the brutal treatment there, he was released sometime later after the personal intervention of the Ukrainian metropolitan Andrei Sheptyts´kyi.

Throughout this time Krawciw was also writing and publishing poetry, while also serving as editor of various Galician Ukrainian newspapers and journals.[6] His first collection, *Doroha* (The Road), appeared in 1929; a second collection, *Promeni* (Rays), came out a year later. While in prison he wrote another collection, *Sonety i strofy* (Sonnets and Strophes), which appeared after his release in 1933. The differences between the first two collections and the latter are evident. In the former Krawciw is still searching for his own voice, testing, among other things, modalities established a decade and more earlier by the two most prominent poets of the powerful new Soviet Ukrainian literature—Pavlo Tychyna and Maksym Ryl´s´kyi. Echoes of these poets (for example, the Ryl´s´kian "Do ostroviv svitlykh" [To the Bright Islands] or "Ja neznanomu navstrich" [Facing the Unknown] or "Rozsmiialys´ temni vulychky" [Dark Streets Laughed Aloud] in *Doroha*, and the Tychynian "Khrystos rodyvsia" [Christ is Born] or "Ratai" [The Plowman] in *Promeni*) are so manifest in their overall style, diction, and topoi of these poems that one can see them as poetic homages to these new bards.[7] At the same time, these styles are so different from each other that Krawciw seems inadvertently to be casting himself as a novice at loose ends, torn in quite opposite and contradictory directions.[8] Even more than the tug of powerful poetic voices, the first two collections reveal the seductiveness—reflecting both an emerging nationalist poetics and the not-so-distant Soviet model—of poetry conceived as a surrogate for political and ideological action, and the poet as tribune and herald of the nation's cause. This became the hallmark of Ievhen Malaniuk, soon the preeminent (and paradigmatic) poet of the interwar period in western Ukraine, a feature of various members of the so-called Prague group of poets, and above all the battle cry of the nationalist ideologue, Dmytro Dontsov, and his journal *Visnyk*.[9] In Krawciw, however, this appears in an attenuated form: both of these collections are free of any overt ideological or political rhetoric or content. What does appear in *Doroha*, however, is a cycle of prose pieces; in effect, poems in prose, the first of which is entitled "Tale of Fire,"[10] which basically present a cycle of campfire talks to a group of scouts (*plastuny*). The tenor of that mode, its symbolism (and unavoidable sentimentalism), its topoi and images, were very much part of Krawciw's real life activities; they are projected here both as an aesthetic recapturing that facet of his world and as a generic projection of the scouting mystique, its love of nature, of the fellowship of the young and innocent—of idealism, as it were, in its pure form. It is, however, consistently unideological and apolitical. The solitary, concluding reference to Ukraine in the last piece, where she is invoked and distanced as a "sublime ideal of an eternal Ukraine," recapitulates again the ubiquitous mode of the idealistic and the symbolic. In a curious reversal of the role of the poet as national leader so powerfully appropriated by Tychyna (and so coveted by Malaniuk), Krawciw projects an almost self-effacing alternative: the poet not even as scout leader, but rather as one who along with the scouts is caught in a transcendent reverie on nature and the scouting way.

The 1933 collection *Sonety i strofy* has a totally different cast, as it recounts his prison experience in fourteen sonnets and eighteen double quatrains. To say "recounts," however, may be misleading, for the formal constraints of the sonnet and the concision of the quatrains, coupled with a consistent, multifaceted removal or bracketing out of the real and sensual, make this a remarkable and somewhat disorienting experience. For in effect this is a poetry of displacement, where the real life sights, smells, sounds, and a myriad of other experiences of prison are indeed at times noted, or alluded to, but seldom if ever depicted in extenso, as experiences in and of themselves. Instead, they are for the most part supplanted, displaced by other moments—images, recollections, fantasies and reveries of the outside, free world, but also by the very texture and process of casting this all as poetry—of a highly formalized and "classicist" kind. In a sense literature, or literariness, supersedes reality; and experience becomes secondary to the text. But this also becomes a form of psychological and existential escape, of structuring and thereby of imagining and creating a separate non-prison reality—precisely foreshadowing the poetry of Vasyl′ Stus written in the gulag in the late Soviet period, the process of reestablishing self-dignity, humanity, indeed one's intrinsic, non-prison identity as such. It is a poetry of self-regeneration—albeit couched in the "impersonal," "cold" structures of a classicist, in effect neoclassicist poetics. A fundamental measure of its authenticity is that it indulges in no political or ideological posturing, no cant, or self-pity.[11] At the same time, a sense of something missing is inescapable—of experience not confronted, of insights into the self not explored. This absence, however, can be apprehended only from a larger perspective, not least of all one that is empowered by Krawciw's later, mature poetry.

The next two collections Bohdan Krawciw published were *Ostannia osin′* (The Last Autumn, 1940), containing poems from the years 1927–38, and *Pid chuzhymy zoriamy* (Under Foreign Stars, 1941), which collected poems written in 1940–41. Both were published in Berlin, where Krawciw had been taken by the German authorities and along with other Ukrainian writers put to work writing journalism for the growing number of Ukrainians in the forced labor brigades. Both collections, especially the second, have a retrospective, at times distinctly elegiac cast and reflect a sense of foreboding and the premonition that he had seen Ukraine for the last time. Much more pronounced—in comparison with his earlier poetry—is his readiness to address political issues of the day. In *Ostannia osin′* this is reflected in several retrospective poems on the political struggle, from the 1 November 1918 dramatic beginning of the Western Ukrainian independence struggle and the seizure of Lviv from the Austrian authorities, which in turn precipitated the Polish-Ukrainian war of 1918–19 ("I, XI, 1918"), to the 1938 proclamation of a free Carpathian Ukraine and the subsequent heroic but doomed struggle of several thousand patriots crushed by the Hungarian army ("Karpats′ka Sich"). In the second collection, the only one in which Krawciw turns to long, narrative, and in part contemplative poems, he devotes some of them to the mode of political commentary and indeed exhortation, particularly "Slovo pro Polkovnyka" (Recollections of the Colonel), an elegiac account of the life and significance for the nationalist movement of its founder Colonel Ievhen Konovalets′ (1891–1938).[12] The topicality here is twofold. Not only is the poem a requiem for a political leader recently assassinated, but the mode in which this is done—a requiem combined with political exhortation and historiosophic musings—echoes the recent appearance of Iurii Klen's highly resonant poem "Prokliati roky" (The Cursed Years, 1937), in which the poet, a member of the Kyiv neoclassicist group[13] who somehow was allowed to emigrate to the West, combines an impassioned lament for the victims of the Soviet terror with a no less passionate indictment of the perpetrators.[14] As with Krawciw's earlier poetry, this again seems to try out another topical role and rhetoric. (That he is cognizant of Klen's model is indicated by his drawing on "Prokljati roky" for an incipit to the first poem of *Pid chuzhymy zoriamy*, "Oleni revut′ 1" [The Stags Roar]).[15] On the whole, the attempt is not sucessful, the voice does not ring true, and it is to Krawciw's credit that he himself apparently comes to the same conclusion, for he was never again to return to this exhortatory mode—although the opportunities and temptations were many.

The following period of Krawciw's poetic activity encompasses the immediate postwar years of 1945–49 and basically coincides with the so-called MUR (Mystets′kyi ukraïns′kyi rukh; Ukrainian Artistic Movement) period where in the various DP (displaced persons) camps scattered across western occupied Germany, principally in Bavaria, Ukrainian literary life experienced a remarkable efflorescence of creativity and organized activity. In a brief period of four or five years, hundreds of literary, critical, and scholarly works were published and dozens of periodicals and organizations came into being (and also quickly disappeared), but in sum the whole constituted a "small renascence" of Ukrainian cultural life after the depradations of World War II and Stalinism.[16] Krawciw actively participated in the literary life of this period, but above all he found what he never could enjoy either before or after: the time and the freedom to actually write poetry. In sum, during this period he wrote only two or three slim volumes of poetry and published two of them, but this is a poetry of a very high order, and dramatically different from what he wrote before and would write later.

"Two or three" volumes seems something of a circumlocution (presumably it should be easy to determine whether it was two or three), but the imprecision flows from Krawciw's own indirection as he revisits, rewrites, and edits his earlier poetry, and in light of his intensive translating from Rainer Maria Rilke, overwrites his voice and poetic persona with the insights gained from that difficult task. In effect, he crafts a new poetic voice, and with it an altogether original and inimitable style. The volumes in question are: (1) the largely retrospective collection *Korabli* (Ships), subtitled "vybrani poeziï (selected poems), 1922–1947," which appeared in 1948;[17] (2) *Hlosarii*, with the baroque subtitle "abo tlumachnyi slovnyk taiemnykh, pryzabutykh i ne zavzhdy zrozumilykh sliv, zibrav u sonetakh Bohdan Kravtsiv," (Glossarium, or Dictionary of Obscure, Forgotten and Often Misunderstood Words, Sonnets by Bohdan Krawciw), which was completed and ready for publication in 1949, although some parts of it, eight sonnets in all, had already been included in *Korabli*, and the whole actually published only twenty-five years later, in 1974;[18] and (3) *Rechi i obrazy* (Dinge und Bilder; Things and Images), a book of translations from Rilke, which appeared in 1947.[19] All three were designed with utter care and attention to detail, and each is graced by covers (and *Korabli* with vignettes and *Hlosarii* by decorative borders) by the soon-to-be-famous Ukrainian painter and woodcutter Jacques Hnizdovsky. Individually and as a set they project an aesthetic standard rarely matched in the entire large corpus of MUR publications. But the MUR period ended in 1949/1950, and *Hlosarii* appeared only in 1974. Hence, too: "two or three."

Korabli, 1948

By all indications the crux lies in the translations; they are not only the first to be published, but are the ineluctable key to the transformations that are occurring in Krawciw's poetry in this period—indeed, they are the watershed itself. In a revealing postscript to *Rechi i obrazy* written in both German and Ukrainian, Krawciw recounts how he came to Rilke, first as a gymnasium student sometime in the early 1920s. In a Lviv antique bookshop he finds a volume of *Das Buch der Bilder* (Book of Images, 1898–1906) and discovers a poet of rare complexity and difficulty who will remain a challenge for him and a frame of reference for years to come.[20] A decade or so later, as a university student (and an ex-political prisoner), he again encounters Rilke when someone presents him with *The Notebooks of Malte Laurids Brigge*, and in this autobiographical novel he comes upon—and, as he writes, "remembers to this day"—the insights into how poetry is to be written: "Ah, there is little value in verses written too early," he cites Rilke, "one needs to wait for their maturation, collecting for a whole lifetime their sweetness…[21]

Translating Rilke was to come later. His first efforts were "Khlopets′" (Der Knabe; The Boy, which first appeared in the collection *Ostannia osin′*, 1940) and "Osinnii den′" (Herbsttag; Autumn Day, which first appeared in *Rechi i obrazy*), which he altogether too modestly considers here a failure.[22] When circumstances brought him to Germany, his first purchases, he says, "were two volumes of Rainer Maria Rilke's poetry, as if purposefully bought so that wandering the squares and autumn parks of foreign cities I could ponder the wise and bitter words of the poet from the poem 'Herbsttag,' 'Wer jetzt kein Haus hat baut sich keines mehr' (Who has no house now, will never build one)."[23] The end of the war found Krawciw in a tiny mountain village of the Frankenwald with some books he had managed to save from the bombing raids, among them Rilke's *Selected* and *Late* poems and his *Sonnets to Orpheus*, and one day in August 1945, he began—unexpectedly for himself—his translating: "What had once been dark and incomprehensible opened to me like the flowers of early morning, and by winter of that year I had some thirty translations ready."[24] In the course of the next year he expanded it by another twenty or so poems, and with the support of Iurii Klen (Oswald Burghardt) as well as Anton Kippenberg, who had been Rilke's publisher and curator of his poetic legacy, succeeded in publishing it in 1947. He drew for its title, as he says, "perhaps somewhat immodestly, on two mysterious but so characteristically Rilkean terms—*things* and *images*."[25]

Krawciw's sense of Rilke's poetry—and of its deeper resonance with his own poetic development—is suggested by his epigraph for the book, which he does not identify, other than with the initials "R.M.R.":

> …doch wir wollen reifen
> Und das heisst dunkel sein und sich bemühn.

These are, in fact, the closing lines from the poem "Im Saal" (In the Hall, 1906) from the *Neue Gedichte* (New Poems), part 1 (1907), where the poet contrasts the life and existential frame of the assembled dignitaries—or their portraits in a museum—with those of the onlooker(s):

> Sie lassen, voller Takt, uns ungestört
> das Leben leben wie wir es begreifen
> und wie sie′s nicht verstehn. Sie wollen blühn,
> und blühn ist schön sein; doch wir wollen reifen,
> und das heißt dunkel sein und sich bemühn.

> With exquisite decorum they allow us
> A life of whose dimensions we seem sure
> And which they cannot grasp. They were alive
> To bloom, that is be fair; we to mature,
> That is to be of darkness and to strive.[26]

The notion that the poet's task is to be open to that which is dark, "to be of darkness" (dunkel sein), is a leitmotif in Rilke's poetry and is a theme and image in a number of poems that are translated here. But as central, as constitutive as it is for Rilke's world, for its defining symbolic and existential dimensions, for Krawciw it is a new and not easily assimilated moment. Even (or perhaps especially) in his poetry of prison—that is, the *Sonety i strofy*—it is far more absent than present. In general, his had largely been the poetry of assertion and optimism, often attained through a highlighting of the rhetorical, and in an all-but-canonical cult of deeds, valor, uplift, and so on. As such it was part of the general discourse of western Ukrainian poetry of the interwar period, and in that context was even rather restrained—as compared, for example, with the unabashedly tribunicial, hectoring, and at times vituperative poetry of Malaniuk.[27] Here the poetry is profoundly decentered. The task of poetry, as these short lines of the epigraph and the entire body of the translations make evident, is not to be beautiful or

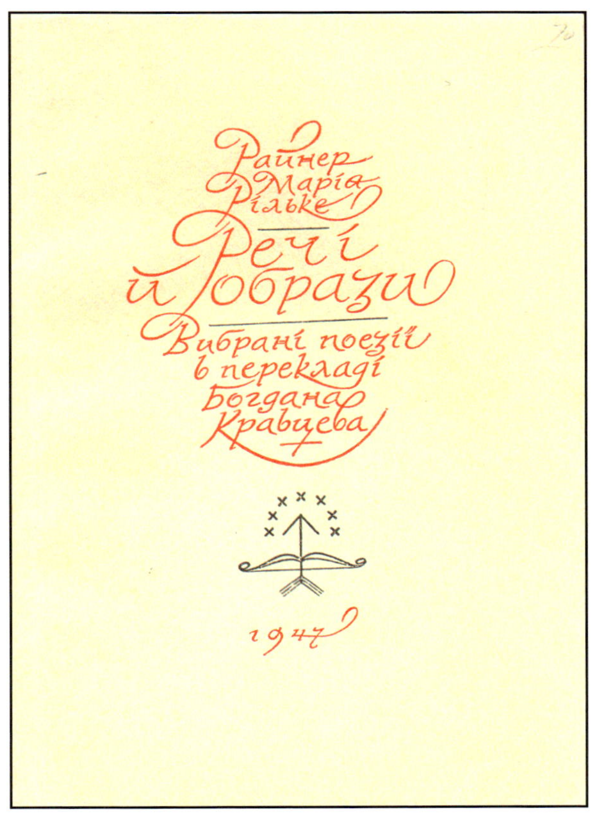

Rechi i obrazy, 1947

pretty ("schön sein"), but to seek that which is deeper and darker—and in this to attain ripeness and with it self-reflection, and self-knowledge. As Krawciw observes about "Herbsttag," Rilke's poetry becomes a model for such re-vision, and as with Rilke, his own poetry begins to focus more deeply on the senses, and emphasizes the task of learning to see ("sehen lernen").[28]

A close reading of these translations is a task for the future, which is all the more pressing and doable now that an anthology of the various Ukrainian translations of Rilke has appeared; in light of it Krawciw's pioneering effort (which in this edition is presented rather skimpily) can be properly framed and contextualized.[29] What is apparent even at a cursory reading, however, is that the poetry that is produced here is somehow both remarkably faithful and remarkably powerful—providing an altogether new and joyful sense of the resources and subtlety of the Ukrainian language and with it a sense of the impressive poetic resources at Krawciw's disposal. At the root of this is the transcendent power of Rilke's poetry, his ability to touch with seeming simplicity and objectivity the very core of our fragile existence, and to persuasively suggest that that core opens itself only to a poetic vision. Krawciw's ability to capture these Rilkean poles of the profound and the simply objective is masterful, as in this well-known poem:

Todeserfahrung

Wir wissen nichts von diesem Hingehn, das
nicht mit uns teilt. Wir haben keinen Grund,
Bewunderung und Liebe oder Haß
dem Tod zu zeigen, den ein Maskenmund

tragischer Klage wunderlich entstellt.
Noch ist die Welt voll Rollen, die wir spielen.
Solang wir sorgen, ob wir auch gefielen,
spielt auch der Tod, obwohl er nicht gefällt.

Doch als du gingst, da brach in diese Bühne
ein Streifen Wirklichkeit durch jenen Spalt,
durch den du hingingst: Grün wirklicher Grüne,
wirklicher Sonnenschein, wirklicher Wald.

Wir spielen weiter. Bang und schwer Erlerntes
hersagend und Gebärden dann und wann
aufhebend; aber dein von uns entferntes,
aus unserm Stück entrücktes Dasein kann

uns manchmal überkommen, wie ein Wissen
von jener Wirklichkeit sich niedersenkend,
so daß wir eine Weile hingerissen
das Leben spielen, nicht an Beifall denkend.

Досвід із смертю

Нам знати не дано про відхід той,
Що не займає нас. Й нема що свій
Являти подив нам, любов, ато й
ненависть смерті, що обличчя ій

страшною скаргою спотворене ж.
Ще світ цей повен роль, що їх ми граєм.
І поки ми подобатись бажаєм,
Хоч і противна всім—смерть грає теж.

Та як ішла ти, то крізь ту з ущелин,
Що нею ти відходила у вись,
Ввірвалась дійсність: зелень, справжня зелень,
і промінь справдешній, і справжній ліс.

Ми ж граєм далі. І слова, і жести
здіймавши іноді—з важким трудом
засвоєні; та вирване до решти
із п'єси нам життя твоє притьмом

нас може часом пошибом пройняти,
постигти дійсности предивним знанням—
і ми на мить в життя забагнем грати,
не думавши про оплески, признання.

Death Experienced

We know nothing of this going-hence
that so excludes us. We have no grounds
for showing Death wonderment and love
or hate, since it wears that age-old mask

of tragedy that hopelessly contorts it.
The world is full of roles—which we still act.
As long as we keep striving for acclaim,
Death also acts a part—though always badly.

But when you went, a streak of reality
broke in upon the stage through that fissure
where you left: green of real green,
real sunshine, real forest.

We go on acting. Fearful and reciting
things difficult to learn and now and then
inventing gestures; but your existence,
withdrawn from us and take from our piece

can sometimes come over us, like a knowledge
of that reality settling in,
so that for a while we act life
transported, not thinking of applause.[30]

A more striking masterpiece, to my mind, is his translation of Rilke's acclaimed "Orpheus. Euridike. Hermes" (1904). In the interest of space, one can only glance at a few fragments here, such as the haunting opening scene of the underworld from which the trio comes and which in the dark is seen with utmost clarity—and which limns the poem's sublime tension of the interpenetration of light and darkness:

Das war der Seelen wunderliches Bergwerk.
Wie stille Silbererze gingen sie
als Adern durch sein Dunkel. Zwischen Wurzeln
entsprang das Blut, das fortgeht zu den Menschen,
und schwer wie Porphyr sah es aus im Dunkel.
Sonst war nichts Rotes.

Felsen waren da
und wesenlose Wälder. Brücken über Leeres
und jener große, graue, blinde Teich,
der über seinem fernen Grunde hing
wie Regenhimmel über einer Landschaft.
Und zwischen Wiesen, sanft und voller Langmut,
erschien des einen Weges blasser Streifen,
wie eine lange Bleiche hingelegt.

Und dieses einen Weges kamen sie.

Це душ було родовище чудове.
Крізь сутінь жилами вони пливли,
мов тихі зложжя срібла. З-між коріння
бистрила кров, що до людей одходить,
і важко—мов порфір—багріла в пітьмі.
То весь був багор.

Скелі там були
І пустка пущ. Мости понад пустинням
І той великий, сірий, темний став,
Що над далеким дном своїм навис,
мов небо дощеве над краєвидом.
І проміж лук терпцю смутного повен
І тиші й миру—смугою блідою
Прослався шлях, як довга біль в далінь.

І шляхом тим надходили вони.

It was the soul's strange mine.
Like silent silver ore they wandered
through its dark like veins. Between roots
the blood welled up that makes its way to men,
and it looked hard as porphyry in the dark.
Nothing else was red.

Rocks were there
and unreal forests. Bridges spanning voids
and that huge gray blind unmoving lake
that hung above its distant bed
like rainy sky above a landscape.
And between meadows, soft and full of patience,
the pale strip of the single path,
laid down like a long pallor being bleached.

And on this single path they came.

And then this vision of Eurydice at the moment Orpheus looks back:

Sie war schon nicht mehr diese blonde Frau,
die in des Dichters Liedern manchmal anklang,
nicht mehr des breiten Bettes Duft und Eiland
und jenes Mannes Eigentum nicht mehr.
Sie war schon aufgelöst wie langes Haar
und hingegeben wie gefallner Regen
und ausgeteilt wie hundertfacher Vorrat.
Sie war schon Wurzel.
Und als plötzlich jäh
Der Gott sie anhielt und mit Schmertz im Ausruf
Die Worte sprach: Er hat sich umgewendet—
Begriff sie nichts und sagte leise: Wer?

Це не була вже та русява жінка,
що іноді в піснях співця дзвеніла,
ні пахощі, ні ясен острів ложа,
ані теж власність мужа он того.
Розвіялась вона волосям довгим,
рясним дощовим ливнем віддалася
і запасом стократним дарувалась.
Корінням стала.
І коли притьмом,
враз бог спинив її і з болем в озві
проказував слова: Він озирнувся—
не зрозумівши, прошептала: Хто?

She was no longer the blond wife
who echoed often in the poet's songs,
no longer the vast bed's scent and island,
and that man's property no longer.

She was already loosened like long hair,
and given over like fallen rain
and handed out like a limitless supply.
She was already root.
And when without warning
the god stopped her and with pain in his voice
spoke the words: he has turned around—,
she was puzzled, and answered softly: Who?

and finally the piercingly muted finale:

Fern aber, dunkel vor dem klaren Ausgang,
stand irgend jemand, dessen Angesicht
nicht zu erkennen war. Er stand und sah,
wie auf dem Streifen eines Wiesenpfades
mit trauerfollen Blick der Gott der Botschaft
sich shweigend wandte, der Gestalt zu folgen,
die schon zurückging dieses selben Weges,
den Schritt beschränkt von langen Leichen-
bändern,
unsicher, sanft und ohne ungeduld.

Далеко ж перед виходом яснющим
стояв померкло хтось, чиє обличчя
непізнанням взялось. Стояв, дивився,
як ген на смузі стежки лукової
бог вісти смутен мовчки одвернувся
і німо йшов за постаттю тією,
що вже верталася тим самим шляхом,
в'язавши крок в довжінь полотен смертних,
непевна, тиха, терплива вкрай.

But there in the distance, dark against
the bright exit, stood someone whose features
could not be recognized. He stood and saw
how on the stripe of the meadow path
with mournful look the god of messages
turned silently around, to follow the shape
already returning on that same path,
its steps, constrained by long winding-sheets,
uncertain, gentle and without impatience.[31]

The impact of these translations on Krawciw himself and on his own poetry was profound. This is reflected not only by external circumstances and in the visible realm—the care that went into the production of *Rechi i obrazy*, the recollection of his encounters with the master, the cathexis that surrounds the themes of this poetry (solitude, resignation, the muted presence of the transcendent, death itself), and even the fact that he so stresses the Ukrainian themes in Rilke, his sympathy for the poetic topoi of Ukrainian history and culture, and his, the eternal wanderer's, import for the Ukrainian exiles.[32] It also reveals itself in a more immanent way—in the very way poetry comes to be defined for him—and in deeper processes and effects that become apparent only after much time. To be sure, one of these effects emerges already in his next collection of poetry, *Korabli* (1948), which is primarily a distillation of Krawciw's earlier work that winnows out a number of weaker poems, primarily the more rhetorical and overtly political, and ultimately reduces his earlier corpus by about a half. At the same time he includes in *Korabli* a number of new works—primarily sonnets from the ongoing *Hlosarii* (eight in all, written in the fall of 1945) and alexandrines (six in all, written in 1945–47). The difference between these and the earlier poetry is striking. In effect, within the confines of this small book of selected poetry (some 120 pages in all) one sees the traces of a fundamental divide.

Much more striking, however, is the subsequent trajectory of Krawciw's poetry and the way this divide is articulated in the course of his remaining years, in effect, of his entire émigré experience in the United States. The key moment here is the fate of *Hlosarii*, a collection of sixty-seven sonnets that was ready for publication in 1949 (with eight, as noted, already published in *Korabli*), but which actually appeared—in the very design that Hnizdovsky had prepared already in the MUR period—only in 1974, a year before Krawciw's death. As he puts it in the postscript to the collection (the only instance I know of where the much abused self-designation "zamist´ pisliamovy," [instead of an afterword], is entirely appropriate by reason of its brevity and succinctness),

> [*Hlosarii*] appears now after twenty-five years. The author had hoped to publish these sonnets with separate commentaries and notes to each of them, but working in the gristmills of newspapers, journals and encyclopedias he never found the time to do so. He gives them over now to the judgement of his readers as they were written.[33]

Hlosarii, presumably, was delayed by the force majeure of the poet-as-critic, with visions of a full (historical? biographical?) commentary overriding the voice of the poet. Fortunately, as if intuiting the end and with it various editorial incursions that in effect did take place, the poet finally prevailed. The *Hlosarii* that thus appeared had both dates and both locations on its title page—1949–1974; Regensburg-Rutherford—and the intervening twenty-five years could, if one so wished, be elided as but a contingency, a small detour in the flow of

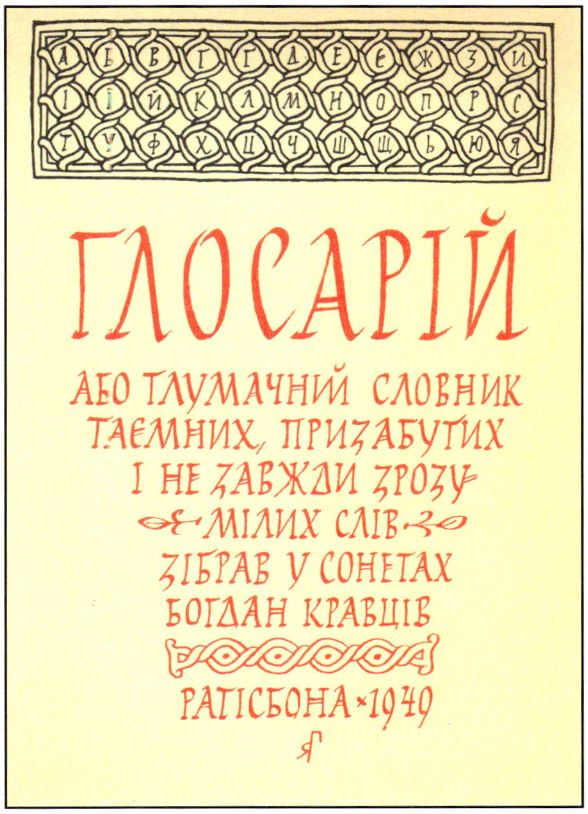

Hlosarii, 1974

poetic time. In the real world, of course, it was hardly that, but the fact that *Hlosarii* emerges as the culmination of Krawciw's poetic effort is also unaffected by that eddy of time. In the larger order of things it returns him to that brief period of extraordinary poetic activity from 1945 to 1948 and within it reveals the true tenor and achievement of his poetry.

Hlosarii is a collection of sixty-seven sonnets styled as a dictionary of mysterious, misunderstood, and forgotten words, with the poems themselves positioned as glosses—in effect, as comments to, illustrations of, elaborations, or meditations on the terms-words that serve as the titles-topoi here. That interplay of title and text might appear at first as an echo of the early Tychyna cycle "Enharmoniine" (Enharmonics) in *Soniashni klarnety* (Clarinets of the Sun, 1918) where the impressionistic opaqueness and indirection of the text is counterposed to the basic reality of the titles: "tuman" (mist), "sonce" (sun), "viter" (wind), "doshch" (rain). As such, Tychyna's minimalist and purposefully obscure or hermetic poems reactivate the form of a riddle, with the text the question and the title the answer. Little if any of that is seen in *Hlosarii*: the texts here are invariably complete and focused (as befits a sonnet), clear in their narrative, and the connection of title and text semantically direct. For the most part (perhaps entirely, depending on the reader) the terms are familiar, at times obviously so; what, might one ask, is obscure about "maty" (mother), "vid´ma" (witch), "bis" (devil), "rid" (ancestry), "dolia" (fate), "dusha" (soul), "zhovten´" (October), "khlopets´" (boy), and many other such words? Other terms: "Argo," "Medea," "lisovyk" (forest-devil), "chuhaistyr" (also a forest-spirit),[34] "Lada" (figure or goddess of spring in Slavic folklore), "vovkulaka" (werewolf), and others like it are quite accessible to those who know Greek mythology and Ukrainian folklore. And "Solovky," "Gonta," "orda" (the horde), even "chaika" (gull) are accessible to anyone who knows Ukrainian history. (To make the connection plain, the epigraph to "Chaika" notes that it refers to a poem purportedly written by Hetman Mazepa.) At the same time, there are terms (words, refererences, names) that are not generally accessible even to the sophisticated reader—"irytsia," "Teklia," "kotora," "ran´," "shumyl´tse," "tyrlych," and so on—and only the poem, either through a direct bibliographic reference as in "Chaika," or in "Teklia," or through the text itself, amplifies the meaning.[35]

What unites the seemingly obvious with the less than obvious, and the fully obscure and mysterious, is the poet's implicit belief that all of these terms/topoi have a hidden, deeper meaning. Nothing in this glossarium, the poet seems to be saying, not even "maty" and "rid," is simple and directly given, and it is thus his higher, ordained task to seek and discern hidden meaning; for only poetry, through its evocations or indeed incantations (or songs, as he will call them), can reveal the sense of these words-that-are-more-than-words. In this process the high calling of poetry is itself laid bare and the collection becomes in part an exercise in metapoetry and in part a celebration of poetry as such. (The two, of course, are always linked.) As he puts it in the dedication to his wife, which also serves as the epigraph to the whole collection:

> З яких же слів мережати, плести
> Тобі пісні мої сьогодні, Ладо?
> Котрого ж не торкнуся, заблестить
> клейнодом завороженого кладу,
> живоцвітом розквітне—і мости
> в минуле прокладе… навмань, без ладу…
> І завжди, завжди гомоном дібров,
> диханням рідних нив нам палить кров.

> Out of what words am I to weave
> my songs for you today, my Lada?
> Whichever one I touch will shine
> like a jewel from a mysterious treasure trove
> will bloom and cast bridges to the past
> haphazardly, every which way
> and always, always stir the blood

with the rustle of forest depths
and the breath of native meadows.[36]

Inescapably, however, the role of the author is blurred here—or rather doubled, or even tripled. He is not only a creator, but a collector, a cataloguer, and through the epigraphs that introduce each of these sonnets and point to a textual source he is also a lexicographer and bibliographer; and, finally, a critic-exegete (and this role, in the guise of the compleat commentator is what may have delayed the publication of *Hlosarii* for twenty-five years). This self-inscription in multiple roles, and precisely in these roles, foretells, of course, the direction of Krawciw's future path, but also shows that already then (and indeed earlier) it was part of his practice and self-image. At the same time, it is only intimated here, all but concealed in the text (with the footnote, for example, transposed into the epigraph). The dedication, however, intimates yet another moment, both in the cited opening lines ("Out of what words am I to weave/ my songs for you today, my Lado?") and the concluding dedication-proper: "Ладі моїй—усе ще під чужими зорями—на Різдво 1949 року" (To my Lada—still under foreign stars—Christmas 1949).[37] In effect, it inscribes the whole collection into the mode of intimate poetry: the collection becomes a gift, a Christmas present, from husband to wife. This, too, is a harbinger of things to come.

To all appearances, the dominant tone and ambience of *Hlosarii* is folkloric, with a particular focus on folk demonology and the magic of the night world and the forest world;[38] it inheres in the topoi already noted, as well as those like "Vodianyk" (Water Spirit), "Domovyk" (House Spirit), "Troizillia" (Magic Herbs), "Rusallia" (celebrations of the mermaids), "Perelesnyk" (incubus), "Kupala" (midsummer night revelry), and others. Along with the range of the documented folkloric, Krawciw is also elaborating or coining his own; for example "Moroka" (the spirit of troubles, akin to the "zlydni") or "Mavs´ka nich" (akin to the "kupala" and echoing "mais´ka nich"). In all, and in its time and place—that is, the period of MUR and of the intense rediscovery of all things Ukrainian after the dark night of the 1930s and the war years—this may well be seen, as Iurii Sherekh (Shevelov) observed, as yet another instance of a general reaching out for organicity (see his notion of an "organic" or "national" Ukrainian style) and simply authenticity.[39]

Krawciw's "folklorism" has a very specific cast, however. For one, it is strongly marked by the erotic. In some poems this inheres in the very term/topos, such as "Barvinok" (periwinkle, with its connotation of betrothal and marriage), "Perelesnyk," "Mavs´ka nich," "Kupala," "Lada," "Uroky" (Enchantments), and others; in other instances, even a neutral or innocent topos, such as "vesnianka" (spring song), or "Pliasanka" (dance), or a topos from the night world (lisovyk) leads to the erotic embrace. Thus the conclusion of "Pliasanka":

І схилится розплетеним розмаєм
на руки мужа зморена і трудна,
в солодкий радо здаючися бран.[40]

And she will lean, with disheveled hair,
tired and heavy, on the man's arms,
gladly submitting to the sweet captivity;

or "Lisovyk":

Обмарений, лягаєш поміж квіттям,
вбачавши прибажно в березій кожній
уроду гожу, білі груди мавок...

Згори ж, з-під брів хвилястих верховіття
підморгує тобі зеленокожий
кудрявий бог спокусливо й лукаво.[41]

Dazed, you lie among the flowers,
seeing in each birch nubile beauty,
the white breasts of mermaids...

And beneath the undulating tree tops
the green-skinned, curly-haired god
bestows his sly and lustful winks.

But from the first, this eroticism is interlaced with the demonic and the world of darkness. Here, again, it may be immanent and stem from the thematically given, as in the mythical story of Jason and the Argonauts ("Argo"), or the balladic tale of "Teklia," who (echoing Scott's *The Bride of Lammermoor* or Donizetti's operatic heroine) kills the man she is forced to marry and prefers a horrible death (being buried alive) as long as she is buried with her lover. More often, however, it appears as a general, nonthematic openness to the darkness of the heart, of human passions, and of fate—and here the impact of Rilke is most telling and adumbrated by the fact that Krawciw was writing some of these poems of at the very time—that is, October–November 1945—that he was intensively translating Rilke. This openness to the darkness of things defines the ontology of *Hlosarii*, the sense that as much as light and beauty, sexuality and fecundity, love and family exist, all of them are shadowed by darkness.

The poem "Blud" ("Error"—literally, the state of being totally lost) stands out in the whole collection by placing the darkness at the center of the self. The folkloric merely contextualizes the trope; that is, in the epigraph: "Як кого нападе блуд, нехай згадає в який день було Різдво, візьме землі з-під ніг і посипле собі на голову. —«З українських приказок» Номиса." (If someone were to fall into Error let him remember on what day Christmas came, take some earth from under his feet and sprinkle it on his head. —"Ukrainian sayings" by Nomys.)

Взялися блудом дні мої й дороги,
все глибше у чужу вгрузавши твань,
лиш пригадом далеких поривань
князь-місяць сновидає круторогий.

Йдуть вибертом ворота і пороги
і думи розтікаються навмань:
за круг заклятий за незриму грань
не виведуть колеса ні остроги.

Даремне, сповнений тривог до краю,
вгадати рвусь: вівторок чи сибота
святилися різдвяними святками.

Дарма навпомацки сліпий шукаю
за крихтою хоч глини хоч болота—
то ж всюди, скрізь тут—не земля, а камінь.[42]

My days and roads are lost in error
as I sink still deeper into foreign muck
and as he lurks, the horned moon
only reminds me of my distant goals.

Gates and thresholds pass in a blur
and thoughts fly out in all directions
and neither wheels nor spurs will breach
the cursed circle and the invisible divide.

Filled to the brim with fear, I seek
in vain, to know: was it on Saturday or Tuesday
that we were sanctified by Christmas day.

In vain, I blindly grope
for but a clod of clay, a clod of mud
But everywhere, around me I find not earth, but
stone.

The existential angst that informs the collection is projected for the most part by the collective, not the personal, frame—and its signposts are the traumas and long stretches of darkness that make up Ukrainian history. Through such poems as "Gonta" (echoing the bloody time of the haidamak uprising of 1768 and the atrocious cruelty of its leader's execution by the Polish authorities), as "Orda" and "Iasyr" (the centuries of Tatar raids and its thousands upon thousands of victims), "Kotora" (on the curse of fratricidal strife going back to early Rus´), "Neryst´" ("enmity" or "rancor," which is cast here as a Shevchenkian meditation on the morass of émigré life—which also closely echoes "Blud"), and "Solovky," a meditation on the fate of Ukraine's leaders perishing on that prison island at the hands of Russian despotism from the time of Kalnyshevs´kyi in the late eighteenth century to Stalin's time, that history is presented—rather perhaps intuited—as a fate that accompanies that people, that defines the Ukrainian essence of things. The opening poem "Argo," which in *Korabli* was still "Argonavty" (the Argonauts) in fact encapsulates the *fatum* of that land and sets the tone for the whole collection:

В ясну далінь Кольхіди-України
ти через море негостинне й чорне
синів Геллади не зряджай Язоне,
за руном золотим! У тій даліні

тебе обмарять ночі солов'їні,
данням любовним зможуть гожі жони
і шлях заступить племя необорне
залізнім проростом зубів зміїних.

Ти може й цілий вернешся в Гелладу,
здолавши чарами догонь чужинну,
і грім і зудар зловорожих скель.

Та замість руна привезеш ти зраду:
вродлива бранка вб'є дітей, дружину,
тебе ж розчавить власний корабель.[43]

Into the shimmering distance of Colchis-Ukraine
do not send forth, O Jason, across the black
and uninviting sea the sons of Hellas
for the golden fleece. In that distant land

you will be dazzled by the nights of nightingales,
comely women will sap you with gifts of love
and a resilient tribe will block your way
with iron sproutings of the dragon's teeth.

You may perchance return to Hellas in one piece
and overcome with magic the pursuing horde,
and the thunder and clashes of malevolent rocks.

Instead of a golden fleece you'll bring back treachery:
the lovely slave will kill your children and your
wife
and you, you will be flattened by your own ship.

Hlosarii, the height of Krawciw's poetic achievement, was not repeated: few poets are fated to repeat their chefs d'oeuvre. The combination of classical elegance and self-restraint so characteristic of Mykola Zerov and the neo-classicism that he was then cultivating on the one hand, and on the other the world of Ukrainian folkloric and regional (largely Carpathian) linguistic inspiration and behind that the collective voice of the people, the *narod* (echoing such precursors as Volodymyr Svidzins′kyi and before him Taras Shevchenko), was not fated to recur—certainly not with such intensity and organic melding. As one looks at the larger context—both the synchronous translating that he was then doing from Rilke, and the nature of the poetry that he wrote subsequently—one may even postulate that this collection also marks the beginning of a partially conscious, but far-reaching retreat from poetry, that is, the high poetry that was synonymous with Rilke. At the very least, this is reflected in the way the message of its darkness, its fatedness, and early demise is encoded into its fabric, and becomes its core feature. Thus the lead poem of the collection makes clear that the search for the golden fleece will end in disaster, and that Jason is fated to be crushed by his own ship. And this shadow, one should note, falls on the still seemingly optimistic ambience of *Korabli*, giving that romantically charged topos a very different cast. For in fact "Argo," ("Argonavty") first appeared in that collection as a prefiguring of *Hlosarii* ("Iz Hlosariia") and was dated October–November 1945. In so doing, Krawciw also rewrites the recent acmeist and triumphalist legacy of Iurii Klen's *Karavely* (1943). The perspective of implacable fate and defeat in fact informs much of the *Hlosarii*, from "Argo," "Gonta," and "Orda" to "Solovky," "Chaika," and "Iasyr," and serves as a constant counterweight to the poet's efforts to recapture in language, customs, and collective memory the life force of the collective. Thus when we encounter despondency and despair in such poems as "Blud" and "Neryst′," there is no surprise; they, too, seem fated. Not least of all, the poet's sense of self is constantly tugged in several directions as he parcels out the creative act with the critic and the bibliographer. The symbolic recurrence of the topos of being torn, like Orpheus, into many pieces, and of images of brutal execution ("Chuhaistyr, "Gonta," "Medea," "Vid′ma," "Teklia") is also not coincidental.[44]

What becomes evident from the larger trajectory of Krawciw's life is that this poetry was accompanied and enabled by the inspiring afterglow of Rilke's great poetry, with which for a while the poet had became one; it was a poetry sheltered, so to speak, in the lee of that great ship.[45] With it gone, when no longer in that company—more concretely, perhaps, without the awesome, all-demanding task of meeting its sublime standards and, above all, its universal addressee—the world of "Blud" and "Neryst′" took over. Poetry was left to fend for itself, to survive as best it could in an ever more inhospitable climate. The *Hlosarii* of the time of its writing did, of course, steel the author for much of the work that was still to come—it was, after all, a *profession de foi*—but not for poetry; it could not maintain that temper under new circumstances, especially not if the poet himself had conceded his despondence, in effect, his defeat. The following collections only confirm this.

A year or two into his stay in the United States, Krawciw published a collection of poetry, *Zymozelen′* (Wintergreen; Philadelphia, 1951), that was partially an echo of *Korabli* (of its fifteeen sonnets, one had appeared in *Korabli* as well as six of its ten alexandrines) and partially a brief attempt at accessing the poetry of *Hlosarii* that was to see light only some twenty-three years later. Notably, too, the poems "Blud" and "Neryst′" are both first published here—and nothing so reflects the mean-spirited and tawdry circumstances of émigré life, and the poet's despondency within it, as do these baleful pictures. That both poems, and the collection itself, were published by the very press (the émigré Catholic and clericalist newspaper *Ameryka*) where Krawciw was then working for coolie wages also provides some measure of poetic justice (although one can hardly assume that the publishers read the allusion). Thus, "Neryst′":

Усотаний у твань і баговиння
мізерних слів і мислей цвіль і слизь,
в безвивідь буднів грузнеш. Глухо скрізь
і меркнуть блідо крижаним промінням

Колишні зорі. Сміття і плавиння
збирається на плесах і в розріз
йдуть ночі з днями і недобра гризь
нам роз'їдає дружбу і сумління.

Пручаєшся, в обридлій рвешся сіті,
як визволу, немов рятунку ждавши
ковтка ясного, чистого повітря.

Та всюди глуш і тиша мертва в світі,
бридня зненависти й підозри завжди—
і ні звідкіль «Ні хвилі ані вітру»...[46]

Sucked by the swamp and mire
of empty words and moldy, sleazy thoughts
you sink into the daily grind's dead end.
No voice is heard and shining on with dim and icy
rays

are former stars. Rubbish and flotsam
gathers on the surface and night opposes day
and vicious squabbles eat away
at trust and conscience.

You claw and struggle at the loathsome nets
And long as if for freedom
for a gulp of air that's bright and pure.

But deadly silence rules the world
and ugly spite and always the mistrust
and from no quarter can be heard "the sound of
waves or wind…"[47]

Krawciw's next collection, *Dzvenyslava* (Jersey City, 1962), takes a very different tack—although the course itself was already signaled in 1949; that is, in the dedication to *Hlosarii*. In effect, it is a full retreat from a harsh external reality into the intimacy and safety of family poetry as the fifteen sonnets of the wreath (sonnet redoublé) are dedicated to his daughter on the occasion of her wedding.[48] While technically accomplished, the tenor and fabric of this poetry, its circumscribed circle of addressees, its essentially laudatory and sentimental mode, sets it fully apart from the earlier public, "high" poetry—and to judge it by the criteria of the latter is to misunderstand both genres and modes. To place them side by side (and indeed to place the family poetry before *Hlosarii*, on the basis simply of publication dates—as both the editor of *Zibrani tvory* and his Lviv imitator do) is to misread and distort the poet's legacy.[49]

The verse that Krawciw will continue to write will basically conform to this new, circumscribed key—some more obviously, some less so; some of it will appear still during his lifetime, for example, the selection "Osinni stansy" (Autumn Stanzas), dated Fall 1972–Spring 1973 and published by *Suchasnist´* in its July/August 1974 issue, or the "Novi stansy" (New Stanzas) dated July–November 1974 and published by the same journal (of which Krawciw was also editor for a time) in its January 1975 issue;[50] some will await inclusion in the posthumous first volume of *Zibrani tvory*. The collection now titled *Kvitolit* (Blossom Year) on the basis of a conjecture by Bohdan Boychuk, the editor of *Zibrani tvory*,[51] was named "Strofy i stansy" (Strophes and Stanzas) by Krawciw (a more modest and traditional, or perhaps simply retrospective title, but also hardly as precious as "kvitolit") and was submitted to *Suchasnist´* for publication in 1974, shortly before his death.[52] The other collections were posthumously reconstructed by Boychuk from Krawciw's archive: "Tavtorymy" (Tautological Rhymes), a second part of "Stansy" (Stanzas) written in July–November 1974, a collection entitled by Krawciw "Osinni strofy" (Autumn Strophes—not the same as the published "Autumn Stanzas," but also not essentially different), "Virshi dlia Nil´" (Verses for Nil´—that is, for his wife), several pages of fragments, and a small selection of explicitly family verses ("Rodynni virshi").[53]

This not inconsiderable late corpus, from the sonnet wreath *Dzvenyslava* of 1962 to the very last poetry he wrote in 1975—in effect, *more than half* of Krawciw's total poetic output (!)—does call for a few concluding comments. For one, as already noted, it is framed in its preponderance as family poetry or verse, and the category Boychuk entered only at the very end of the first volume of *Zibrani tvory*—that is, that of "Rodynni virshi," could easily and with good conscience, have included all of it; the benefit of such a discrimination would have only worked to the advantage of the corpus and the reputation of Krawciw as poet. Some further qualifications are also in order. The "Tavtorymy," for example, are not expressly family verse. These are distichs whose sole point and effect is either the rhyming of the same word with a different meaning, or a play with homophony. Their effect is mostly ironic and comical, at times caustic and expressly satiric, with émigré reality and pretensions the butt of the joke—but they do not purport to be serious poetry. For example, "Parubok" (The Bachelor):

Жінок мав безліч у забав бездонні—
І старість мав: без сина і без доні.[54]

He had innumerable women and bottomless
pleasures—
And an old age without son or daughter.

or "Kholuistvo" (Sycophancy):

У виструнчені рівно ставши ряди
Ревли вождеві враз: Наказуй і ряди.[55]

Lining up stiffly in straight ranks
They roared as a man to their leader: Command us
and rule.

This is the humor that one expects to find on the last page of the official paper of the Union of Ukrainian Writers in Kyiv, *Literaturna Ukraïna* (Literary Ukraine); arguably, this is also the kind of humor one can share around the family dinner table. Krawciw wrote over two hundred of such *tavtorymy*; clearly, he found the wordplay interesting. And yet the avatar of this play can also be found in the seminal Rilkean period, as in his translation of Rilke's concluding "Schluss Stück" (Coda) from the cycle *Die Stimmen* (Voices) in *Das Buch der Bilder*, 1906:

Der Tod ist gross.
Wir sind die Seinen
Lachenden Munds.
Wenn wir uns mitten im Leben meinen,
wagt er zu weinen
mitten in uns.

Death is great
We are in his keep
Laughing and whole.
When we feel deep
In life, he dares to weep
Deep in our soul.[56]

Thus Krawciw's "Kintsivka":

Смерть є велика.
В неї в полоні
ми і наш сміх.
В час, як живемо, здається нам—вповні,
смерть плачем повнить
нас же самих.[57]

Very different questions are posed by the strophes and stanzas of what is now called *Kvitolit* and in the "Stansy."[58] They basically constitute a kind of a diary, where, for example, the "Stansy" of *Kvitolit* are dated day by day for long stretches of time (i.e., from 1 October 1972 to 2 May 1973, and the "Stansy" (1974) are dated 21 June 1974 to 26 November 1974, with here and there some pauses or lacunae. This exercise of writing a poem a day that seems to echo Stendhal's injunction to write "twenty lines a day, genius or not" was realized to high effect by the American Oulipian writer Harry Mathews, who saw it "as a method for overcoming the anxiety of the blank page."[59] For Krawciw it may have had a similar purpose, given his silence for long periods, and his complaint (as in the postscript to *Hlosarii*) of never having the time to write poetry. The poems themselves are not merely rote—some are indeed interesting—but they do not contain the high poetry that Ostap Tarnavs´kyi espies there.[60] Tellingly, Krawciw himself calls attention to their retrospective, autumnal, and sentimental nature, for example in "Uroky" (Lessons):

В далекий ранок, що будив нас півнями,
Поривом юности виводив до воріт
І мчав на хвилях річкової піняви—
Керуєм, без упину пізніх мислей літ.

У спогади вникаємо знеможено,
Сантиментальні вірші пишемо й листи,
Пожовклим листям бродимо—й неможемо
У круг той чарівливий стежки ізнайти.[61]

15 листопада 1972

Into that distant morning that awoke us with
cockcrows
and with youth's stirrings led us to the gates
and sped us on the waves of foaming rivers
we ceaselessly direct the flight of our late thoughts.

Enervated we sink into our memories
and write our sentimental verse and letters
and traipse through withered leaves—and never
find the magic center of our path.

15 November 1972

Although articulated near the end of his life, this positioning of his poetry within a space of limits, doubt, and fatedness (*pryrechenist´*) harkens back to feelings expressed decades earlier in "Blud" and "Neryst´." Presumably, it is precisely to counter them that he finally decides to resurrect *Hlosarii*—fully commented or not. External causes may have also cast a shadow. One of them may have been the already noted article by the leading critic of that period and in the late 1940s, the de facto ideologue and guru of MUR, Iurii Sherekh (Shevelov). In a retrospective overview and assessment of the literature written in the emigration in 1945–49—which is also the first history of MUR and one of the several apologias for it that he was to write—he discusses the various poets of the period and at the very end also turns his attention to Krawciw.[62] His assessment is not expressly negative—he calls Krawciw's poetry both interesting and promising; and even as he notes that Krawciw is influenced by Maksym Ryl´s´kyi (for many, an exemplary neoclassicist), he also claims that Krawciw surpasses him in this modality.[63] But Sherekh's larger assessment of neoclassicism as a spent phase, as something invariably imitative and epigonic, is also prominently asserted in this and various other articles of this time.[64] Probably more discouraging in this overview was Sherekh's readiness to pair Krawciw with Iar Slavutych—in fact to discuss him only after he deals with the latter. And yet for anyone who knew poetry, and Sherekh certainly did (and indeed even tried his hand at translating Rilke), it was clear that Slavutych was a hack, a *grafoman*. The implied dismissal of Krawciw's poetry is hardly concealed.

The other external cause was more objective and harder to ignore. This was the poetic context itself: in effect, the emergence in the mid-1950s of the so-called New York Group of poets—Bohdan Boychuk, Bohdan Rubchak, Yuriy Tarnawsky, Emma Andiyevska, and a few others—all of whom Krawciw knew, and supported in his various capacities as editor and literary elder, and whose poetry, by all indications, he genuinely admired. Their reciprocal esteem was reflected (implicitly) in their collaboration in the journal *Suchasnist'*, and later, more directly, in the anthology *Koordynaty*, and then explicitly in the posthumous *Zibrani tvory*. Mutual esteem and affection aside, neither he nor his younger fellow poets could ignore the sea change occurring in Ukrainian émigré poetry that dramatically separated the past from the present. In the face of their unalloyed modernism, their openness to urban and erotic themes, a new heady intertextuality, and a cosmopolitanism free of émigré complexes—and indeed, for the most part, of the experience of trauma—not only his poetry, but that of more established colleagues like Malaniuk, with their poetics of interwar neoclassicism or of (*Visnyk*-style) national pathos, was rapidly being consigned to history. Becoming an anachronism, suffering Teklia's fate (to harken back to his and Franko's phobias), was not something that Krawciw was disposed to accept. With a decisiveness that from the distance of time may appear prescient, but which in reality was more a function of survival and a sense of duty mixed with unending demands on him, he was already embracing a different role.

In the interwar period Krawciw had been no stranger to journalistic and editorial work. His reentry into this mode in the last third of his life, from the 1950s to the early 1970s, is accompanied by both new range and depth, as well as some new departures. While virtually none of his prewar critical and journalistic writings are now readily available, many (if not all) of his later writings are, and while they show, as does his poetry, eddies, detours, and recapitulation, they also demonstrate a progressive movement from journalism and various quotidian publications to ever more insightful criticism and finally moments of significant scholarship. This facet of his life's work—which already occupies two volumes of *Zibrani tvory*, but could probably add two more—has received much less attention than the poetry, and the commentary that has appeared has been superficial.[65] The result is not only ignorance about a major facet of Krawciw's life's work, but also a blank spot in the cultural history of the diaspora—which, as future events would show, became a major factor in Ukraine's political revival in the late twentieth century. Krawciw's criticism was a major voice in this camp and a clear influence on the discourse of the revival of the 1970s and 1980s, and of post-independence Ukraine; and it is impressive both for its range and its intense pace.

Just a few months after arriving in the United States in October 1949, Krawciw was at work on an article for a commemorative edition of the *Igor Tale*, the *Slovo o polku Igoreve*, on the occasion of the 150th anniversary of its publication in St. Petersburg in 1800.[66] In keeping with the tenor of the whole, his "Mitolohichnyi svit *Slova o Polku Ihorevi*" (The Mythological World of the Tale of Prince Ihor's Campaign) presents a popular, but cogent exposition of the topic, reflecting the perspectives and the state of knowledge at the time. The problem here is not just that this edition, one of the first such projects of the "new emigration," had a political agenda: to show the flag, and present the "Ukrainian epic of the twelfth century" in defiance of the Russocentric and Soviet-backed world of slavistics where the *Igor Tale*, by the very nature of things, was simply a "Russian" or at best an "Old Russian" monument. The deeper, all-but-metaphysical problem was that from the outset, whether for the Russians or the Ukrainians, the *Igor Tale* was first and foremost a national project, indeed a paradigmatic "national text" (and thus, by its nature, a national mystification), where collective validation and teleology—and with them antiquity, heroism, the canonically weighty Epos, and so on—are the desired value and effect, and the traditional scholarship that surrounds it is invariably contingent, that is, designed to validate these goals and values. In short, its task was to assert and not skeptically assess the claim of authenticity; in essence it was to be a form of window dressing.[67] In this respect this jubilee edition is comparable to the various other popular-scholarly editions published by the Soviet and then the post-Soviet Russian and Ukrainian literary establishments. As in all of them, the *Igor Tale* is presented *sub speciae* of its purported patriotic message—as is explicitly argued by the editor of the project, the artist, poet, and translator Sviatoslav Hordynsky.[68] Krawciw's own effort at placing the various, often murky and disputed references to pagan deities in the context of Indo-European mythology draws on a broad range of sources. As a popular and somewhat belletristic account it is as informative as it is entertaining; its synthesizing and speculative thrust follows in the footsteps of such prominent scholars as Golubinskii, Hrushevs´kyi, Vernads´kyi, Jakobson, and others (many of whom were also in thrall to the *Igor Tale*). And the overall frame and pathos of the mythological draws, of course, on Krawciw's earlier work on *Hlosarii*.[69]

Two major contributions that Krawciw undertakes in the course of the following decade move in a very different direction—the reconstruction not of an aestheticized, very distant, and largely imagined poetic past, but one that was real, recent, and wrenchingly traumatic; in effect, the period of totalitarian terror in the first half of the twentieth century. The anthology *Obirvani struny* (Broken Strings), published in 1955 jointly by the Shevchenko Scientific Society and the Ukrainian Free Academy of Sciences in the United States,[70] constitutes perhaps the first such anthology commemorating the Ukrainian poets who perished in this period, principally at the hands of the Soviet regime.[71] It precedes by four years the more famous and influential anthology *Rozstriliane vidrodzhennia* (Executed Renaissance) of Iurii Lavrinenko[72] and by just one year the groundbreaking study of George S. N. Luckyj (Iurii Luts´kyi), *Literary Politics in the Soviet Ukraine, 1917–1934* (1956), and is one of several works that in the 1950s and 1960s initiate the process of studying and documenting the Soviet depradations against Ukrainian literature and culture and which in turn played a major role in stimulating the birth of modern Ukrainian studies in the U.S. and Canada in the late 1960s and 1970s.[73] In comparison with *Rozstriliane vidrodzhennia*, Krawciw's anthology is clearly the more rudimentary: its scope is narrower (focusing only on poetry), while at the same time its time frame is more diffuse (covering the Soviet period up to the end of World War II); its bibliographic apparatus and its notes are more modest. Its conceptualization of its task and purpose is more openly martyrological (although that mode also defines Lavrinenko's work).[74] At the same time, its broader focus allows the author to include writers who might have otherwise remained forgotten.

Bearing witness to the past and to the victims of unprecedented oppression, and then ceaselessly responding to Soviet efforts to conceal the crimes and rewrite history becomes, in short, a central and defining task for Krawciw for the next decade and more. In *Na bahrianomu koni revoliutsiï* (On the Crimson Horse of Revolution, 1960), a small book published by the publishing house Proloh and intended for dissemination within Soviet Ukraine, he provides in a series of biographical and statistical sketches an overview of the hundreds of writers murdered or imprisoned and a whole literature radically censored and altered, and ultimately eviscerated and lobotomized. But bearing witness and documenting crimes and depradations soon evolves into a larger concern for taking stock of cultural and historical processes and preparing the groundwork for future efforts. Beginning thus with *Obirvani struny* and *Na bahrianomu koni revoliutsiï*, Krawciw continues throughout the 1960s and early 1970s to write article after article, sometimes two or more a year, where virtually every major facet of Ukrainian literary and intellectual life under the Soviet regime is examined and discussed—for example, the state of Ukrainian studies relating to the early period of Ukrainian literature ("Udar v osnovu," 1961); the state of Ukrainian bibliography ("Bibliohrafiia u vykryvlenomu dzerkali," also 1961); Ukrainian literature in the Soviet Ukrainian encyclopedia, 1962; the Fifth Convention of the Ukrainian Union of Writers, 1967; and the Sixth, in 1971; the state of Russification in Ukraine, 1972; the rolling-back of the period of the Thaw ("Pid praporom pohromnytstva," 1973); and many others.[75]

As much as they draw their polemical edge from the cold-war setting and the collective experience and trauma of the Ukrainian emigration, Krawciw's writings are also progressively moving toward a new understanding of the tasks ahead. Significant in this connection is his focus on those developments in Ukrainian society under the Soviets that serve as harbingers of change and the promise of rebirth: de-Stalinization and the Khrushchev Thaw, and above all the new Ukrainian poetry of the 1960s, the so-called *shistdesiatnyky*. To the latter he devotes both a number of articles and two anthologies: *Poety chumats´koho shliakhu* (1962), and *Shistdesiat´ poetiv shistdesiatykh rokiv* (1967).[76] Individual poets also draw his attention, prominently among them Ivan Drach, the then leading (most often discussed, and "most promising") young Ukrainian poet (see "Protuberantsi sertsia i kredo Ivana Dracha"). At times, the very fact of reprinting writers who were destroyed and works that were censored—as in the case of Hryhorii Kosynka and particularly the poem "Sliptsi" (The Blind Bards) of the outstanding twentieth-century Ukrainian poet Mykola Bazhan—becomes in itself a contribution to the new, emerging canon of Ukrainian literature.[77]

Throughout this period Krawciw was also contributing numerous articles on Ukrainian literature and culture to the *Entsyklopediia ukraïnoznavstva* edited by Volodymyr Kubiiovych and published by the Shevchenko Scientific Society (1949–), and later also contributing to and serving as editor of the English language *Ukraine: A Concise Encyclopaedia* (1963–71).[78]

A major focus of Bohdan Krawciw's literary criticism was the national poet, Taras Shevchenko (1814–61). He addressed this topic with great intensity in the early 1960s, as he worked on a major émigré publication project—the fourteen-volume complete edition of Shevchenko's works that appeared in the U.S. in 1959–63.[79] While based largely on the prewar Warsaw edition of Shevchenko's works (1934–39) under the editorship of Pavlo Zaitsev, along with Oleksandr Lotots´kyi and Roman Smal´-Stots´kyi, and following its overall conceptualization of Shevchenko's legacy and of Shevchenko studies, its methodology, and even design and layout,[80] the new edition did attempt to include new studies and to update its research—and, above all, sought to counter the accumulating and aggressive distortions, censorship, and propaganda in Soviet editions of Shevchenko, including academic editions. In a larger sense, of course, it was also a project for maintaining and projecting national identity—even in the face of, indeed especially because of, massive political adversity. Krawciw's role in this effort was significant. Apart from contributing a number of articles,[81] he was editor of three of the edition's fourteen volumes: volume 4, the poetry of the years 1857–61 as well as variants of earlier poems (1962); volume 12, which is an anthology of Shevchenko's poetry translated into various languages, much of which goes beyond the original Warsaw edition of 1938 (1963); volume 13, which is an anthology of critical writings on Shevchenko, from Panteleimon Kulish to contemporary non-Soviet criticism (1963); and, finally, he authored the introduction to volume 11 (also 1963), which includes studies by Dmytro Antonovych and Pavlo Zajtsev on Shevchenko's art. During this time Krawciw also published in the journal *Suchasnist´* several articles on the Soviet treatment of Shevchenko and Shevchenko studies that combine, as in his other critical writings in this period, fundamental bibliographic and source studies with an indictment of the Soviet destruction of Ukrainian scholarship in general and of this field in particular.[82]

What remains unknown, however, is that Krawciw also worked on Shevchenko with the same intensity in the interwar and into the war period, from the late 1920s to the early 1940s. In a large manuscript folder (actually the book bindings of a German-Polish dictionary) that is part of the Krawciw archive at the Ukrainian Research Institute there is material, all of it handwritten in an extremely dense hand on longitudinal (4" x 13") folded strips of poor (now totally yellowed and brittle) paper, most often utilizing the back of various official blanks (most often of the Polish railroad office in Stanisławów)[83] that shows work on a large Shevchenko project, by all indications a book, which includes a biographical sketch of Shevchenko's life, "Zhyttia Tarasa Shevchenka" (which in one of its two drafts is subtitled "U 120-litni rokovyny narodzhennia"—that is, referring to 1934); sections and voluminous commentaries on many of the major poems, numerous extracts (presumably for future citation) from critical studies; material for an index of historical figures, places, and so on; drafts and plans for dealing with subsections (e.g., Shevchenko's historical poems) and so on. Included among this material are notes for a talk or speech on Shevchenko that are not dated, but which from internal evidence, that is, quotations from a well-known poem, "Shevchenko" by Ievhen Malaniuk (1930), also suggests the period of the 1930s. The book or monograph on Shevchenko never materialized, not only because of the vagaries and instability of the times during which it was being worked on, but also, one may assume, from a shifting sense of the addressee: from an initial popular exposition and an emphasis on national pathos to ever greater scholarly contextualization and nuance. One may also assume that much of the pasion and many of the insights accumulated there were accessed again in Krawciw's Shevchenkiana of the early 1960s.

Bohdan Krawciw's criticism during the last period of his life, from the late 1960s to his death in 1975, shows a clear turn to scholarship as against publicistic work and to a fundamentally analytical, intellectual, and ultimately academic frame of reference—even while not abandoning his basic engagement with crucial, national issues. This also corresponds, as was noted at the outset, to his growing involvement with the Ukrainian Research Institute established at Harvard in 1972 (HURI), and specifically his work there as a research fellow in 1973–74. To be sure, an analytical and scholarly frame and a dispassionate, or at least not principally polemical, cast to his criticism was evident already in a number of earlier studies—in the various Shevchenko articles of the early 1960s, in his overview of the critical literature devoted to Iurii Fed´kovych, or his introduction to Leonid Mosendz's novel *Ostannii prorok* (The Last Prophet; 1961) in which he directly addressed the issue of the moral vacuum at the core of the nationalist leadership.[84] But in that period (the 1950s and early 1960s), and even more so before that, his writings were principally polemical and often drew on pathos and on rhetorical effect. Now there is a palpable shift as the latter mode is basically abandoned—suggesting the emergence of a different discourse and a different reader. In effect, in the space of a decade, beginning with the appearance of *Suchasnist´* in 1961, or slightly more, if one focuses on the student-led campaign that was initiated in the mid-1950s to fund a chair of Ukrainian studies at a leading American university, the tenor of Ukrainian émigré life was changing and an audience attuned to new intellectual standards and a much more sophisticated and optimistic understanding of the issues facing Ukrainians both in Ukraine and the diaspora was rapidly asserting itself. The enabling sociological and demographic factors aside, there was also the basic ability and readiness on the part of Krawciw to engage in this new intellectual discourse. The works produced in this relatively short space of time—the already mentioned publication of Mykola Bazhan's magnificent and all-but-forgotten poem *Sliptsi* (1969); a year earlier his retrospective study of the poetry of Ievhen Malaniuk, where the lyrical and not the civic and bombastic are at center stage; a similarly focused study of the lyrical poetry of Ivan Franko (which was published posthumously, but which by all indications is also from this period); his excellent study of the Renaissance in Ukrainian literature (1974), and (anticipating our final frame and the subject of the present book) his equally excellent overview of Guillaume le Vasseur, Sieur de Beauplan in Ukrainian historiography (also published posthumously in 1977), and last, but certainly not least, his editing (with a final note) of Volodymyr Svidzins′kyi's *Medobir* (1975)[85]—are among the best criticism that Krawciw was to write and fit comfortably within any short list of the better work done in the area of Ukrainian studies of that entire period.

Krawciw's critical strength then and before was in scrupulous source studies and bibliographic work entirely open to the comparative frame—to which he added a fine ear for literature, particularly poetry. His last works, however, are also informed by an openness to revisionism, the first promise of which we can already see in his 1961 introduction to Mosendz's *Ostannii prorok*. Its sequel, so to speak, is his introduction to Mykhailo Sosnovs′kyi's fundamentally revisionist *Dmytro Dontsov: politychnyi portret* (Dmytro Donzow [*sic*]: A Political Portrait), where the study of the chief ideologue of Ukrainian interwar nationalism is presented as something that must be subjected to objective scrutiny and not obscurantism or mythmaking.[86] His essay on literary criticism in the Ukrainian diaspora, "Literaturnoznavstvo i literaturna krytyka v diiaspori," which was first presented as a paper at the Seminar in Ukrainian Studies at HURI in December 1972,[87] is openly critical of the remnants of the earlier nationalist establishment and actively anticipates and welcomes a new beginning, from new quarters.

For Krawciw, too, this remained not only a hope, but a course of action, an investment in the future—as he willed his remarkable library and archive to Harvard's library and to HURI. Several generations of students were to become beneficiaries of his foresight and generosity. I remember to this day (a phrase used before) the summer day in 1974, in Rutherford, New Jersey, when Bohdan Krawciw presented me with a rare 1940 copy of the Lviv edition of the poetry of Svidzins′kyi—which despite the book's minute size (32 mo) immediately expanded for me severalfold the available poetic corpus of this rare poet. (*Medobir* came a year later.) This, too, was an investment in the future, for the recent complete edition of the poetry of Volodymyr Svidzins′kyi was in no small measure generated by that generous gesture.[88]

That Krawciw was also a passionate, driven map collector, and that his unique, and basically priceless collection, assembled over the course of his entire life in the New World, would now find a fitting home at Harvard University should also come as no surprise. The perspectives provided by his poetry and his criticism bring that commitment—to tirelessly search and collect, and ultimately to will it to serve a larger audience and purpose—into clear focus. Collecting the multiform cartographic representations of Ukraine became a form of naming and reconstituting his homeland, image by image. The bibliographer and scholar in him saw further references to the object of his life's study. But they were also things of beauty, complete in themselves, both signs and icons. Like the rare words in his masterful *Hlosarii*, researched with much effort but also contemplated with a sense of aesthetic distance, like Rilke's transcendent images and things, the maps of Ukraine, from so many different periods and perspectives, became emanations, constantly recurring objective correlatives of a transcendent vision of his country, reconstituted in discrete images and texts, always recurring and never exhausted. There were as many maps as there were perspectives—and history had seen to it that these would be many. Even such that seemed to deny that Ukraine was there. But it was there, and the activity of collecting was a constant search for a lost homeland for which the discovery and acquisition of each map was visible, ineffable evidence that it existed—and gave tangible proof to his poetic and critical intuitions. As with the poet's drive to create and the critic's need to analyze and redress the existing state of affairs, the collector in Krawciw could not but reach out and acquire each new and surprising map of his Ukraine—and by making it his own, to reinforce the bond between them. At the end, cognizant of his impermanence, his fatedness, he made sure that it would be passed on so that in the sweep of the whole collection others could also share his vision.

Milton, Mass.
9 May 2010

Notes

1. Bohdan Krawciw, *Poeziia*, vol. 1 of *Zibrani tvory* (New York: The New York Group, 1978); *Statti*; vol. 2 of *Zibrani tvory* (New York: The New York Group, 1980); *Publitsystychna proza*, vol. 3 of *Zibrani tvory* (Kyiv: Svitovyd, 1994).
2. See Rainier Maria Rilke, *Rechi i obrazy: vybrani poezii*, trans. Bohdan Krawciw (Nuremberg: Chas, 1947).
3. Bohdan Krawciw, *Poezii* (Lviv: Feniks, 1993), compiled by Taras Salyha. The collection also includes Krawciw's translation of the *Song of Songs*, which had first appeared in full in 1935; see 405–29. Given his later aggressive championing of the nationalist slant of interwar Ukrainian literature, Salyha's treatment of Krawciw is still quite restrained here, presumably reflecting the slow dissipation of Soviet attitudes among the newly converted literati. In recent Ukrainian writings on Krawciw, particularly on the internet, the nationalist claim on him is quite evident.
4. Bohdan Boychuk, ed., *Koordynaty: antolohiia suchasnoï ukraïns′koï literatury na Zakhodi*, vol. 1 (Suchasnist′, 1969–), 265–76. See also Ostap Tarnavs′kyi, "Poetychnyi shliakh Bohdana Kravtseva," *Suchasnist′*, no. 11 (1977): 18–39.
5. His output is certainly much smaller than the prodigious production of a Vasyl′ Barka or an Emma Andiyevska, but is certainly considerably larger than that of such contemporaries like Oleh Olzhych, Olena Teliha, or Oksana Liaturyns′ka.
6. Tarnavs′kyi ("Poetychnyi shliakh Bohdana Kravtseva," 28) notes that he was editor of such nationalist newspapers and journals as *Visti*, *Holos natsiï*, and *Holos*.
7. Krawciw, *Poeziia*, 27–28 and 49–50.
8. Just a few years earlier, in 1925, in the diptych "Suchasnykam" (To my contemporaries) the soon to become preeminent émigré poet Ievhen Malaniuk had cast these two, Ryl′s′kyi and Tychyna, as not just opposite, but totally incommensurate entities: with Ryl′s′kyi a poet of culture who finds the inner strength to resist the Soviet system, and Tychyna the poet of genius, indeed the national poet, who betrays his calling and allows himself to become a witless tool of deadly totalitarianism. That Krawciw invokes both poets without any apparent recourse to this judgment is perhaps revealing; whether it signals his belief in poetry free of such ideological judgments and strictures is another matter.
9. Dontsov was editor of the Lviv-based *Literaturno-naukovyi vistnyk* from 1922 to 1932, when it became the programmatically nationalist *Visnyk,* which he then edited until 1939—that is, until the outbreak of the war.
10. In the Salyha edition, *Poezii*, the whole cycle is so named; in *Poeziia* the cycle is not named.
11. Krawciw may have also felt that basic understanding, let alone sympathy, would be in short supply. Malaniuk's brief (one-and-a-half page) review of *Sonety i strofy* (*Visnyk* 2, no. 1 [January 1934]: 71–72) is a case in point. It has virtually nothing to say of the merits, strategies, or texture of this poetry. The existential fact that it was, after all, prison poetry is glossed by the comment that this is "a high tradition in our literature," adding then the parenthetical "(Franko)" (p. 71). All of the concrete, textual observations are quibbles about particular words—and the rest are generalities about "internal energy" and "self-restraint." The purportedly positive is couched in equivocation ("The intended perfection [*navmysna doskonalist′*] of Krawciw's last collection might give rise to certain reservations"; 71) and the recurring notion that the poet "does not rest on his laurels" basically damns with faint praise. For Salyha this constitutes perceptive and positive criticism—coming as it does from the mouthpiece of the nationalist camp (Krawciw, *Poezii*, 17–18). What is plain, however, is not only that this is niggardly praise, but that it cloaks what cannot but be a gnawing aporia: how is it that "minor" nationalist poets spend time in prison while the "major" ones, the all-but-official mouthpieces, remain untouched? (The charge that Dontsov actually had a working relationship, or "nonaggression pact," with the Polish security services—the avowed enemy of the nationalist camp—was indeed later voiced by nationalist spokesmen.)
12. Ievhen Konovalets′ was assassinated by a Soviet agent in Rotterdam in 1938.
13. In terms of its organization the group was altogether tenuous, and the poetry of its "members" differed widely; in terms of their impact the neoclassicists—Mykola Zerov, Maksym Ryl′s′kyi, Pavlo Fylypovych, Mykhailo Drai-Khmara, and Klen himself—had a massive impact on twentieth-century Ukrainian poetry, especially in the non- and then the post-Soviet frames; see below.
14. See Iurii Klen, *Prokliati roky* (Lviv: Nakladom "Vistnyka," 1937).
15. Krawciw, *Poeziia*, 107. Krawciw's contacts with Klen and his reliance on Klen's help in publishing the Rilke translations is attested in the postscript to them; see below.
16. See George G. Grabowicz, "A Great Literature," in *The Refugee Experience: Ukrainian Displaced Persons after World War II*, ed. Wsevolod W. Isajiw, Yury Boshyk, and Roman Senkus, 240–68 (Edmonton: Canadian Institute of Ukrainian Studies, 1992).
17. Bohdan Krawciw, *Korabli: vybrani poezii, 1922–1947* (Beyreuth: n.p., 1948).

18. Bohdan Krawciw, *Hlosarii: abo, Tlumachnyi slovnyk taiiemnykh, pryzabutykh i ne zavzhdy zrozumilykh sliv* (Rutherford, N.J.: Suchasnist′, 1974).
19. Rilke, *Rechi i obrazy.*
20. Ibid., 119.
21. Ibid.
22. Ibid., 44 and 119.
23. Rainer Maria Rilke, *Buch der Bilder*, trans. Edward Snow (San Francisco: North Point Press, 1991).
24. Rilke, *Rechi i obrazy*, 123.
25. Ibid.
26. Rainer Maria Rilke, *The Best of Rilke: 72 Form-true Verse Translations with Facing Originals, Commentary, and Compact Biography*, trans. Walter Arndt (Hanover, N.H.: Dartmouth College Press, 1989), 83.
27. See, e.g., his condemnations of Tychyna in "Suchasnykam" (1924).
28. Cf. Silke Pasewalck, *"Die fünffingrige Hand": die Bedeutung der sinnlichen Wahrnehmung beim späten Rilke* (Berlin: De Gruyter, 2002).
29. Cf. Rainer Maria Rilke, *Temni plachi: poetychni tvory u dvokh tomakh* (Ternopil: Navchal′na knyha—Bohdan, 2007).
30. Rilke, *Rechi i obrazy*, 24–25. Rainer Maria Rilke, *New Poems*, sel. and trans. Edward Snow (New York: North Point Press, 2001), 95.
31. Rilke, *Rechi i obrazy*, 50–51, 54–55. Rilke, *New Poems*, 141, 145, 147.
32. See *Rainer Mariia Ril′ke i Ukraïna: naukovi studiï pro Rainera Mariiu Ril′ke* (Drohobych: Kolo, 2003–), where the theme of Rilke and Ukraine is revisited through several (early and mid-twentieth century) articles and where some older and more recent Ukrainian translations are also included. Paradoxically, while Krawciw's *Rechi i obrazy* is mentioned, it was not available to the editors and his translations are thus neither included nor discussed.
33. Krawciw, *Hlosarii*, [75].
34. Known in Hutsul folklore and appearing in Mykhailo Kotsiubyns′kyi's *Tini zabutykh predkiv* (Shadows of Forgotten Ancestors; 1911).
35. These references may be spurious. Thus, the conjecture (in the epigraph to "Chaika") that the author of the poem "Oi hore tii chaitsi, hore tii nebozi" [Woe to the poor seagull] was Hetman Ivan Mazepa is not contained in the *Istoriia Rusov* as Krawciw claims (what the *Istoriia* does contain is his speech to his countrymen on Ukraine torn between Muscovy and Sweden). The reference to Żegota Pauli, presumably his *Pieśni ludu ruskiego w Galicji* (Lviv: Nakł. Kajetana Jabłońskiego, 1839–40), is also spurious: the lines cited in the epigraph are from the song beginning "Uzhe Ivasenkovi v namastyren′ku zvoniat..." included as no. 17 in the "Dumy i dumky" section of *Rusalka dnistrovaia* (Budym, 1837), 32. (Cf. the facsimile edition, Kyiv, 1972.)
36. Krawciw, *Hlosarii*, [5]. The English translations of Krawciw's poems are my own—GGG.
37. There is also something of a tension here, for the inscription of the "intimate" Christmas present also invokes the altogether public name of his earlier collection of poetry, *Pid chuzhymy zoriamy* (Under Foreign Stars, 1940). See below.
38. As a leading member of Plast, the Ukrainian scouting organization in the 1920s (and beyond), Krawciw was probably present at the founding (in 1922) and certainly from the early years was a member of the *kurin′* (company) of older scouts (*starshi plastuny* and *seniory*) called the *Lisovi chorty* (Forest Devils).
39. Iurii Sherekh [George Shevelov], "Ukraïns′ka emigratsiina v Evropi, 1945–1949," in idem, *Porohy i zaporizhzhia: Literatura, mystetstvo, ideolohiï*, 1:196–235 (Kharkiv: Folio, 1998), 232–33. Sherekh is referring here to the collection *Korabli* (1948), which does have eight sonnets from the future *Hlosarii* (the cycle "Iz Hlosariia")—and these are, in fact, the only such "folkloric sonnets" in the collection. The whole of *Hlosarii* would only substantiate this judgment.
40. *Hlosarii*, 51.
41. Ibid., 40.
42. *Hlosarii*, 11.
43. Ibid., 7.
44. In the case of "Teklja" and "Medea" this is the horror of being buried alive. One may note here that Ivan Franko, whose immense involvement in civic and community matters tended to overwhelm his ability to find the time and psychic space to create poetry was also given to depicting this customary state of affairs as one of being buried alive; see my "Vozhdivstvo i rozdvoiennia: 'valenrodyzm' Franka," in *Teksty i masky*, 95–139 (Kyiv: Krytyka, 2005).
45. Krawciw's translation of Rilke's "Werkleute sind wir..." may serve as an emblem of this apprenticeship and inspiration. But Ostap Tarnavs′kyi is wrong (in effect, encomiastic) when he claims that all of Krawciw's poetry was basically Rilkean (see his "Poetychnyi shliakh Bohdana Kravtseva," 18–19 and passim). As we see, it was not all cut from that cloth.
46. Bohdan Krawciw, *Zymozelen′: sonety i oleksandryny* (Philadelphia: Amerika, 1951), 15.
47. As the obligatory epigraph makes clear, the citation, "The sound of waves or wind," i.e., "Ой нема, нема ні вітру, ні хвилі / Із нашої України" are the opening lines from Shevchenko's "Hamalia" (1842), which depicts the lament

of Cossack prisoners in Turkish captivity.

48. Note that *Dzvenyslava* (Jersey City, 1962) is a private edition, published by Krawciw himself ("накладом автора").
49. To be sure, in purely technical terms the form of the sonnet redoublé is of some interest. In this recherché mode Ostap Tarnavs´kyi preceded Krawciw by some ten years: see his *Zhyttia (vinok sonetiv)* (Philadelphia: Kyïv, 1952).
50. See *Suchasnist´*, no. 7–8 (163–64) (July–August 1974): 3–6, and no. 1 (169) (January 1975): 3–6.
51. Boychuk, to be sure, chooses "Kvitolit" from an earlier, draft version in Krawciw's manuscripts; Krawciw, *Poeziia*, 346. This practice is repeated in the 1993 Lviv edition (Krawciw, *Poezii*̈), thus making it altogether likely that this name will remain.
52. Krawciw, *Poeziia*.
53. Ibid., *passim*.
54. Ibid., 254.
55. Ibid., 268.
56. Rilke, *Best of Rilke*, 53.
57. Krawciw, *Rechi i obrazy*, 87.
58. Krawciw, *Poeziia*, 273–300.
59. Harry Mathews, *20 Lines a Day* (Elmwood Park, Ill.: Dalkey Archive Press, 1988), 3.
60. Tarnavs´kyi, "Poetychnyi shliakh Bohdana Kravtseva," 36–37.
61. Krawciw, *Poeziia*, 208.
62. Sherekh, "Ukraïns´ka emigratsiina literatura."
63. Thus: "Bohdan Krawciw (in the collection "Ships") developed this manner not without the influence, perhaps, of Maksym Ryl´s´kyi, but in this he is more consistent than his putative teacher"; ibid., 271.
64. See especially his "Legenda pro ukraïns´kyi neokliasytsyzm," in *Ne dlia ditei: literaturno-krytychni statti i eseï*, 97–156 ([New York]: Proloh, 1964), here 98–99 and passim.
65. Krawciw, *Statti*; Krawciw, *Publitsystychna proza*.
66. Sviatoslav Hordynsky, ed., *Slovo o polku Ihorevi: heroïchnyi epos XII viku* ([Philadelphia]: Kyïv, 1950).
67. See my "National Poets and National Mystifications," in *Literární mystifikace, etnické mýty a jejich úloha při formování národního vědomí: sborník příspěvků z mezinárodní konference konané ve dnech 20.–21.10.2001*, ed. Blanka Rašticová, 7–24. Studie Slováckého Muzea 6/2001 (Uherské Hradiště: Slovácké muzeum, 2001).
68. Thus see Sviatoslav Hordynsky's lead article "Slovo o polku Ihorevi iak literaturna pam'iatka," in *Slovo o polku Ihorevi*, 45–57, and its English version, 58–62.
69. Unfortunately, too, the editors of the *Zibrani tvory* give Krawciw's writings on this topic more prominence than they deserve in the context of his criticism, while possibly leaving out more important essays.
70. *Obirvani struny: Antolohiia poezii poliahlykh, rozstrilianykh, zamuchenykh i zaslanykh 1920–1945* (New York: Naukove t-vo im. Shevchenka v Amerytsi, 1955).
71. Of the forty-one poets included here, all but two were victims of the Soviets; the two exceptions, who died at the hands of the Nazis, are Oleh Olzhych and Olena Teliha.
72. Iurii Lavrinenko, *Rozstriliane vidrodzhennia: Antolohiia 1917–1933, Poeziia—Proza—Drama—Esei* (Paris: Instytut Literacki, 1959).
73. In his basically well-researched overview of the historiography of this topic, Serhii Bilokin´ does not mention *Obirvani struny* and in fact begins with Lavrinenko's *Rozstriliane vidrodzhennia*; Bilokin´, "Istoriohrafiia derzhavnoho teroru i teroryzmu v SRSR: problemy systematyzatsiï i periodyzatsiï. Povoienni zakordonni vydannia (1945–1991)," http://bilokin.myslenedrevo.com.ua/terror/idtt3.html (accessed 23 September 2010), n. 111. In fact, chronologically the first to broach this topic was Viktor Petrov (*Ukraïns´ki kul´turni diiachi 1920–1940: zhertvy bil´shovyts´koho teroru*, written in the late 1940s, but published only in 1959 by Proloh).
74. The fact that Krawciw does not attempt to provide a theoretical and historiographic frame for his work is all to the good: the requisite concepts and distance were not in place yet. Lavrinenko's ad hoc choice and conjectures to the contrary ("vitaïzm," "klarnetyzm," "neo-Baroque," etc., and above all the very paradigm of "rozstriliane vidrodzhennia") present more problems in the long run than they provide answers or insights.
75. Krawciw, *Publitsystychna proza*.
76. The introductory articles for each are included respectively in vol. 2 (*Statti*) and 3 (*Publitsystychna proza*) of his *Zibrani tvory*.
77. Mykola Bazhan, *Sliptsi: peredmova Bohdana Kravtseva* (Munich: Suchasnist´, 1969).
78. *Entsyklopediia ukraïnoznavstva* (Munich: Naukove t-vo im. Shevchenka, 1949–); *Ukraine: A Concise Encyclopaedia*, vols. 1–2 (Toronto: Ukrainian National Association, University of Toronto Press, 1963–71).
79. Taras Shevchenko, *Povne vydannia tvoriv*, 2nd ed., 14 vols. (Chicago: Mykola Denysiuk, 1959–63).
80. Taras Shevchenko, *Povne vydannia tvoriv Tarasa Shevchenka* (Warsaw: Ukraïns´kyi naukovyi instytut, 1934–39).
81. See Krawciw, *Statti*.
82. "'Kobzar' pid tsenzuroiu" (1961), "Shevchenkoznavstvo v sotsrealistychnii diisnosti" (1961), "Ostrakizm u shevchenkoznavchii bibliohrafiï" (1964), and "Dolia ukraïns´koho shevchenkoznavstva v URSR" in Krawciw, *Statti*, 151–56, 157–72, 173–90, and 191–205, respectively.
83. That is, "Polskie Koleje Państwowe—Dyrekcja w Stanisławowie."

84. Krawciw, *Statti*, 268–80 and 316–45.
85. Volodymyr Svidzins´kyi, *Medobir: poezii̇̈* ([Munich]: Suchasnist´, 1975).
86. Mykhailo Sosnovs´kyi, *Dmytro Dontsov: politychnyi portret* (New York: Trident, 1974), 111–13.
87. Krawciw, *Statti*, 467–88.
88. Volodymyr Svidzins´kyi, *Tvory u dvokh tomakh* (Kyiv: Krytyka, 2004).

Ukraine under Western Eyes

1 *European Renaissance Cartography and the Framing of Ukraine*

WELL AWARE OF THE REALITIES of political discontinuity in Ukraine's past, and that the ongoing contests for the naming and claiming of Rus´ had more than symbolic or academic significance, Bohdan Krawciw collected maps of the present-day (late twentieth-century) lands of the region dating back to the Renaissance in the fifteenth and sixteenth centuries.[1] His notebooks on the history of cartography indicate a detailed knowledge of variegated historical labels for Ukrainian lands—Rus´, Ruthenia, and Little Russia, for instance—and an attempt to offer an alternative to the Cold War geopolitical gradient of a Soviet bloc or Russia and Eastern Europe. Krawciw shared with other map enthusiasts and proponents of national self-determination in Central and Eastern Europe the view that historical maps validated the existence of polities.

Yet it is worth noting that European Renaissance cartographers treated Ukrainian borderlands as peripheral, if not altogether marginal, until after 1648. Not until Beauplan were European cartographers directly concerned with the mapping of Ukrainian lands for geopolitical purposes. Located between other empires, the Ukraine Krawciw traced in various Renaissance maps was imagined as one which, in a European and North American context, deserved cultural recognition and, ultimately, political independence. The early European maps in his collection reflected his ambition as an émigré scholar to "discover his country" in many languages and through differing historical trajectories, and thereby to achieve these twin goals.

Relative to Ukraine, European cartography during the fifteenth and sixteenth centuries was stimulated by two major events. First, the commercial and political competition among Italian city-states sparked a growing interest in the scientific-technical importance of maps for charting navigation routes around the Mediterranean Sea (Mare Mediterraneum, Mare Nostrum) and Black Sea (Pontus Euxinus). Second, the rise of the Ottomans along the northern Black Sea coast and its hinterland, and, ironically, the capture of Constantinople, had a surprisingly beneficial effect on humanistic scholarship in Europe. Byzantine manuscripts made their way to Italy in the fifteenth century, including most famously Ptolemy's *Geographia*.[2] With the revival of ancient Greek and Latin scholarship, and due largely to the geographical position of waterway passages for commerce, Renaissance maps generally followed along a north-south gradient, not one divided into halves of "Western" and "Eastern," or even of a landlocked "Central" Europe.[3] Medieval cosmographical conceptions of space such as Dante's *Commedia* yielded gradually to mathematical considerations of territory on land and by sea, based on astronomical points and scaled measurement. Nevertheless, medieval *mappae mundi*, or world maps, offered new and fascinating spatiotemporal compendia for Christian and secular cartographic arrangements of knowledge, the origins of which have been reconsidered recently by a series of monographs.[4] With perspectives shifting toward more worldly considerations such as the collection of taxes and commercial and military expansion, the Ukrainian lands became situated between at least five areas with changing borders: (1) the Habsburg lands and the Kingdom of Hungary; (2) Polish-Lithuanian Commonwealth; (3) Ottoman Empire; (4) Muscovy; and (5) the lands of the Golden Horde (often referred to in maps as "Tartary").

Framing the "Founding Fathers" of European Cartography

In the fifteenth century, the rediscovery of Ptolemy's *Geographia* and the works of Sebastian Münster, Abraham Ortelius, and Gerard Mercator played a key part in structuring a political and territorial framework for Renaissance and early modern Ukrainian lands.[5] As late medieval monastic scholars flocked from Byzantium carrying manuscripts, Renaissance humanists translated classical Greek and Latin works in astronomy, geography, and cartography. The most important of these was by Greek astronomer and mathematician Claudius Ptolemy, or Ptolemaeus (AD 127–80), who had labored in the library of Alexandria.[6] A surviving manuscript was brought to Italy by the Byzantine scholar Manuel Chrysoloras, and translated from Greek into Latin by his pupil Jacopo d'Angiolo (Jacobus Angelus) of Tuscany in 1406. A first edition was finally printed in Bologna 1475, and the first one with maps was crafted in 1477. Engraved atlases of the world were circulated in cities such as Florence, Basel, and Rome. Initially they were of mainly academic interest: five hundred copies of the *Geographia* were sold in Bologna in 1477. Some of the most famous editions with maps modeled on the *Geographia* were Martin Waldseemüller's edition with twenty new maps in 1513, Jacopo Gastaldi's miniature edition with sixty maps, printed in Venice in 1548, and Gerard Mercator's classical edition in 1578. This map, *ASIAE // SARMATIAM // ASIATICAM // REPRAESENTANS* (**figure 1, #403**), is based on Ptolemy's second table in the *Geographia*, reprinted in a later edition (1730) in Amsterdam.[7] It offers a vision of Ptolemy's conception of Sarmatia between Europe and Asia; lands encompassing Ukraine are included under the place names Sarmatiae Europae pars (Sarmatian Europe), Taurica Chersonesus (Crimea), Maeotis palus (Sea of Azov), and Ponti Euxini pars (Black Sea). The Sarmatian idea, and cultural and political interpretations of it, remains as a key legacy of the Ptolemaic revival in Renaissance cartography.[8]

The earliest of the "founding fathers" in the collection, Sebastian Münster (1488–1552), was a Dutch German theologian, born in Hesse and originally from a Dutch family. He became Professor of Hebrew in Freiburg, where he lectured on mathematics and cosmography. Münster is probably best known for his editions of Ptolemy's *Geographia*, and maps he made to accompany them while he lived in Basel. His new map of Poland and Hungary, *POLONIA ET VNGARIA XX -- NOVA TABVLA* (**figure 2, #374**), was based on Bernard Wapowski's map of Southern Sarmatia (1st ed. 1526, 2nd ed., 1528), and published in 1545 as part of his revised edition of Ptolemy's *Geographia*.[9] The Ukrainian borderlands are designated on the map as Rus´ ("Russia"), Volyn ("Volhinia"), Podillia ("Podolia"), Kodyma ("Codimia"), Pokutia ("Pocutze"), Besarabia ("Bessarabia"), White Rus ("Rvssia Alba"), Little Tartary ("Tartaria Minor"), and Perekop Tartary ("Tartaria Precopien[sis]").[10] Münster's *Cosmographia* was first produced in Basel in 1544, and continually expanded until his death in 1552, when it had an impressive 1,162 folio pages. The edition included old maps from Ptolemy's version in addition to new maps, including, for instance, sketches of Muscovy (Moscoviters Land) and Transylvania (Siebenberg). Although expensive to produce because of all the engravings, the *Cosmographia* became so popular that it appeared in over thirty separate German editions, also in Latin, French, Italian, German, English, and Czech editions. Münster's work was the final stage of the Ptolemaic revival; after the success of Gerard Mercator, older ancient and medieval non-scaled models of the European and Asian lands were generally regarded as outdated and inaccurate.[11]

In the mid- to late sixteenth century, the first comprehensive atlas compilation, *Theatrum Orbis Terrarum*, was arranged by Abraham Ortelius (1527–98). The monumental work was the first systematic collection of maps in uniform size, scale, and format made by nearly ninety different European cartographers.[12] Ortelius was a Dutch cartographer, cosmographer, and publisher who worked his whole life in Antwerp. His first original cartographic work was printed in 1567; he was known for maps of Asia and North Africa (Egypt) in addition to Europe. King Philip II of Spain appointed him to the newly created official position of Royal Geographer in 1575, placing him in charge of hydrographic (waterway) charts and colonial mapping projects. One of Ortelius' original maps from 1572, *Europae* (**figure 3, #387**), was printed in Antwerp as part of his *Théâtre de l'univers contenant les cartes de tout le monde*. Relative to Ukrainian lands, the map design includes a toponymy that designates "Russia" and "Podolia" in the west, and "Zevera," "Cumani," "Cabardi," and "Gazaria" (Crimea) in the east. In a later map of the Black Sea, *PONTVS // EVXINVS. // AEQUOR IASONIO PULSA- // TUM REMIGE PRIMUM* (**figure 4, #388**), which also appeared in a 1590 edition of the *Theatrum Orbis* published in Antwerp, Ortelius followed old Ptolemaic labels to designate the lands of southern Ukraine around the Black Sea: "Scythjae, sive Sarmatjae Evropaeae Pars," with "Tavrica, quae et Scythica Chersonesvs," and "Maeotis Palvs."

Around the same time, Gerard Mercator (1512–94)

invented what would become commonly known as the Mercator projection, an innovative idea for mapmaking that allowed navigators to plot the same constant compass bearing on a flat map. Mercator also developed a model for the production of a world atlas. Even his chief rival, the cartographer Abraham Ortelius, called him "the Ptolemy of our time." Born in East Flanders, Mercator was a polymath Dutch cartographer, publisher, engraver, and a land surveyor who studied mathematics and astronomy.[13] His first original work was a six-sheet map of Palestine, drawn in 1558. Among the most important maps he drew and engraved were maps of Flanders and the British Isles, in addition to large maps of Europe. In 1563, the German Duke Wilhelm V ("the Rich") of Jülich-Kleve-Berg, whose domains in the Netherlands and Central Europe were under the Habsburg Holy Roman Emperor Charles V after 1542, appointed him to the position of Royal Cosmographer. In 1569, Mercator produced his great world chart in eighteen sheets. Mercator spent the last three decades of his life at his workshop in Duisburg. In 1578 he conceived of a grandiose atlas, *Atlas sive Cosmographiae Meditationes de Fabrica Mundi et Fabricati Figura*, with one hundred original maps, as part of a five-part treatise on cosmography that would cover the entire world. Due in part to the shortage of engravers, he began to publish the maps in piecemeal form, and did not live to see the completion of the project.[14] The plates covering lands around Ukraine, including Poland-Lithuania, Muscovy/Rus′, and the Ottoman Empire, were published posthumously by successors such as Jodocus Hondius and Nicholaes Visscher (the younger) in various European languages. Mercator's 1670 map of the Ottoman Empire, *TVRCICVM // IMPERIVM* (**figure 5, #358**), was printed in Amsterdam and published in an atlas by Nicholaes Visscher. It includes lands in western Ukraine, labeled as "Russia," "Volinia," and "Podolia," and lands in southern Ukraine, Crimea, and along the Black Sea basin, designated as "Gazaria."

The rediscovery of Ptolemy's *Geographia* and the works of Renaissance Europe's "founding fathers" of cartography—Münster, Ortelius, and Mercator—would make the sixteenth, seventeenth, and eighteenth centuries an age of a scientific-technical revolution for cartography, characterized by engraving of large-scale maps and atlases for military, commercial, and administrative purposes by European imperial and colonial powers. From Krawciw's perspective, these were the humble beginnings of a more systematic cartography of Ukraine as a strategic geopolitical pivot point.

Terrae Incognitae and European Cartographic Ventures

For most European state sovereigns in the sixteenth century, the geographical area between the Habsburg lands, Poland-Lithuania, Ottoman Empire, and Muscovy was an empty space at best, and at worst, relatively unknown. Cosmographical depictions were researched by European travelers and ambassadors, who were also Renaissance gentlemen of letters, steeped in classical antiquity and humanistic learning. Based on their descriptions and sketches, maps were made by cartographic artisans who gained a strong reputation by working in places such as Amsterdam, Venice, Nuremberg, and Vienna. Maps and descriptions of the lands and peoples in these areas of supposed "Eastern Europe" were made by Sigismund von Herberstein, Augustin Hirschvogel, Wolfgang Lazius, Cornelius de Jode, Tomasz Makowski and Hessel Gerritsz, Wacław Grodecki, Johannes Janssonius, Jacopo Gastaldi, Anthony Jenkinson, Isaac Massa, and Jodocus Hondius, all of whom are represented in the collection.[15] We turn now to some examples of their work.

Habsburg Dynastic Lands and the Kingdom of Hungary. A Renaissance humanist writer, traveler, and ambassador of the Habsburg monarchy to Moscow, Sigismund von Herberstein (1486–1566) came from a prominent German family in Wippach/Krain, part of the Styria region. There he had learned some Slovenian, a South Slavic language, which enabled him to do later research in Moscow. In 1506 he became an officer in the imperial army. Lauded and rewarded for his service, he was knighted by the Holy Roman Emperor in 1508, and entered the Imperial Council in 1515. One of most celebrated diplomats of the sixteenth century, he traveled on behalf of the Habsburg monarchs throughout Europe, on nearly seventy different missions from 1515 to the end of his career in 1553. He is best known for his pioneering description of Muscovy, *Rerum moscoviticarum commentarii* (Notes on the Muscovites), a manuscript he began around the early 1520s and printed in full in 1549.[16] His traveler's description was what Marshall Poe has aptly called a Renaissance traveler's "ethnography," a work offering "a general description of Russia according to a standard template of topics—geography, the royal court, the administration, the army, social classes, popular customs and mores, the economy, religion, and so on."[17] The information from Herberstein's *Notes* was borrowed frequently by other European travelers to Muscovy, and used repeatedly as a source for subsequent works. Herberstein's *Notes* were published with a map drawn by Augustin Hirschvogel (Hirschfogel, Hirsvogel, Hirssfogel) (1503–53), a form-cutting artist based at first in Nuremberg, and later commissioned to draw maps for the Habsburg monarchy. By the early 1540s, Hirschvogel himself had earned quite a good reputation, and after the ambassador Herberstein had completed his work on Muscovy, he summoned Hirschvogel to Vienna, the intellectual and political center of the Holy Roman Empire, where he would receive Habsburg patronage for completing cartographic work. Information was gathered by Herberstein and based on the sketches of Ivan Liatskoi, a boyar who had distinguished himself during the Lithuanian campaigns, and produced an outline map of Muscovy for him. In 1546, Hirschvogel produced a fine woodcut of the map, *Moscovia Sigismundi Liberi Baronis in Herberstein, Nieperg, et Gutenhag* (**figure 6, #234**), and published it in 1549.[18] Hirschvogel's other works included maps of Ottoman-Habsburg borders, Upper Austria, Carinthia, Saxony, and Hungary.[19]

Another Renaissance humanist who gathered information in his travels was Wolfgang Lazius (1514–65), a Hungarian physician who became a geographer and cartographer also based in Vienna. Many of the cartographers in Vienna came from Nuremberg, a center for artisanship and the study of chorography, or regional description.[20] A professor of medicine at the University of Vienna, Lazius traveled throughout Central Europe, documenting geographical areas and place names, and jotting down observations about local inhabitants. In Vienna he compiled a history of Austria, which forms the basis for his maps of Austria, Hungary, and Bavaria. His cartographic works include maps of Austria, Hungary, Bavaria, and Greece. He etched the works himself, since he could find no craftsman to do it for him in Vienna. The maps were widely influential, used by Ortelius, Jode, and others.[21] Lazius' 1556 ten-sheet map of Hungary, *Hungariae Descriptio, Wolfgango Lazio Auct[ore]* (**figure 7, #301**), was used by Abraham Ortelius for his atlases, and by successors such as Gerard de Jode. In the map, Hungary is depicted in its borders with Austria, Bulgaria, Wallachia, Moldova, Poland, and Transylvania. Ukrainian lands are represented within Rus′ ("Rvthenie Pars") and the Transcarpathian region. Cities include Rakhiv ("Rana"), Uzhhorod ("Vngvar"), Nevytske ("Newykze"), Mukacheve ("Munkacz"), Perechyn ("Peregzaz"), Iasynia ("Ieczenew"), Humenne ("Humel"), Makovytsia ("Makowice"), and Stropkiv ("Stropsko"). The national emblem of Hungary is in the upper left corner.[22]

The Kingdom of Poland and Grand Duchy of Lithuania. Renaissance cartographers also mapped the Kingdom of Poland and Grand Duchy of Lithuania, which were joined together by the Union of Lublin in 1569. Gerard de Jode (1515–91) and Cornelius de Jode (1568–1600) were a family of Dutch map publishers based in Amsterdam who produced some original works and reproductions.[23] Among the maps they reproduced were Ortelius' eight-sheet world map (1544), and the world map of Gastaldi (1555). Gerard de Jode started his own work in the early 1570s, and compiled the atlas *Speculum Orbis Terrarum* (1578; rev. ed., 1593). Jode's 1593 map of the Kingdom of Poland, *POLONIAE // AMPLISSIMI / REGNI...TYPVS // GEOGRAPHICVS* (**figure 8, #175**), appears in the newer folio. It was modeled on the 1565 nine-sheet map of Germany by Christian Schrots (Sgrooten, Shrooten), a prolific cartographer of maps of Germany and the Spanish Netherlands, and a Royal Cartographer to King Philip II of Spain.[24] The left-hand corner of the map shows a portrait of King Sigismund III (Vasa), King of Poland-Lithuania and Sweden. Place names for Ukrainian lands include "Rvssia," "Volhinia," and the "Sulcz flu[vis]" (Sluch River).

Wenceslaus Godreccius, or Wacław Grodecki (ca. 1535–91), was a Polish cartographer and the son of a Silesian gentleman, who studied mathematics and philosophy at Jagiellonian University in Kraków. He is probably best known for his 1557 map of Poland.[25] The map was republished in Basel in 1562, and used by Ortelius (1570) in subsequent Latin and Dutch editions of his *Theatrum Orbis Terrarum*, and by many others for atlases of the world. The only known example of the original Godreccius -Pograbius work (1570) was destroyed in Munich in 1945. Grodecki's map of Poland-Lithuania, *POLONIAE, // LITVANIAEQ. // DESCRIPTIO* (**figure 9, #216**), was coauthored with Andrzej Pograbka (Pograbius) (d. 1602), whose earlier map of Poland (1570) was printed in Venice. Krawciw's collection has one reproduced engraving with updated place names, published in Antwerp in 1601.[26]

The project of mapping the crown lands of Poland and lands of the Grand Duchy of Lithuania was carried out by kings and nobles, and therefore not entirely centralized as an enterprise.[27] In attempting to define a Ukrainian cultural and territorial identity, Krawciw collected many copied editions of a Polish-Lithuanian map which included "Ukraina" as a place name. The map was commissioned by Prince Mikołaj Krzysztof Radziwiłł (1549–1616), a magnate and grand marshal of Lithuania and a collector and major patron of cartography. He commissioned Tomasz Makowski (c. 1575–1630), a Polish draftsman, engraver, and printer, to make the 1613 map.[28] The 1613 original, reproduced in 1631, *MAGNI DVCATVS // LITHVANIAE* (**figure 10, #412**), appeared in Blaeu's *Theatrvm orbis terrarvm, sive Atlas novvs...Amsterdami, Apud Guiljelmum Blaeuw, 1635*.[29] In 1613, this Radziwiłł-Makowski map of Lithuania was published for the first time. The map in the collection is its third printing, made without the inset map of the Dnipro River. Later, it was printed on four sheets and included in Blaeu's atlas. Notably, the map is the first on which a part of Ukrainian territory is designated with the name "Ukraina." The map also includes western Ukrainian lands, labeled as "Podlachia," "Russia Rubra," "Polesia," "Wolynia Citerior," and "Podolia," while Right-Bank Ukraine is labeled as "Volynia Vlterior" with cities including "Kijouia," "Biallacerkiew," and "Czijrkassy." The map itself went through many reproductions. An influential source of information about Poland-Lithuania and the Ruthenian parts of Ukraine, the authorship of the map was falsely attributed for a long time to Radziwiłł, but it was actually drawn by Makowski and engraved by Hessel Gerritsz (1581–1652), a famous Dutch cartographer for the East India Company, maker of maps covering the Iberian Peninsula, and engraver of some of the first maps of the Australian coast.[30]

THE OTTOMAN EMPIRE AND OTTOMAN TURKS. The Ottoman Empire and its European/Asian borderlands also make an appearance on Renaissance maps during the early seventeenth century, and these maps usually reflect a Eurocentric, and often Orientalist, view of the Porte. Johannes (Jan, Joannes) Janssonius (Jansson, Jansz. II, Janszoon) (1588–1664), the son-in-law of the Flemish map and globe engraver Jodocus Hondius the Elder (1563–1612), was part of a Dutch printing and bookselling family in Amsterdam.[31] A prolific cartographer, publisher, and eventual heir to the Hondius firm, he assisted in 1612 with the production of the important Mercator-Hondius atlas, which was revised and reprinted in many languages. In 1637, he printed the *Novus Atlas* with the assistance of his brother-in-law Henricus Hondius. The work was intended to rival the published work of Willem Janszoon Blaeu (Blauw, or Blaeuw, 1571–1638), a meticulous student of the Danish astronomer Tycho Brahe. Blaeu was a Dutch cartographer who adopted a serious and scientific approach to cartography, and with whose name Jansonius' was commonly confused.[32] Influenced by Mercator, Janssonius and Hondius added more leaves to revised versions of their atlas, having comprehensiveness of relatively unknown lands as their shared aim. As one part of a revised version of the atlas, Janssonius' map of the Ottoman Empire, *Turcicum // Imperium // concordia res parvae crescunt // Discordia maximae dilabuntur* (**figure 11, #264**), was printed as part of Janssonius' *Nouvel Atlas ou Theatre du Monde* in Amsterdam in 1649. Offering an image of Oriental despotism, the map includes a cartouche depicting the sultan on a throne flanked by armed figures.

MUSCOVY AND TARTARY. Renaissance and early modern depictions of Muscovy relied heavily on the work of Herberstein and Hirschvogel, and many new works were copies of the information they had presented. Culturally and politically, tsarist and Orthodox conceptions of space and place interfaced with mathematical models imported from Europe.[33] Jacopo (Giacomo) Gastaldi (ca. 1500–66) was a Venetian astronomer, engineer, and cartographer from Piedmont. A prolific publisher, by 1539 he was active in Venice, holding the position of Cosmographer of the Venetian Republic. He produced general maps, engineering and hydrographic maps for Venice, and over thirty new maps for Italian editions of Ptolemy's *Geographia*. By the mid-1540s he had developed a distinctive style of copper engraving and acquired a good reputation.[34] Gastaldi's 1550 map, part of his *Descriptione de la Mo // scovia* (**figure 12, #202**) printed in Venice, was taken directly from the work of Herberstein and Hirschvogel.[35]

In the second half of the sixteenth century, an interesting original map of Muscovy and the borderlands of Rus´ was made by Anthony Jenkinson (c. 1525–1611). Jenkinson was an English merchant, navigator, astronomer, and cartographer of the Mercer's Company of London. He served as chief factor of the London-based Muscovy Company from 1556, and in the 1550s and 1560s traveled in the Ottoman Empire, Russia, and Persia, including a journey from Muscovy to Bukhara in 1558. Upon his return to London he drew a map of Muscovy (1562) that for a long time was thought to survive only in duplicate form, in the atlases of Ortelius and Jode.[36] Jenkinson's 1562 map of Muscovy, Russia, and "Tartaria," *RVSSIAE, MOSCO- // VIAE ET TARTA- // RIAE DESCRIP- // TIO* (**figure 13, #272**), was published in Amsterdam in 1570 as part of Ortelius' *Theatrum Orbis Terrarum*. The map has many historical and geographical inscriptions, and includes Right-Bank Ukraine labeled as "Litvania" with "Kiou," and Left-Bank Ukraine with "Stara dowha" and Crimea above "Asou."[37]

Some other plans of Muscovy were drawn by Isaac Massa (1586–1643), a Dutch merchant and cartographer who spent nearly ten years in Muscovy (c. 1600–1609), and wrote a historical description of Muscovy in Dutch in 1610.[38] From his station in the "German colony" (*nemetskaia sloboda*) designated by the tsars for foreign visitors to Moscow, he followed the Herberstein template. Massa apparently obtained native geographical sketches and produced plans of Muscovy in the 1610s. Reporting back in Dutch, he spoke with some authority on the geography, history, and political features and court rituals of Muscovite tsars and their loyalist boyars.[39] His maps were copied and used by Blaeu and Janssonius in their competing atlases. Massa's map, *Moscoviae // Pars Avstralis* (**figure 14, #339**), was published in Amsterdam in 1647 as part of Janssonius' *Novus Atlas Sive Theatrum orbis Terrarum*. The map includes three cartouches in color, surrounded by figures of soldiers, royalty, and traders, and decorated with all kinds of animals. One cartouche shows a crown, and has a coat of arms with a crowned two-headed eagle at its center, with a medallion of St. George.

Jodocus Hondius the Elder (Josse de Hondt) (1563–1612) was a Flemish engraver and globe and instrument maker from Ghent who worked at first in London, then settled in Amsterdam and set up a map publishing business in 1593.[40] His son, Jodocus the Younger (1594 or 1595–1629), and son-in-law Johannes Janssonius continued the business with great success. In 1604 he acquired the publishing rights and plans for Mercator's atlas and in 1606 printed a first edition of it with thirty-seven new maps. His youngest son, Willem, would go on to publish the maps of Ukraine by Guillaume le Vasseur Sieur de Beauplan. Hondius' 1606 map, *Tartaria* (**figure 15, #254**), was an imperial map of the Crimean Khanate. The black and white version includes cartouches of the khans and "Tartaria" as a place name covering lands ranging across Central Asia. The first version of the map was printed in the edition of the Mercator-Hondius Atlas in 1606; an updated work appeared in the 1649 edition of Janssonius' atlas, *Nouvel Atlas ou Theatre du Monde*, published in Amsterdam.

Conclusion

Much literature in the history of cartography has concentrated on European transatlantic empires—Spain, Portugal, England, and France, for instance—but European cartographers looked to the south and east for passages and territories as well, even as the Mediterranean was displaced as a center for commercial traffic.[41] Still, it would be fair to say that as a general observation, Ukraine was "off the map" in European Renaissance cartography, treated as a borderland to other empires. "Ukraine" itself was not often used as a geographical term, and labels for its territories were multiple. Yet it is not possible to understand the mapping of Ukraine without taking into account the scientific and political framework for how Renaissance cartographers measured the bordering regions of Ukraine, namely the lands of the Habsburg, and, by extension, Holy Roman Empire, the Polish-Lithuanian Commonwealth, the Ottoman Empire, tsarist Muscovy after Ivan III, and Tartary. Ukraine's strategic importance along the Black and Azov Seas, and the place of the Cossack Hetmanate and steppes between these loosely construed and fairly decentralized polities, would make it increasingly important to early modern European conceptions of autocracy and territorial dominion, not to mention statehood and sovereignty. We turn now to the influential work of Guillaume le Vasseur de Beauplan, the French cartographer who would, once and for all, place early modern Ukraine on various post-Westphalian Great Power maps.

2 *Attempts at Mapping Early Modern Ukraine*

In early modern Europe, maps acquired a new geostrategic importance for sovereign rulers as tools and emblems of statecraft. Cartographers participated in overseas commerce and colonialism, working mostly on commission for European monarchs, ministers, trading companies, and, on occasion, the nobility.[1] The signing of the Treaty of Westphalia brought an end to the Thirty Years' War across the European continent, and signaled a new era for cartography. Although dynastic borders were mutable, if not always negotiable, and changed hands frequently through shifting alliances and at treaty settlements, new state interests in large-scale topographical cartography reflected emerging Great Power concerns with centralized administration, dynastic claiming, and territorial surveying by military means, on land and by sea.[2] As a reflection of these trends, European centers for the production of maps shifted northward from Venice and Florence to Amsterdam, Paris, and London.[3]

Beauplan's Maps and Their Legacy

The period of the Thirty Years' War and Vasa rule in Sweden and Poland-Lithuania paved the way for more militarily oriented cartography in Central and Eastern Europe. With practically all the major European powers engaged in the war, Władysław IV (r. 1632–48) recruited engineers from abroad to work for him. He commissioned the artillery officers Jan Pleitner, Fryderyk Getkant, Sebastian Aders, and Guillaume le Vasseur de Beauplan (1595–1685), a French military engineer and hydrographer.[4] Beauplan was directed by the kings of Poland to search out suitable sites for fortifications in Ukraine, and in particular, to draw a map of Left-Bank Ukraine. Starting in 1648 in Gdańsk, he published his general and specific delineations of Ukraine in eight sheets, engraved by the Dutch cartographer Jodocus Hondius. At the time, the maps were important for survey purposes of a relatively unknown region, and offered the first European military coverage of such a large area. Author of the *General Map of Ukraine* (1652), Beauplan was instrumental in the founding of more than fifty settlements and colonies after the Khmel′nyts′kyi Uprising and Treaty of Pereiaslav (1654).[5] He also authored a new map of the Kingdom of Poland in 1652, a manuscript map of Brittany in 1666, and a twelve-sheet map of Normandy in 1667.

Three examples from the collection illustrate Beauplan's tremendous attention to detail as well as Krawciw's interest in the political recognition of Ukrainian territory and iconographic representation of Ukraine, for instance in the image of the Cossacks. The 1660 map by Beauplan, *Carte d'ukranie contenant plusieurs Provinces comprises entre les Confins de Moscouie et les Limittes de Transilvanie* (**figure 16, #33**), delineates the borders of Ukraine and surrounding territories, and includes a cartouche depicting Cossack rule (detail shown on next page). Beauplan's map of Ukraine, *Ukrainae Pars, que Kiovia Palatinatus Vulgo dicitur. Per Guilhelmum le Vasseur de Beauplan S.R.M. Poloniae Architectum militarem etc* (**figure 17, #32**), includes the territory of the Kyivan palatinate. It appeared in a special reprinted edition of the *Atlas major* (1659–72) by Johannes Blaeu. Finally, Beauplan's general map of Ukraine showing the palatinates of Podolia, Kyiv, and Bratslav, *Typus Generalis Ukrainae sive Palatinatuum Podoliae, Kiovensis et Braczlaviensis terras nova delineatione exhibens* (**figure 18, #45**), appeared in influential

Detail of cartouche from figure 16.

atlases printed in 1680 by Johannes Janssonius and Moses Pitt. These maps originally appeared in Beauplan's major compilation, *Description d'Ukranie* (1652). In addition to these works, the collection features a number of cartouches in color representing other Ukrainian regions such as Podolia, Volhynia, and "Braclaw," many of which have engraved images of the Cossacks in iconographic form.[6]

Why would an apparently obscure late seventeenth-century series of maps acquire such importance later on? Beauplan's maps were particularly important "national treasures" to Krawciw because they offered an alternative representation of Ukraine under Western, and in this case, European, eyes. Beauplan's maps visually codified for future reference the usage of the term "Ukraine" as an umbrella term to designate a specific territory inscribed by historical, ethnographic, and territorial borders. Initially, "Ukraine" was used in the late sixteenth century to designate the western Ukrainian lands under Polish rule, but the place name was adopted and maintained by the Cossacks in the seventeenth century, and employed by Beauplan and subsequent European observers. From the Muscovite and then Russian imperial perspective, the earlier ecclesiastical concept of Mala Rus´ was revived as Little Russia (Malorossiia) after the Treaty of Andrusovo in 1667, in part in order to "reunite" Kyiv to tsarist Russian domains following Polish-Lithuanian conflict with Muscovy and the rule of Khmel´nyts´kyi and the Cossack state there (1648–54).[7] The collection also has an edition of the work from 1959, published in New York by the Organization for the Defense of Four Freedoms of Ukraine, for which Krawciw himself wrote a short preface (see chapter 5 and figure 77, below). For Krawciw, "Ukraine" in this context came to assume a historical continuity with the early modern Hetmanate, which he treated, like Taras Shevchenko and other Ukrainian romantics, as a polity separate from Poland-Lithuania. As a Ukrainian nation-builder, he sought a historical continuity on European terms for early modern Ukrainian history, and a means to validate Ukraine as a separate state, standing apart from other forms of Polish and Russian claims to its regions.

Beauplan's maps also were influential for more pragmatic reasons. Throughout the early modern period, a time when legally structured notions of intellectual property were absent, cartographers tended to draw heavily on each other's work as "sources" for their own maps, often to the point of brute plagiarism. Thus cartographers often treated Beauplan's work as authoritative, reproducing his maps for the next century and a half in atlas compilations depicting the *terra incognita* to the east. As a result of this continual reliance on Beauplan's works, from the Treaty of Pereiaslav (1654) to the settlements after the Vienna Congress (1815) Ukraine became commonly labeled in many European maps as the "Land of the Cossacks."[8]

Capitals of European Mapmaking

In early modern Europe, maps were needed for successful commerce, administrative statecraft, and military intelligence. Since the engraving of maps was expensive and required substantial financial resources and trained craftsmen, European maps were generally produced in major capitals. Polish-Lithuanian cartography based on scaled grids preceded Muscovite and later Russian cartography of the same genre. In a larger geopolitical context, cartographic production in terms of quality and quantity was contingent on the post-Westphalian fate of European states on the rise or in decline. Amsterdam became the center of the map- and atlas-producing world throughout the seventeenth century. Venice, like the other Italian city-states, was in decline due in large part to the displacement of commerce from the Mediterranean to the Atlantic. Vienna, the heart of the Habsburg monarchy and Holy Roman Empire, remained an important site for map and globe production, but not for overseas imperial or colonial maps. Some maps were produced in Stockholm, reflecting a last gasp for Swedish state interests in Poland-Lithuania and along the Baltic littoral. The ascendancy of Louis XIV and the end of the English Civil War and Glorious Revolution in England would portend growth for English and French military and colonial cartographic production in the late seventeenth and early eighteenth centuries.

AMSTERDAM. Since at least the early sixteenth century, cartographic production in the Netherlands was concentrated in Antwerp and Amsterdam.[9] Around the 1590s, Amsterdam supplanted Antwerp because of the reputation of the former's craftsmen and its suitable geographical position for maritime, coastline, and commercial maps. Cartographers there became known for hydrographical and nautical cartography. Form-cutting shops produced a substantial number of woodcut maps for the market. As copperplate engraving became more common for printed maps, Dutch craftsmen earned a reputation and created a burgeoning industry for map production.[10] Starting in the early seventeenth century, small publishing houses emerged and produced sea charts and large-scale atlas compilations in the style of Mercator-Hondius and Ortelius, and even some of the first miniature pocket atlases. Firms for sea atlases included P. van Alphen (1600), Jacob Colom (1632–33), Arnold Colom (c. 1640), Anthonie Jacobszoon (1653), Pieter Goos (1654), H. Doncker (1659), Arent Roggeveen (1675), and Jacob Robijn (1683). Dutch publishing houses owned by Blaeu, Hondius, Witsen, Jansson (Janssonius), and Mortier ventured into the printing of wall maps as well as hydrographic works. It was not until the start of the eighteenth century that Dutch maps experienced a relative decline in production, when the center for European cartographic output shifted to London and Paris.

Claes Jansz[oon] Visscher [Piscator] (1587–1652), Claes II (Nicholas Ioannis Piscator) Visscher (1618–79), and Nicholas Jansz. Visscher (1649–1709) were a Dutch family of engravers and publishers based in Amsterdam. Most famously, one of Visscher's wall maps appeared in Johannes Vermeer's *The Art of Painting* (1667).[11] Claes II worked in his youth for the influential firm of Blaeu, and traveled throughout Europe learning the craft of cartography. His map compilations eventually competed with the firms of Blaeu and Hondius-Janssonius. In 1664, Visscher printed his first atlas with sixty maps, the *Atlas Contractus*, which was widely distributed for at least the next fifty years. After the death of Nicholas Janszoon Visscher in 1709, the business folded and the plates for *Atlas Contractus* were sold by his widow to the firm of Pieter Schenk.[12] The 1651 map of Muscovy-Rus´ (**figure 19, #534**) is an exact copy of Hessel Gerritsz's *Tabula Russia* (1614). The copy was printed in a 1632 atlas by Claes Jansz. Visscher. Nicolaes Piscator Visscher updated the version and reissued the map in 1651. The map itself includes views of the Narva and Kremlin ("Kremelenagorod") as supplements to Muscovy ("Celeberrima ingens et imperialis civitas Moscua"). The map also includes all of modern Ukraine with "Polesia," "Wolynia," and "Podolia" in the west, and "Crimea, seu Tartaria Perecopensis" in the east. The Black Sea is labeled as "Pontus Euxinus qua Zorno More Rhuteni vocant."[13]

Another prolific Dutch map publisher was Frederick de Wit (1629 or 1630–1706), an engraver and seller whose work encompassed the full range of cartographic production, including world atlases, wall maps, and plans and views of Dutch and other European towns and cities. In 1688 he obtained a privilege from the States General to publish his maps; atlases after this date are marked "cum privilegio." His geographical maps established his fame, and were still in demand many years after the closing of the family's three-generation business. Frederick de Wit's total output stood at about 124 land maps and 27 sea charts, in addition to at least one map of the world. His son and grandson produced an additional 130 maps, but the business did not survive for long after the father's death, and it folded in 1712. The plates were acquired by the firm of Mortier, an engraver based in Amsterdam.[14] Wit's 1654 map of Poland, Lithuania, Ukraine, Prussia, and Livonia, *REGNI // POLONIAE // et Ducatus // LITHVANIAE //*

VOLINIAE, PODOLIAE // VCRANIAE // PRVSSIAE, LIVONIAE // et CVRLANDIAE descriptio (**figure 20, #561**), a likely borrowing from Beauplan's maps, appeared in one of his atlases published in Amsterdam from 1654 to 1672. The Ukrainian lands are named as "Volhynia," "Podolia," "Palatinatus," "Kiovensis," and so on.

The Dutchman Pieter van der Aa was a publisher in Leyden who ran a successful business from about 1713 to 1730. He printed a cartographic sketch by François Grata, *PLAN EXACT DES FORTIFICATIONS DU CHÂTEAU DE CHOCZIM SUR LE NIESTER, AVEC LES PORTS ET RETRANCHEMENS DE L'ARMÉE DES TURCS DEFAITE EN 1673 PAR JEAN SOBIESKI, // GÉNÉRAL EN CHEF DES POLONOIS ENSUITE ROI DE POLOGNE, SOUS LE NOM DE JEAN III* (**figure 21, #210**). The plan of battle appeared in Aa's major atlas, *La Galerie Agréable du Monde* (1729), an edition which sold one hundred copies.[15] The engraving depicts the Battle of Khotyn in 1673, and inset maps indicate the battles along the Dnister between Jan Sobieski and Ottoman forces.

Justus Danckerts the Elder (1635–1701) and his three sons Cornelis Danckerts (1664–1717), Theodore Danckerts (Danchers) (1663–1727), and Justus Danckerts the Younger (d. 1692) were a Dutch family of publishers, engravers, and map sellers based in Amsterdam. Although their works had some errors in the measurement of territory, their maps were important in that they paid close attention to changing and extant political boundaries. In sum they produced over three hundred sheets of maps for their atlases, which were printed until about 1710.[16] Justus Danckerts' 1685 map, *Exactissima // Asiae // Delineatio. // In Praecipuas Regiones // Caeteraiq. Parte Divisa. // et denuo in lucem edita // Per // T. Dankerts // Amsterdami cum Privilegio* (**figure 22, #139**), shows Muscovy and Asia with Orientalist imagery of a caravan at rest (detail below). The map is taken from his *Atlas Tot Amsterdam* (1670–87).

Carolus (Karel) Allard (Allardt) (1648–c. 1706) was a Dutch engraver and publisher based in Amsterdam. His maps were known among contemporaries for their scientific accuracy. Initially, he did not intend to include his maps in larger atlases, but he eventually gave way and his sons, Karel and Abraham, went into the profitable atlas business and included some of their father's maps in larger compilations and productions of wall maps. The engraved atlases of the family firm include works of Jansonius, Wit, and others. Allard's 1696 map, *Regni Poloniae, Magni Ducatus Lithuaniae Coeterarumq, Regi Poloniae Subditarum Regionum Tabula; In Omnes suos Ducatus, Palatinatus* (**figure 23, #24**), depicts Poland-Lithuania and bordering areas. Latinate place names include "Lithuaniae Magnus Ducatus," "Polonia Major," "Polonia Minor," "Russia Rubra," "Ukraina," "Podolia," "Walachia," "Moldavia," "Transylvania," "Hungariae Regni Pars," "Volhynia," "Podlachia," and "Podolia." Allard's other maps in the collection include maps of Hungary, the Tatar lands, and Russia's expanses in Asia and Siberia. By this time, the name "Ukraine" or "Ukraina" was used with greater frequency, even when included within other geospatial entities.

VENICE. The craft of mapmaking and the map trade had also become highly developed in Venice since the early sixteenth century. Venetian cartography flourished thanks to accomplished craftsmen working in the arts of woodcutting, copper engraving, and printing. As with Amsterdam, the physical geography of Venice also contributed to its importance as a site for cartographic production. Venetian

Detail from figure 22.

commercial ships traveled to the Levant and around the Black Sea, to the ports of Spain and Portugal, and along the coasts and ports of Western Europe. Following the Ptolemaic resurgence and commercial development along the Mediterranean, Italian regional cartography of Tuscany, Piedmont, Sicily, Malta, Corsica, and Sardinia made great strides during the sixteenth century, but declined substantially in the seventeenth century beginning with the Thirty Years' War. Nevertheless, a number of atlasmakers remained in these areas well into the seventeenth century. For one example, Marco Vincenzo Coronelli (1650–1718) was a Venetian friar who set up his own map- and globemaking workshop at a convent in his native city. He was appointed to the position of Cosmographer of the Venetian Republic, and later founded the Gli Argonauti, a geographical society. Coronelli made a number of interesting terrestrial and celestial globes.[17] His 1695 map of Europe, *Parte Orientale Dell' Europa, Descritta, e Dedicata Dal P. Cosmografo Coronelli All' Illustrissimo, et Eccellentissimo Signore Giovanni da Mula, Senatore amplissimo Nella Serenissima Republica di Venetia* (**figure 24, #118**), was engraved with a cartouche decorated with angels, a crown, and plants. The original work appeared in his twelve-volume edition, *Atlante Veneto*, published in Venice from 1697 to 1701.

STOCKHOLM. Erik Jonsson Dahlberg (1625–1703) was a count, military engineer, and cartographer in Stockholm, Sweden. His atlas of Sweden was published in 1698 and presented to Charles XII. His corrected *Tabula chorographica Regni Poloniae* was done posthumously in c. 1729.[18] A chorographical map of the Kingdom of Poland (**figure 25, #134**) was probably made around 1660 after the Battle of Brest-Litovsk. This version appeared in Samuel Pufendorfs' *De rebus a Carolo Gustavo Sveciae Rege gestis commentariorum libri septem*, published in Nuremberg in 1696–97. The map includes a part of "Russia Rubra," "Palatinatus Chelmensis," and "Palatinatus Belsensis." All Polish, Lithuanian, and Ukrainian lands are designated by their coats-of-arms. An interesting map of Vasa rule, it includes heraldic seals for provincial lands, and is dedicated to the Vasa king of Sweden, who was also the elected king of Poland-Lithuania.[19]

PARIS. Starting around the 1640s, Paris began to rival Amsterdam as the epicenter of European atlasmaking. Cartography was mostly regional, and in France as in the Netherlands, chorography (regional cartography) emerged around the mid-sixteenth century. Parisian map sellers started circulating maps relating to local French lands—for instance, Calais, Bourges, Boulonnais, Vermandois, Languedoc, Provence, and Burgundy.

From the 1640s until the end of the seventeenth century, the maps and atlas compilations of Nicolas Sanson d'Abbeville (1600–1667) and his sons dominated the French cartographic scene. Sanson began as a military engineer but turned to cartography with the encouragement of his publisher, Melchior Tavernier, who was greatly impressed with his map of ancient Gaul. The French map trade blossomed in the 1660s, thanks largely to Sanson.[20] As his fame grew, Sanson was appointed by Louis XIV to the position of Géographie ordinaire du Roy. Incredibly prolific, Sanson produced over three hundred maps during his career. In 1695, Alexis-Hubert Jaillot (c. 1632–1712), a sculptor and cartographer who produced his *Atlas nouveau* in 1689 based on the materials of Sanson and his sons Adrien and Guillaume, took over his father's family printing and publishing business and produced large atlases. Sanson and Jaillot strongly influenced a number of important French royal cartographers, geographers, and atlasmakers, including Guillaume Delisle and Jean Baptiste Bourguignon d'Anville. They made Paris the main European center for cartographic production in the eighteenth century. Sanson and Jaillot's 1692 map of the Kingdom of Poland and Grand Duchy of Lithuania, *ESTATS DE // POLOGNE // Subdivisés suivant l'estendue des // PALATINATS...* (**figure 26, #458**), includes the duchies of Prussia, Mazovia, Samogitia, Volhynia, and Podolia, and also "Ukraine or the Land of the Cossacks." It was published in Paris in 1692 as a slightly reduced version of a 1675 map, and appeared in Sanson and Jaillot's updated *Atlas Nouveau contenant toutes les parties du Monde*. This map is accompanied by a chart indicating regional divisions of Poland, printed on one large folio leaf. The map also depicts western and central Ukrainian territories, designated as "Russie Noire" with "Pokutie" in the west, and "Ukraine ou Pays des Cosaques" with "Volhynie" and "Podolie" in the east, on both sides of the Dnipro River. Eastern and southern Ukraine are labeled as "Frontières des Petits Tartares," and the northeastern lands are labeled as "Frontières des Estats du Czar de Moscovie."[21]

LONDON. The geopolitical scene gradually shifted away from the commercial centers in Amsterdam and Venice as London became a center for British imperial and colonial cartographic pursuits.[22] The first large-scale mapping enterprise in English was undertaken at the start of the seventeenth century, resulting in the printing in London in 1611–12 of *The Theatre of the Empire of Great Britaine*, a very influential work that was used throughout the late seventeenth and early eighteenth centuries. A grand project, the maps were made by John Speed (1552–1629), engraved by Jodocus Hondius in Amsterdam, and based on the work of John Norden and William Smith, two late-sixteenth- and early seventeenth-century estate surveyors who sought to produce a comprehensive atlas for all the counties of England. Such ambitions were echoed later in the century in the hapless story of Moses Pitt (1641–97), an English map publisher and bookseller based in London.[23] Pitt's cartographic production is a saga of complex continuities and borrowings. He adapted the printed works of the Dutch Janssonius firm, based on the Mercator-Hondius atlas, which itself had been adapted to rival the works of Blaeu starting in the 1630s. Pitt's name was added to 1660s editions of the Janssonius-Waesbergius firm's 1666 *Atlas Contractus*. An ambitious character and a Freeman of the Haberdashers' Company from 1667–88, Pitt planned a massive twelve-volume world atlas. Only four volumes were published, as *The English Atlas* (1680–82), and he was imprisoned for debt in the Fleet Prison from 1689–91. Pitt's 1680 map bore the title *The // English // Atlas. // Volume I. // containing a // Description // of the // Places next the North Pole; // as also of // Muscovy, Poland, Sweden, Denmark, // and their several Dependances* (**figure 27, #398**).[24]

Also based in London, Edward Wells (1667–1727) was an atlasmaker who produced a popular compilation, *A New Sett of Maps* (London, 1700). Wells' 1700 *New Map of EUROPE // according to the Present // General Divisions and Names // of its Countries together with // their Capital Cities Chief Rivers / Mountains &c // Dedicated // To His Highness William // Duke of Gloucester* (**figure 28, #554**), offers a vision of Europe on the cusp of a new century. The Eurocentrism is evident, as is the tendency to omit entire states or regions. Europe is defined continentally, with the British Isles, Norway, and Sweden left out. On the continent: Portugal, Spain, France, Netherlands, Switzerland, Italy, Germany, Poland, Hungary, Transylvania, Wallachia, Moldova, Turkey in Europe, "Lesser or Crimea Tartary," "Great Russia," and "Moscovie" are indicated. European peripheries across imagined continental divides are vaguely labeled as "Part of Africa" and "Part of Asia," and Ukraine is not included as a separate entity.

Conclusion

Krawciw's commemoration of Beauplan in his collection, and his attempt to situate Beauplan's maps of Ukraine in a European context, may be interpreted as a major signpost in his quest to preserve Ukraine's cultural monuments and to suggest concrete geographical evidence for Ukrainian national self-determination. In a broad historical context, Beauplan's attempts at mapping early modern Ukraine revealed the purposes and practices of European cartography, including a major shift in large-scale cartographic production away from Mediterranean to transatlantic cities. After the signing of the Treaty of Westphalia in 1648 and the dominance of Khmel´nyts´kyi in the Zaporozhian Cossack Hetmanate, the deployment of "Ukraine" as an umbrella term became more commonplace. It was also reflective of a set of larger political mapping projects within the new state system. The cultural and geopolitical milieu was not culturally reducible to an Orient and Occident, or categories of Eastern and Western Europe. The mapping of Ukraine reflected military and commercial interests aimed at strengthening state sovereignty and administration. States such as England and France were on the rise as imperial powers, and Spain, the Netherlands, Sweden, and Poland-Lithuania were declining as state powers in terms of their commercial and military capacities. Cartographers and engravers were technical craftsmen who needed expensive resources, and relied upon both their reputation and recognition by rulers and other elites. As highly skilled artisans, they made a living primarily by working on royal and state-imperial commissions. To the end of the seventeenth century, urban markets began to develop for the map trade and production of large-scale world atlases, especially in cities such as London and Paris. In this context, Ukraine had to be envisioned geopolitically, with a ready eye toward shifting early modern alliances and loyalties amid European imperial and colonial pursuits.

3 *Enlightenment Cartography and Ukraine as "East"*

DURING THE EIGHTEENTH CENTURY, European geographical studies and cartography flourished. Beauplan's maps contributed greatly to this flowering, and were reproduced in European capitals and became authoritative as sources of aesthetic and technical information. Ukraine was generally perceived through the lens of the Enlightenment as "the land of the Cossacks," at a time when philosophes such as Voltaire and Diderot supplanted an earlier north-south axis from the Mediterranean to the Baltic and North Seas with one that imagined the West as the locus of "civilization," in opposition to "barbaric" lands and peoples living within a vaguely defined East.[1]

Cartographers advanced Eurocentric notions based on biological conceptions of race and the assumed primacy of French or English. The maps they produced thus excluded not only continents other than Europe from the map of civilization, but also regarded supposedly primitive peoples, such as the Cossacks, as ruling over, or simply inhabiting, "non-historical" borderlands in European popular imaginations.[2]

The acquisition by force or negotiation of the many historical territories of Ukraine, including Galicia, Volhynia, Podolia, and Crimea, became a desideratum for both the Russian Empire and Poland-Lithuania because of Ukraine's strategic location along the Black and Azov Seas. Understanding the prerogatives of European state-to-state politics in a staunchly realist manner, Tsar Peter I (r. 1682–1725) and Tsarina Catherine II (r. 1762–96) became participants in the "progressive" mode of European state cartography; namely, the territorial expansion and scientific-technical administration of empire.[3] The maps of Ukraine that were produced during the age of Enlightenment are reflective of the increasingly frequent European attempts at expansion and state centralization in a Great Power world.

The maps discussed in this chapter demonstrate the extent to which the production of maps flourished in such cities as Amsterdam, London, Paris, Augsburg, Vienna, and the new Russian imperial capital of St. Petersburg, where equipment and the trained personnel necessary to produce them were scarce. Dutch and English cartographers continued to make maps and atlases: in Amsterdam, Witsen, Schenken and Visscher, and Châtelain; in London, Senex and Moll. German-language maps were produced in Vienna and Nuremberg. Cartographers and family-run printing firms made maps and atlases, often on commission for state sovereigns. Four factors contributed to a new concentration on the acquisition and administration of territory in Central and Eastern Europe: (1) the conflictual economic interests of England, France, Spain, and Portugal in the New World; (2) the decline of Sweden and the instability of Poland-Lithuania; (3) the decline of the multiconfessional and multilingual Ottoman Empire in relative terms of military power; and (4) the emergence of Prussia and the Russian Empire. Between Europe's imperial powers, Ukraine's strategic situation as a southern coastal borderland placed it in a volatile position.

Imperatives of Enlightenment Cartographers

In France, cartographers and publishers such as Delisle, Fer, Crépy, Bellin, and Robert de Vaugondy embraced the paradigm of Enlightenment with its confidence in science and progress. Guillaume Delisle (1675–1726) came from a French family dynasty of cartographers. His father Claude (1644–1720) was a French geographer and historian, and pupil of Nicolas Sanson the elder. He himself had studied with Giovanni Domenico (Jean Dominique) Cassini

(1625–1712), an Italian astronomer and geodesist who came to Paris in 1669, measured the Paris meridian, and directed the royal astronomical observatory. Delisle's first map of the world appeared in 1700, and generated a great deal of public interest. In 1702, he became a member of the French Académie Royale des Sciences. For his efforts as a scientist, he earned acclaim throughout Europe, and he and his sons were sought after by a number of state sovereigns. Delisle was appointed to the esteemed position of Premier géographe du roi (First Geographer to the King) in 1718.

In terms of scientific quality, Delisle made great strides in astronomy and geography. In his maps of France and of Europe, he stressed the importance of accuracy in the calculation of astronomical points; for instance, he corrected the longitudinal width of the Mediterranean to 42 degrees, which had not been determinable previously. One of the giants of European cartography, he produced over one hundred maps and globes during his lifetime. He thought of his maps in the spirit of the Enlightenment, as tools that would improve knowledge of the world. His work was collected posthumously into the *Atlas nouveau* by the firm of Covens & Mortier (Amsterdam, 1735). Delisle's son-in-law, Phillippe Buache, worked with J. A. Dezauche to revise Delisle's plates. The two men compiled an updated atlas published in Paris in 1789.[4] The 1700 map, *L'EUROPE // DRESSÉE SUR LES OBSERVATIONS DE M.rs // DE L'ACADEMIE ROYALE DES SCIENCES // ET QUELQUES AUTRES; & SUR LES MEMOI- // RES LES PLUS RECENS* (**figure 29, #176**), appeared variously in the *Atlas de Géographie* (1700–26) compiled by Guillaume Delisle in Paris. The map includes Ukrainian territories within the limits of "Pologne," "Moscovie Européene," and "Petite Tartarie." Mirroring Beauplan, the region on both banks of the Dnipro River is designated in rather typical form for the time as "Ukraine ou Pays des Cosaques."

Nicolas de Fer (1646–1720) was another prolific French geographer based in Paris. He took over the family business after his father Antoine's death, and transformed it into a flourishing map publishing company. Louis XIV appointed him to the position of "Géographe ordinaire de Sa Majesté catholique." Fer produced many atlases of France, and maps of Europe. His 1703 map of Europe, *L'EUROPE // PAR DE FER* (**figure 30, #151**), was published in a manual for the study of Enlightenment geography by Robbé, *Methode pour apprendre facilement la Geographie.*

English cartographic and encyclopedic pursuits were equally important in the eighteenth century. John Senex (c.1678–1740), for one, was an English surveyor, engraver, bookseller, and publisher of maps and atlases. The son of a Shropshire gentleman, in 1702 he set up his own publishing business producing maps of England and continental Europe. He is perhaps best known for his many antique globes. Ephraim Chambers, who was apprenticed to Senex when the latter was making globes, developed his plan for the influential two-volume *Cyclopaedia, or Universal Dictionary of Arts and Sciences*, first printed in London in 1728, while working at Senex's workshop in London from 1714 to 1721. The *Cyclopaedia* proved to be a source of inspiration for Diderot and D'Alembert's *Encyclopédie*, which started as a French translation of the work in 1743. Senex was elected as a fellow to the Royal Society of London in 1728, where he delivered lectures on astronomy and served as the royal geographer.[5] Senex's 1708 map of Europe, *A NEW MAP // OF // EUROPE // From the latest Observations* (**figure 31, #494**), envisioned European states as discrete political units. In this map, Ukraine was again "off the map" of civilization, at least in the Western European imagination. It was seen as a stateless borderland, and therefore ill-suited for inclusion in a general "modernizing" or progressive political map.

In another line of continuity in the business of producing maps, Gilles Robert de Vaugondy (1688–1766) and his son Didier Robert de Vaugondy (1723–86) ran a flourishing globe business in Paris from 1731–78. They produced over five hundred maps and geographical works, including two major atlas compilations in 1748 and 1752. In 1731, Gilles inherited Sanson's copperplates, globes, manuscripts, geographical papers, and rolling press, and also bought out the shares and cartographic materials of Jaillot, who had been Sanson's partner. Gilles became a professor of mathematics in Paris and was appointed to the position of Géographe du Roi (Geographer to the King) in 1734. Didier, a teacher of mathematics, royal censor, and globemaker, became the Géographe du Roi to Louis XV in 1751. In 1778, the business was taken over by Jean Baptiste Fortin, and then by Charles François Delamarche (1740–1817), a geographer, publisher, and globemaker in Paris.[6] The 1749 map *Partie Orientale // de la Russie Polonoise // où se trouvent // la Basse Volhynie et la // Basse Podolie. // Par l S.r Robert de Vaugondy // fils de M.r Robert Geog. du Roi. // Avec Privilege 1749* (**figure 32, #433**) appeared in the *Atlas portative*, published in Paris in 1749. It includes "Palatinat de Kiovie" and "Cosaques Zaporoski" as place names.

Jacques Nicolas Bellin (1703–72), a major contributor to the *Encyclopédie* project, spent over fifty years in the French Hydrographic Service. Bellin was the first ever appointee in France to the position of Ingénieur hydrographe de la Marine (Hydrographic Engineer of the

Navy), and later became the king's personally appointed hydrographer, in charge of making and compiling sea charts of high scientific quality. Bellin carried out major surveys of the coasts of France, and eventually of all known coasts across the world. A general success story and representative of scientific pursuits in the high Enlightenment, Bellin was elected fellow of the Royal Society of London.[7] Bellin's 1750 global and marine map *An Essay of a New and Compact Map Containing the known Parts of the Terrestrial Globe* (**figure 33, #52**), with translations of the title in French and Dutch and a cartouche depicting a compass, includes Poland and Ukraine as parts of the known world.

Ukraine between Empires

Krawciw's collection of eighteenth-century maps represents the continuation of his attempt to promote the recognition of Ukraine and Ukrainians in imperial borderlands, and, ultimately, to carve out an independent political homeland that encompasses Ukraine's many historical and territorial legacies. Whereas most of the sixteenth- and seventeenth-century maps in the collection were printed in Latin and French, the eighteenth-century maps appear principally in French, English, and German, although there are still some Latin examples.

Habsburg Lands and the Kingdom of Hungary. Habsburg cartography in the age of Enlightenment reflected state interests and a growing popular interest in science and education.[8] Johann (Hans) Georg Bodenehr (1631–1704) and his sons Gabriel (1664–1758), Morritz (d. 1749), and Georg Conrad Bodenehr were cartographers who owned a publishing and engraving business in Augsburg. They printed maps mainly in French and German. The patriarch, Hans Georg, published a small atlas of German lands in 1677. They acquired the business of Johann Stridbeck (1665–1714), another Augsburg map publisher who was known for his maps of Bavaria. The Bodenehr family made some new maps, and printed revised editions of Stridbeck's general compilation, *Atlas curieux* (1704–37).[9] The 1704 map of Hungary by Johann Bodenehr, *Ungarn samt angräntzenden Ländern Croatien, Dalmatien, Sclavonien, Siebenbürgen, Moldau, Wallacheu, Bulgarien u. Romanien, bis an das Schwartze Meer und Constantinopel* (**figure 34, #71**), covers most of southeastern Europe including the Ottoman and Habsburg dynastic lands, ranging from the Black Sea to Constantinople.

Despite the decline of the Netherlands relative to the Great Powers after 1700, Dutch cartography remained important. Its influence extended into France, German, and Habsburg lands, and as far as tsarist Russia. Petrus (Pieter) Schenk I (1661–1711), Petrus (Peter) Schenk (Schenken) II (1693–1775), and Petrus Schenk (Junior) III (1728–1803) were an influential German-Dutch family of publishers. Petrus Schenk the Elder came to Amsterdam, studied engraving, and became a seller of art and books. Schenk the elder became a partner with Gerard Valck (1650 or 1651–1720), another Dutch engraver and publisher in Amsterdam. In 1685 Schenk the elder and Valck bought the firm of Blaeu and printed a general atlas that covered many German lands. With a strong connection to Dutch cartography, Petrus Schenk II reprinted in Latin, French, and German some of the works of the first and second generations of the Sanson and Visscher firm. The Schenk-Valck firm became quite successful, and a branch of the shop opened in Leipzig in 1695 to export some of Valck's globes. They obtained a charter to publish some of Nicolas Sanson's maps. Valck himself parted ways with the Schenks after 1700 and opened a rival business.[10] The 1705 map of Hungary and the Habsburg Empire by Schenk and Visscher, *REGNORUM // HUNGARIAE, // DALMATIAE, CROATIAE, SCLA // VONIAE, BOSNIAE, SERVIAE, // ET PRINCIPATUS // TRANSYLVANIAE // TYPUS ACCURATIOR, ET PRINCIPATUS // TRANSYLVANIAE // Typus accuratio, et plane novus* (**figure 35, #482**), was published in Amsterdam in the atlases by Schenk-Valck. Seen from a German vantage point, with Ukrainian lands depicted along the northeastern borders of the Habsburg Länder, the map includes all of the Transcarpathian area and the southern part of Galicia, labeled "Stat Polonicus" and "Russia Rubra."

Imperial Interests in the "Easts." Enlightenment ideals about the overall improvement of human knowledge, confidence in modernization, and the ability to "civilize" uncivilized lands and peoples through education and literacy easily became entangled in the geostrategic realities of Europe's many "Easts." Cartography, in short, was as much a political and military exercise in the context of imperial great games, where charts of naval fortifications and commercial trade routes were as valuable as ethnographic or encyclopedic compendia. Crimea, the Tatar lands, the Black and Azov Sea regions, and other lands between the Ottoman Empire and Muscovy (after 1721, the Russian Empire) were all territorial parts of the Ukraine for which Krawciw sought political recognition beyond Russian and Soviet claims. The cartographic works in his collection indicate how these areas became mapped in the eighteenth century as contentious geopolitical spheres of influence among European powers.[11]

In Paris, the topographical maps of Delisle and Fer were made for use by French kings and Russian tsars. The 1714 map by Fer, *La journée de // Poltawa // en Ukraine, // le 8.e juillet 1709. // Entre l'Armée de sa Majesté // Suedoise Charles XII, et celle de // sa Majesté Csarienne Pierre I. // Empereur de la grande Russie, // dit vulgairement Grand Duc de // Moscovie // de Fer. 1714* (**figure 36, #160**), which appeared in the 1725 *Suite de l'Atlas Curieux* by Fer, shows the results of the Battle of Poltava. Made in Paris, Delisle's 1723 map, *Carte // des Pays Voisins de la // Mer Caspiene // dressée pour l'usage du Roy // sur la Carte de cete Mer faite par l'ordre du Czar* (**figure 37, #622**), depicts coastal lands and routes around the Black Sea and Caspian Sea.

Dutch maps were also dual products of Enlightenment science and colonial pursuits. Nicolas Witsen (1641–1717) was a Dutch mayor of Amsterdam and an avid map collector and mapmaker. He served as an ambassador to Russia, and was a consultant to the court of King William III. In 1687, he printed a map of Siberia, and became known for manuscript charts of Dutch colonial coastlands in Indonesia and Asia.[12] Witsen's 1700 map of the Russian Empire, *IMPERII // RUSSICI // sive // MOSCOVIAE. // Status Generalis, in ejus // Regna. Ducatus, Provincias, Populosq. Subjacentes Divisus, et ex Tabula* (**figure 38, #566**), was published by the Dutch firm of Covens and Mortier. Again following the Beauplan model, the 1735 map of Muscovite/Russian expansion toward the southern coasts by Schenk and Visscher, *NOUVELLE CARTE GEOGRAPHIQUE DU GRAND ROYAUME DE MOSCOVIE REPRESENTANT LA PARTIE MERIDIONALE & C.* (**figure 39, #541**), depicts Ukrainian territory as a borderland and designated it as the "Pays de Cosaques" (Land of the Cossacks).

Especially important for early eighteenth-century cartography was the intellectual exchange between Dutch and other European cartographers. The Frenchman Henri Abraham Châtelain (1684–1743) was a member of a Dutch-French family of booksellers and publishers in Amsterdam. He became best known for his *Atlas Historique*, which appeared under various imprints from 1705 to 1739.[13] Châtelain's 1714 map of Muscovy, *Nouvelle Carte de Moscovie où Sont Representés les Diferents Etats de Sa Maiesté Czarienne en Europe et en Asie et le Chemin d'un de ses Ambassadeurs a Peking Ville Capitale de L'Empereur de la Chine et son Sejour Ordinaire* (**figure 40, #101**), was taken from his *Atlas Historique*. The map of Muscovy and its environs is based on the work of the cartographer Ides Evert Isbrand (1642–94). The historical maps appearing in his atlas often include heraldic seals, medallions, and other insignia as proxies for Russian imperial claims and a means of organizing dynastic territory.[14]

Eighteenth-century cartographers in the Habsburg lands were also interested in the mapping of Ukraine, but since the state lacked a strong navy to compete with England or Russia, or an army to compete with France, their ambitions on Ukrainian lands would become mainly confessional and dynastic, rather than commercial or military in nature. In German lands, Johann Baptist (Baptista) Homann (1663–1724) was a self-taught copper engraver from Swabia who founded his own cartographic and publishing establishment in Nuremberg in 1702. In 1707, he began compiling his own large-scale atlases, some of which had over one hundred maps. In 1715 he became Geographer to the (Habsburg) Holy Roman Emperor, and also served as a member of the Prussian Royal Academy of Sciences. After the death of his son Johann Christoph Homann (1703–30), the company became known as Homännische Erben (Homann Heirs).[15] Homann's 1700 German-language map of Constantinople, *Accurate Vorstellung der Orientalisch-Kayserliehen Haupt- und Residenz-Stadt CONSTANTINOPEL samt ihrer Gegend und zweyen berühmten Meer-Engen, Bosphoro Thracio, und Hellesponto oder dem freto der Dardanellen* (**figure 41, #237**), shows a view of the Dardanelles Straits and Constantinople (Istanbul), and is decorated in Enlightenment style with coats of arms, figures in native garb, and weaponry. Homann's 1710 map, *Generalis Totius // Imperii Moscovitici // Novissima Tabula // Magnam Orbis terrarum partem à Polo Arctica usq. // ad mare Japonicum, et Chinae Septentrionalis // confinia exhibens // cum via Czaricae nuper Legationis ex urbe Mo // scua per universam Tartariam ad magnae Chinae Imperatorem // ex conatibus // Johannis Baptistae Homanni // Norimbergae* (**figure 42, #240**), includes "Ukrania" and "Zaparovie." The map is also decorated with royal, religious, mythical, and laurel-wreathed figures as well as a shipbuilder and cupid measuring a map. Finally, Homann's finely crafted 1730 map of Ukraine, *Ukrania // quae et // Terra Cosaccorum // cum vicinis // Walachiae, Moldaviae, Minorisq. Tartariae Provinciis // exhibita // à Joh. Baptista Homanno // Noribergae* (**figure 43, #248**), is engraved with Polish and Cossack soldiers and flags.

Also from Augsburg, Matthäus Seutter (1678–1757) was from a German family of artists, painters, goldsmiths, engravers, and map publishers. He was a cartographer, engraver, globemaker, and map publisher, first apprenticed to Johann Baptist Homann in Nuremberg. He returned to Augsburg in 1707 and set up his own publishing house, which rivaled the Homann firm. Seutter was awarded the title of Imperial Geographer by Charles VI in 1731. Although he crafted about forty original maps, in the end most were copied from other works.[16] Seutter's 1740 map of Russian imperial expanses, *SPATIOSISSIMUM // IMPERIUM // RUSSIAE MAGNAE // juxta recentissimas Observationes // MAPPA GEOGRAPHICA // accuratissime delineatum* (**figure 44, #508**), includes the lands of the Russian Empire, the Ottoman Empire, and so-called Tartary. Seutter tended to see Ukrainian lands as incorporated into other polities, in this case as part of the Russian Empire.

An interesting cross-cultural case in European cartography can be seen in the work of Herman Moll (ca.1654–1732), an English cartographer of German and Dutch origin who came to England in 1670s. He worked in London as an independent cartographer and geographer, and founded his own map trade business and publishing house. Moll's 1736 map of the Ottoman Empire and Kingdom of Hungary, *A General Map of // Turky in Europe, // Hungary &c.* (**figure 45, #366**), appeared in his *Atlas Minor*. The map includes views of Belgrade and "Temeswar" (Timişoara) in the lower right-hand corner, as well as lands of the Ottoman Empire and the southeastern parts of Ukraine. Krawciw took an interest in such maps even though Ukraine is not singled out as the focal point of the map, since they recognize Ukrainian borderlands at the crossroads of empires other than Russia and Poland-Lithuania.

Russian Imperial Cartography and the Academy of Sciences. With the proclamation by Peter I of Russia as an empire in 1721 after the end of the Great Northern War, the Academy of Sciences (Academia Scientiarum Petropolitana) in St. Petersburg was founded by Senate decree on January 28 (February 8), 1724. Joseph Nicolas Delisle (1688–1768), son of the eminent French geographer Guillaume, worked closely with the secretary of the Senate Ivan Kirilovich Kirilov (1695–1737), who in 1721 had been appointed by Peter to collect, register, and store all maps and surveys for the empire. Joseph Nicholas Delisle was one of three of Guillaume's twelve children who became eminent geographers in France, and was known throughout Europe. He studied with the Cassini family of cartographers in France as an astronomer, and worked at the court of Louis XV. His brother Louis (1687–1741) became an astronomer, and later joined Joseph Nicolas in Russia in 1726, where he worked until his death while on an expedition in 1741.

Joseph Nicholas Delisle occupied the chair of astronomy, geography, and navigation at the Academy of Sciences in St. Petersburg from the time of its establishment. He dreamt of enormous projects for which the Russian Empire, especially under subsequent tsars after Peter I, had few resources or trained personnel. In 1729, Joseph Nicholas Delisle made a proposal for a general atlas spanning the entirety of the empire, and in 1735 he suggested the formation of a Geographical Department. It took ten years for the piecemeal *Atlas Russicus: mappa una generali et undeviginti specialibus vastissimum imperium russicum adiacentibus regionibus* to be printed by the Academy in St. Petersburg. The work would serve as the first general territorial map of Russia. Delisle returned to Paris in 1747, and brought with him important collections of maps detailing Russian imperial spheres of influence.[17]

There are a few maps printed by the Academy of Sciences in Krawciw's collection. The 1736 map of the Russian Empire in conflict with the Ottoman Empire and the Tatars, *CHARTE // derer von der// RUSSISCH-KEYSER: ARMEE/ im Iahr 1736 // zwischen und an dem Dnieper und Donn // wieder die Turcken // und Tartarn // siegreich unternommenen // Kriegs-Operationen* (**figure 46, #12**), is the first printed Russian map (in German) of central and eastern Ukrainian territory including the Right and Left Banks, designated in the upper center as "Ein Theil der Ukraine." Place names include "Wolhinien," "Podolien," "Die Crim," and "Die Kuban."[18] Also made by the Academy of Sciences in St. Petersburg, the 1745 map of the Russian Empire, *MAPPA // GENERALIS // TOTIUS IMPERII // RUSSICI* (**figure 47, #13**), includes only the lands of Left-Bank Ukraine, named in Russian as "Kiewskaia Guberniia" and "Bielogorodskaia Guberniia," and parts of southern Ukraine, labeled as "Krym" (Crimea).[19] The 1787 map of the Russian Empire, *Carte Générale de l'Empire de Russie* (**figure 48, #287**), engraved by Pierre Tardieu, was produced in French at the Academy in St. Petersburg for Empress Catherine II. An iconic imperial map emphasizing Russia's territorial vastness, the map represents Russia stretching across Europe and Asia, and to the Pacific coast of North America.

Conclusion

Krawciw was especially interested in European maps of Ukraine during the Enlightenment. In fact, nearly 25 percent of the total map collection dates to the eighteenth century. The maps generally reveal a west-to-east European, or "Western," template for imagining the boundaries of civilization, a convention that followed from Beauplan in the depiction of Ukraine as "the land of the Cossacks." In a practical, and geostrategic, sense, maps were also used by European monarchs to advance imperial and colonial interests in the post-Westphalian state system. Even while cartographic production shifted to Paris and London, there were strong cultural and political continuities with the early modern and even Renaissance periods. On the European continent in an age of religious and dynastic wars, territories still changed hands frequently between state sovereigns, and borders were rarely fixed to exact geographical points, even in treaties. Most maps before 1800 were still produced for state sovereigns in French or English, and sometimes Latin, since language was not yet a marker for nationality or "ethnic" identity in Central and Eastern Europe. By the eighteenth century, cartographic works were disseminated along with Diderot and d'Alembert's famous *Encyclopédie* to a more general reading public in France, England, and other parts of Europe. Universal maps and atlases became a template to view, experiment with, and in some cases construct a template for understanding the history, geography, and politics of Ukraine, about which little of concrete value was actually known.

4 *Nineteenth-Century Great Power Politics and the Ukrainian Question*

IN HIS QUEST FOR UKRAINIAN cultural and political recognition, Krawciw collected over one hundred maps detailing the late-eighteenth-century annexations of Ukrainian lands. Tsarina Catherine II ended Ukrainian autonomy in Sloboda Ukraine (1765), destroyed the Zaporozhian Sich (1775), and liquidated the Cossack Hetmanate (1785) through policies of territorial expansion by land and by sea, and strategies of state co-optation and centralization.[1] The three partitions of 1772, 1793, and 1795 put an end to the Polish-Lithuanian Commonwealth that had been established in 1569 to unite the Polish Crown and Grand Duchy of Lithuania.

Russia's annexation of Crimea in 1783 revealed that Catherine had imperial designs even prior to the outbreak of the French Revolution. Following European technical and aesthetic models, Russian maps became tools of Great Power statecraft, fulfilling military and administrative functions, and serving as iconic emblems of both scientific and governmental progress—even when such progress was lacking. The geopolitical ambitions of Russia, Prussia, and Austria in Central and Eastern Europe led to the partition of not only the Polish-Lithuanian Commonwealth, but also the Ukrainian lands within it.

Representing Territorial Changes

The threat of partition loomed in the eighteenth century when state sovereigns, following the lead of the Bourbon monarchs in France, aimed to rationalize control over borderland territories through maps. European cartographers continued to reprint Beauplan's textual descriptions and visual representations, but many offered more detailed works of lands within Poland-Lithuania. Tobias Conrad Lotter (1717–77) was of a German family of engravers and publishers who produced maps and atlases in Augsburg. He worked with his father-in-law, Matthäus Seutter, and reprinted some of the copperplates of the Seutter firm in 1758.[2] Made prior to the partitions, Lotter's 1759 map of Poland-Lithuania, *Mappa // Geographica, // ex novissimis observationibus // repraesentans // Regnum // Poloniae // et // Magnum Ducatum // Lithuaniae* (**figure 49, #318**), included as place names "Ukraina," "Podolia," "Pokutia," "Volhynia," and "Isles ou le retirent les Cosaques." The cartouche depicts a merchant, a horseman and farm animals, and a coat of arms. Another depiction of Poland before its first partition in 1772 was Gilles Robert de Vaugondy's 1767 *Le Royaume de Pologne* (**figure 50, #440**), engraved in Paris at the firm of Delamarche. Intended for the king's use, this map included a list of all the regions, major cities, and even minor towns within the Polish Crown lands and along their borders.

Partitions were, of course, legitimated in the maps drawn by the victorious expansionists, primarily the Habsburg and Russian empires. Since at least 1768, the Russian imperial army had been present on Polish territory in Right-Bank Ukraine, and in 1769 it set up a garrison in Lviv, the capital of the Rus′ palatinate, which the tsars had claimed as patrimonial lands descended from Kyivan Rus′. For their part, Habsburg monarchs justified the first partition by laying historical claim to some of the lands of Kyivan Rus′, arguing that they had belonged to Hungary in the twelfth century as the kingdom of Galicia and Lodomeria (Königreich Galizien und Lodomerien), or Galicia (Galizien) in shorthand.[3]

The first partition of Poland-Lithuania in 1772 dramatically changed the map of Europe, and also directly affected European conceptualizations of Ukraine.[4] Prussia annexed Polish crown land territory along the Baltic Sea, and Russia annexed lands in northeastern Poland-Lithuania (present-day Belarus). The initially reluctant Maria Theresa of Austria annexed the mainly Ukrainian-inhabited Polish palatinates of Galicia-Rus′ and Belz, territory along the south Vistula (Wisła) River, and the city of Cracow (Kraków, Krakau, Cracovia), former capital of the Jagiellonian dynasty. In 1774, the Habsburgs acquired the mountainous area of Bukovina from Moldavia within the Ottoman Empire, and the Russian army finally withdrew from Lviv. The Homann firm's German-language 1775 map of the Kingdom of Galicia and Lodomeria, *Lubomeriae // et // Galliciae // Regni // Tabula Geographica* (**figure 51, #252**), served to legitimate these partitions and provided a historical justification for Habsburg dynastic claims to the lands of Rus′.

Although European maps of the partitioning of lands in Poland-Lithuania were hardly neutral, and most were made by the victors as a political fait accompli to the annexations, not every map from London, Paris, or elsewhere supported enlightened autocracy by the imperial powers. Many skillful cartographers did work on commission for the last Polish king, Stanisław August II Poniatowski, or for influential members of the Polonized *szlachta* who resisted the partitions during their course and afterward in the nineteenth century.[5] For example, Giovanni Antonio Rizzi-Zannoni (1736–1814) was a Paduan geographer and astronomer who worked in Paris. In 1774, he became the chief hydrographer to the Dépôt des Cartes et Plans de la Marine, directly succeeding the eminent Jacques Nicolas Bellin. Rizzi-Zannoni was extraordinarily prolific as a cartographer. He worked, for instance, on a survey of the Kingdom of Naples in 1778, was a founder and director of the Topographical Office in Naples in 1781, and published series of maps of Poland-Lithuania and Italy from the 1770s to the 1790s.[6] Rizzi-Zannoni's 1781 Italian-language map of the Bratslav and Kyiv palatinates, *LI PALATINATI // di // BRACLAW, // e // KIOWIA* (**figure 52, #423**), printed in Venice, was part of a series of maps depicting palatinates in Poland-Lithuania, including Volhynia and Podolia. The map includes a cartouche of trees and houses, showing a provincial landscape (detail below). Rizzi-Zannoni's place names appear mostly in Polish; they include "Ukraina Russa," written to the east of "Kiiow." The Dnipro River runs through the center, dividing Left- and Right-Bank Ukraine. Other bodies of water include the "Przypek," "Deszna," "Bog," and "Niester" Rivers.

Another map of lands in southern Ukraine was printed by Antonio Zatta (c.1724–1804), a publisher based in Venice who authored a number of atlases in Italian. Zatta's 1783 map of "Little Tartary," *LA PICCOLA // TARTARIA // divisa // NE' SUOI TERRITORI // Di nuova Projezione* (**figure 53, #576**), includes lands comprising much of southern and eastern Ukraine between Poland ("Polonia") and Russia ("Russia") along the Black Sea, and "Tatar" lands: "Tartaria di Oczakow," "Territorio

Detail from figure 52.

dei Zaporowiani," "Territtori di Azof," "Tartari Nogaiesi," "Krimea," "Circassia." The map includes rivers, mountains, and seas, and a cartouche depicts Orientalist images of incense, palm trees, and colored tents.

Erasing Poland-Lithuania as a polity, the Russian imperial division of territory into "European" and "Asiatic" parts would become a more common trope during the course of the partitions. During her reign, Catherine II boldly announced to heads of state that Russia had arrived on the Great Power stage, and was in fact a "European" power ruling over at least two continents, Europe and Asia. With this in mind, she saw maps as she envisioned her empire, expansive and rationally ordered. Catherine used maps both to prove her claims to actual territory in diplomatic settings, and as visual scenarios of power to exhibit her benevolence, intelligence, and "enlightened" system of ordering lands into provinces.[7] Meanwhile, other mapmakers served their own states' political interests. Based in Paris, Rigobert Bonné (1727–95) was a hydrographer to the French king. He depicted Russian imperial gains for the sake of French foreign policy and intelligence. His 1785 map of the Russian Empire, *Partie Occidentale de l'Empire DE RUSSIE, Extrait de l'Atlas Russien et d'autres Cartes Projetée et assujetie aux Observations Astronomiques Posterieures* (**figure 54, #73**), was printed in Paris. It shows the objects of Catherinian territorial expansion and rational organization of acquired lands into provinces (*gubernias*). In the map, Russia is divided only into "Russie d'Europe," and "Siberie." New provinces include "Bielogorod" and "Kiow." There are no place names to mark independent or autonomous status for Poland-Lithuania or the Cossack Hetmanate.

Radically new European political narratives through maps were suggested after the first partition of Poland-Lithuania by Prussia, Russia, and Austria. John Bowles (c. 1701–79) and his brother Thomas Bowles III (c. 1712–62) were from a family of map publishers and sellers in London. In 1723, the Bowles established their own commercial firm selling and trading maps.[8] They printed the works of many influential Anglophone cartographers, including Herman Moll, Emanuel Bowen (c. 1693–1767), Thomas Kitchin (1719–84), and Robert Sayer (1725–94). Bowles' success relied in part on his promotion of their prestigious, and presumably progressive, works. His 1785 map of Europe, *Bowles's New One-Sheet Map of Europe, Divided into Its Empires, Kingdoms, States, and Subdivisions; laid down from Observations of the most Celebrated Geographers* (**figure 55, #82**), was printed in London. A table at the lower right-hand corner of the map, "A Methodical Division of Europe...," illustrates the first partition of Poland-Lithuania.

The annexations were also shown in key works by French cartographers. The celebrated Jean Baptiste Bourguignon d'Anville (1697–1782), another Paris-based geographer and cartographer to the French king and elected member to the Académie des Sciences, drew maps and amassed a huge collection of over two hundred rare and contemporary exemplars, now stored in the Bibliothèque Nationale in Paris. Modeled on his work, a new map of Europe displaying the first partition was printed numerous times after his death. In *The // Third Part of // Europe, // Drawn & Engraved // from D'Anvilles // Two Sheet Map* (**figure 56, #145**), dated 1791, one can see the place name, "Krayn or Ukrane." Russian territorial acquisitions along the Baltic, Black, and Caspian seas are seen as a politically and intellectually progressive extension of Catherine's empire.

With the last of the three partitions of Poland-Lithuania in 1793 and 1795, Russia gained 178,900 square miles (463,200 square kilometers) and about 5,500,000 inhabitants in total; the Habsburgs received 49,000 square miles (128,900 square kilometers) and about 4,150,000 inhabitants; and Prussia gained 54,600 square miles (141,400 square kilometers) and about 2,600,000 inhabitants.

After the army of Catherine II suppressed the uprising of Tadeusz Kościuszko in late 1794 and early 1795, Russian imperial officers looted Warsaw and liquidated the most important map collections in Poland-Lithuania.[9] They took the more than 400,000 volumes of the National Załuski Library as booty and sources of intelligence. State registers and geographical-statistical descriptions were among the stolen items, and the 325 maps in the king's private Crown Register became the nucleus of the Imperial Public Library in St. Petersburg and the Register for the Russian Army's General Staff. In response to this, nineteenth-century Polish nobles and intellectuals assembled private "national" collections of Polonica, in much the same way that Krawciw would respond to Soviet control over cartography and information in the twentieth century by collecting Ucrainica.[10]

For cartographers, the decision to reject the legitimacy of imperial land acquisition could be expressed in their maps, by using an alternative narrative of "dismemberment" rather than Great Power "acquisition." This narrative generally became more commonplace in English and French, rather than German, cartography.[11] The maps of the cartographer William Faden (1749–1836), the geographer to the king of England, followed this trend. His 1799 map of Poland-Lithuania and the Baltic littoral, *A Map of // the Kingdom of Poland // and Grand Dutchy of Lithuania // Including Samogitia and Curland, // Divided according to their Dismemberments // with the Kingdom of Prussia* (**figure 57, #199**), includes an overview of Warsaw (detail below) and the territorial overview of partitions.

In contrast to English and French practice, German cartography overwhelmingly treated the acquisitions by Prussian and Habsburg monarchies as legitimate. For instance, a survey map of Eastern and Western Galicia was made in 1801 by George Rudolph, Freiherr von Schmidburg, of the main German geographical institute in Weimar.[12] The *Charte // von // Ost und West // Galizien // nach den neuesten astronomischen // Ortzbestimungen entworfen und, revidirt // auf der Sternwarte Seeberg bey Gotha* appeared in a Weimar publication of *Allgemeiner Handatlas der Ganzen Erde.*

As territories changed hands during the Napoleonic wars, maps proliferated. State sovereigns created centralized maps and systematized territorial surveys, but copies of maps also entered the popular reading mainstream. The map printing business of the London-based team of Robert Laurie (1755–1836) and James Whittle (1758–1818) flourished from 1794–1812. A cartographer himself, Laurie had been apprenticed to Robert Sayer as an engraver, and purchased Sayer's firm after the latter died. The 1807 map of the European theater of war, *A New Map of // the Seat of War, // Comprehending // Germany; Poland, // With its Dismemberments, // Prussia; Turkey in Europe, // Italy &c.* (**figure 58, #295**), detailed the partitions, and included part of Galicia. A cartouche shows a background of forested mountains topped by a crowned eagle.

Across political borders, maps expressed the Enlightenment's emphasis on progress and the rational organization and administration of territory, and would soon come to serve as sources for proving the history of "non-historical" lands and peoples on the European continent and beyond. Since at least the Treaty of Westphalia, they were created for Great Power purposes, but soon became the property of national movements as they clamored for resources and recognition.

Historical Atlasmaking and Geopolitical Representation

The French Revolution and Napoleonic wars transformed cartography from a purely elite craft of commissioned artisans and scientists into a "modern" amateur pursuit among nation- and state-building intellectuals, collectors, and a growing and increasingly diverse reading public. A case in point is that of the popular historical atlasmaker, Comte de Las Casas, Emmanuel Marie Joseph Auguste Dieudonné, a French aristocrat, dispossessed émigré, and ardent supporter of Napoleon, who used the pseudonym Emanuel Auguste Dieudonné Le Sage (1766–1842). Le Sage's spectacularly influential *Atlas historique, généalogique, chronologique et géographique* was first printed in London (1801) and Paris (1802–4). It was bought by kings and aristocrats and used later in French schools during the period of Bourbon Restoration. The engraved atlas and its map models and tables were reprinted in subsequent decades almost everywhere in major European capitals and in America, in cities all the way from St. Petersburg to Philadelphia. Le Sage's political and historical map, *L'Europe Moderne // avec ses Divisions Politiques* (**figure 59, #307**), appeared in his atlas and was published in Florence in French by the firm of Mollini, Landi and Company in 1805. The map reflects Russian imperialism

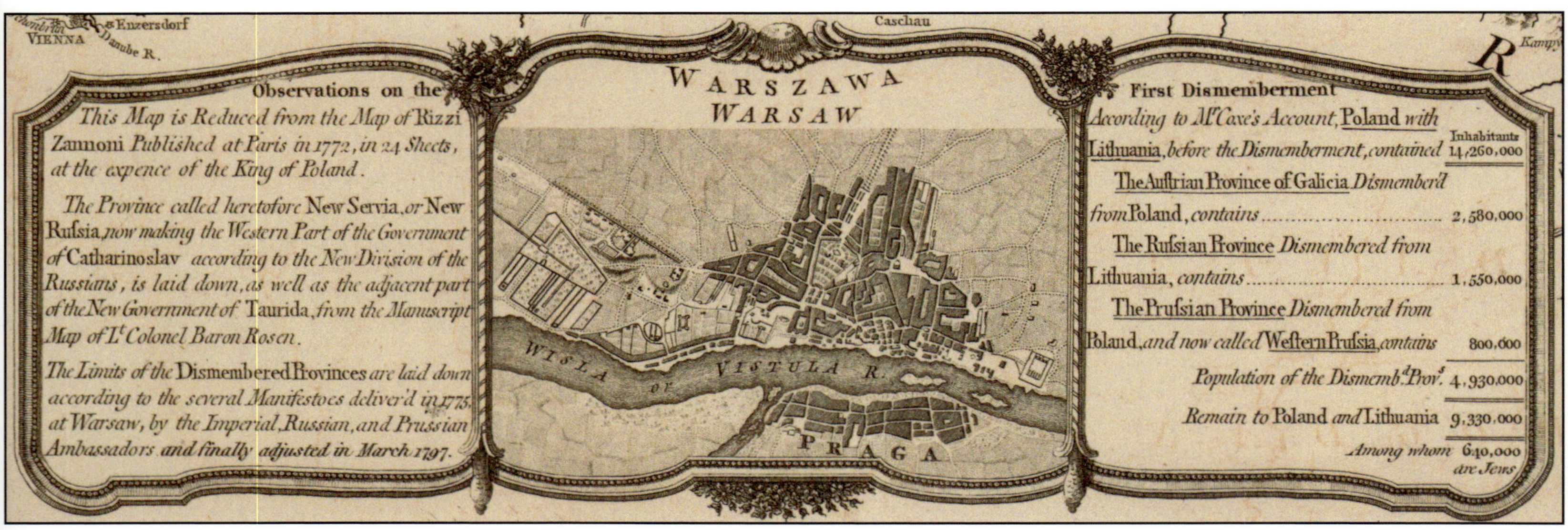

Detail from figure 57.

in Europe and Asia. It also includes maps and texts for the "Plan du Tableau, Contenus, Direction," "Divisions Politiques," "Campagnes de Charles XII, Roi de Suède," "Partage de la Pologne indiqués sur la carte," and "Statistique de L'Europe dans L'année 1805."[13] An aristocrat and French Catholic, Le Sage stressed in his maps the enduring presence of an imagined French golden age in the seventeenth and eighteenth centuries. He also promoted the idea that a continuity existed within European history between classical antiquity and early modern colonialism and imperialism, in contrast to the notion that the arrival of modern Europe marked a rupture from the past.

One fervent admirer of Le Sage's work, an Irish entrepreneur named Mathew Carey (1760–1839), was a native of Dublin who settled in Philadelphia and started a successful family publishing and bookselling business with associated companies. Carey assisted in the making of American versions of Le Sage's work; for instance, C. V. Lavoisne's *Complete Genealogical, Historical, Chronological, and Geographical Atlas*, initially made (without maps) in a first American edition in 1807. Carey's slightly revised version of an 1800 map, *The // RUSSIAN EMPIRE, // in // EUROPE AND ASIA* (**figure 60, #90**), appeared in 1814 in his improved and enlarged atlas, printed in Philadelphia. The map itself includes part of Ukrainian territory, labeled as "Little Tartary" in the south. Representing the American adoption of European cartographic styles and purposes, Carey's atlas encouraged the work of influential American mapmakers such as Samuel Lewis and Amos Doolittle.[14]

Mapping a New Imperial Order: The Congress of Vienna

At the Congress of Vienna, which was convened after the defeat of Napoleon and his Grande Armée in 1814, Tsar Alexander I, to the great befuddlement of Lord Castlereagh, imagined the Russian Empire as part of a Holy Alliance with Prussia and the Habsburg Empire, destined to preserve the political and territorial status quo in Europe. In the first half of the nineteenth century, the tsars suppressed revolts in Poland, and were called in by the Habsburg monarchy to put down the Hungarian Revolution in 1849, an act which stabilized the fragile state and allowed its new emperor, Franz Joseph (r. 1848–1916), to reestablish Habsburg legitimacy. Advancing earlier eighteenth-century policies of Peter I and Catherine II, the Russian Empire soon emerged as the greatest land power in Europe, and its territorial annexations continued practically unabated in Finland (1809), Bessarabia (1812), the Congress Kingdom of Poland (1815), Transcaucasia (1801–28), Transcaspian and Central Asian territories (1822–95), the Amur region (1858–60), and along the Far East, the Pacific coast, and briefly even Alaska.[15]

After the defeat of France and the new diplomatic settlements at the Congress of Vienna in 1815, European and American cartographers struggled to understand and represent such historic changes in hegemony and territory. The French-American C. V. Lavoisne and his British publishers, C. Gros and J. Aspin, published atlases such as the influential 1807 *A New Genealogical, Historical and Chronological Atlas* (1807). Lavoisne published American editions of Le Sage's pro-French work and those of other European cartographers, with additional maps describing American history and geography.[16] Gros' 1813 *MAP OF // POLAND, // PRUSSIA and HUNGARY. // indicating the places // rendered celebrated by // SIEGES and BATTLES // intended for the Elucidation of // LAVOISNE'S // HISTORICAL ATLAS* (**figure 61, #219**) appeared in a new and updated version in Lavoisne's *Complete Genealogical, Historical, Chronological, and Geographical Atlas*, published in London. The map illustrates lands of Galicia-Rus′, the boundaries of Galicia in western Ukraine, and the palatinates of "Brzesc," "Wolhinia," "Podolia," "Braclaw," and "Kiow" in the east. The map is accompanied by English-language text, and includes such general descriptions as: Poland, Austrian Dominions (Galicia), Battles and Sieges in Poland, Battles and Sieges in Hungary, and Hungary.

Between the Vienna Congress and the uprisings of 1848, maps of the former Polish-Lithuanian Commonwealth became contested texts and images for the writing and narration of history. Political maps of the European continent tended toward the black-and-white in considering the national/imperial story of legitimate "acquisition" versus illegitimate "dismemberment" of borderlands. Even cartographers who focused on mainly provincial maps could not escape English, French, and German geopolitics in relation to the many "Easts" imagined initially by Enlightenment concepts of civilization. For instance, John Cary (the elder; 1755–1835) belonged to an English family that engraved and published maps, charts, and globes. Cary is probably best known for his county atlases of England.[17] Cary's 1819 map of Poland-Lithuania, *A // New Map // of // Poland, // and the // Grand Duchy of Lithuania, // Shewing their Dismemberments and Divisions // between // Austria, Russia and Prussia, // According to the Congress of Vienna* (**figure 62, #91**), indicates dates in which divisions took place, and uses colors to indicate to which state they were incorporated ("Galicia...to Austria...1772"). Another cartographer of English provinces, Aaron Arrowsmith (1750–1823) and his son Samuel (1790–1873), were based in Durham. Aaron initially worked as a land surveyor, and then set up his own family publishing business.[18] Samuel Arrowsmith's 1828 map, *Russia in Europe* (**figure 63, #27**), was made in London and shows the expanses of the Russian Empire and its provinces in "European Russia." Finally, Alexander Keith Johnston I (1804–71), an English geographer and fellow of the Royal Geographical Society, published a general atlas in 1832. His maps of European Russia and his *South-West Russia // Showing the extent of // the Kingdom of Poland // previous to its partition in 1772* (**figure 64, #277**) depict the former Poland as a state and Ukraine as a region or borderland.

German-language maps of real or imagined territorial holdings to the east often combined history, statistics, and military intelligence to a greater extent than English or French geopolitical variants. While some were vaguely supportive of Polish independence against the tsarist empire, they did not conceive of Ukraine as a separate political entity. Rudolph Schlicht, a publisher in Mannheim, produced the 1831 *Historisch-statistisch-geographische Darstellung // des // Koenigreichs Polen, 1770, 1773, 1793, 1795, 1807, 1815* (**figure 65, #484**), which depicts the various dismemberments of Poland through the 1830–31 uprising. On the other hand, the 1844 map of Major (Hauptmann) Carl Christian Franz Radefeld (1788–1874), *Europaeisches // Russland* (**figure 66, #404**), which appeared in the German-language edition of the vast *Meyers Handatlas* and was later printed in copy form in Amsterdam, Paris, and Philadelphia, acknowledges Russian imperial claims in "European" territories.

Great Power Conflict and Ukraine in the Crimean War

Krawciw's collection showcases nearly fifty maps of Ukrainian lands during the Crimean War (1853–56), a pivotal example of the collision of Great Power imperialisms. With tsarist Russia on one side and an alliance of Great Britain, France, Piedmont-Sardinia, and the Ottoman Porte on the other, the war took place largely in the areas annexed by Catherine II in 1783. The Crimean War disrupted local economies and caused rising discontent among landowners in Ukraine, who had long called for reform in the Russian Empire.

After his 1851 coup d'état, Napoleon III of France ordered the French ambassador to force the Porte to comply with a statement recognizing his sovereign authority in the Holy Land. In turn, Tsar Nicholas I advanced political and religious counterclaims to Constantinople, and ordered the mobilization of his armies along the Danube. Russian imperialist ambitions also collided with British commercial and naval interests in the Dardanelles Straits, with the threat of the Russian Black Sea Fleet advancing into the Mediterranean. In a peace attempt that failed in 1853, Nicholas I refused to withdraw troops from Moldova and Wallachia, and to abandon what he perceived as the right to interfere in Ottoman affairs on behalf of Orthodox Christians. He calculated, mistakenly, that Austria would remain neutral in the conflict based on his prior assistance given to the Habsburg emperor in suppressing the 1848–49 national uprisings. Following a British-French ultimatum, the powers went to war.

Krawciw's collection of maps depicts some of the most dramatic events of the war, including the siege of Sevastopol (1854) and the Azov campaign (1855). Contrary to conventional wisdom, the theater of war was not limited to Crimea; in 1855, the Russian Baltic Sea Fleet battled the Western Allied Sea Fleet along the Finnish coast, and there were naval skirmishes in the Pacific along the Kamchatka Peninsula. While some of the maps were clearly drawn after the battles were concluded, they reveal the extent to which European Great Powers, including Russia and the Ottomans, had become embroiled in the conflict. The British Major Delafield's 1856 map, *Sebastopol and its Environs showing the Russian Defences and the Approches and Other Works of Attack of the Allied Armies during the Siege of 1854–55 from the Report of Gen. Niel Chief of the French Engrs. Prepared to accompany the Report of Major Delafield on the art of War in Europe in 1854, 55, & 56* (**figure 67, #759**), shows the involvement of British, French, and Russian armed forces at this key moment in the nineteenth-century history of Ukraine.

Other key maps of the Crimean theater of war were made by the geographer and engraver James Wyld, Sr. (1790–1836), and his son James Wyld, Jr. (1812–87). James Wyld, Sr. had been apprenticed to William Faden, and became the Geographer to kings George IV and William IV. He was a founding member of the Royal Geographical Society (est. 1830), and worked in the Quarter Master General's Office. James Wyld, Jr. entered the family business in 1830, and would later become the Geographer to Queen Victoria and Prince Albert. He republished many of his father's works and became a leading London mapmaker in his own right, printing a new series of ordinance maps. Wyld's famous globe, a model of the earth viewed from its interior, was perhaps *the* nineteenth-century emblem of scientific and technological progress. It was exhibited in Leicester Square as part of the Great Exhibition of London from 1851–62. Wyld's 1855 map, *The // Crimea* (**figure 68, #570**), printed in London, shows British imperial interests in conflict with the Russian Empire over the Black Sea and the town and harbor of Sevastopol.

Maps depicting the effects of the Crimean War show a wide range of images of European Great Power militarism and imperialism. From the Russian viewpoint, the commemorative map titled *Vid i plan Sevastopolia (s okrestnostiami)* (Map and Plan of Sebastopol and Environs; **figure 69, #788**) includes a "Map of the Black Sea" and military depiction of the British and French siege. From another perspective, the American publisher Charles Desilver's 1856 *A Complete // Map of the // Crimea // Showing the // Military and Carriage Roads // with Distances // from // Various Points of the Western Coast. // From English & Russian Surveys* (**figure 70, #654**), printed in Philadelphia and Chicago, includes sketches of the theater of war for Sevastopol and all of Crimea.

Russian and European conflict over the "Eastern Question" codified the tendency to view borderlands and waterway passages from a geostrategic vantage, a historical trend that would have a long-term effect on Ukraine and Ukrainians under Western eyes. Amid competing claims to territories within Crimea, the Balkan region, and lands of former Poland-Lithuania, European imperialism left lasting effects on the cartographic representation of Ukraine.

"Modernization" Cartography and Late-Nineteenth-Century Geopolitics

It is particularly interesting that Krawciw's collection features "modernizing" or development maps from the last half of the nineteenth century onward that indicate major administrative, urban/rural, and demographic changes.[19] Following European trends of Haussmannization, named for Baron Georges-Eugène Haussmann (1809–91), who worked on a commission from Napoleon III and drew blueprints that restructured the urban space of Paris, cities in present-day Ukraine also became the subjects of planning and the gathering of population statistics. Planning naturally demanded the creation and application of military, administrative, and transportation and communication maps to modernize cities such as Kyiv (in the Russian Empire) and Lviv (in the Habsburg, then Austro-Hungarian, Empire). For example, the maps of Kyiv ("Kiew") and Lviv ("Lemberg") (**figure 71, #669; figure 72, #670**), show views of the two cities circa 1850.[20]

Between the Habsburg and Russian Empires, ethnographic maps of nationalities made by states and copied by national self-determinationists like Krawciw offered forms of statistical "proof" for claims to territory. The maps, however, had a darker side of which Krawciw seemed nevertheless conscious: the power to exclude entire populations, usually on the basis of language, from belonging to the nation or state. Such maps had an especially destructive potential if and when conjoined to the natural sciences, and to nineteenth-century prejudices about the biological origins of race, ethnicity, and gender. After the 1848 uprisings, the concept of nationality emerged as a way to classify and categorize peoples by language within the borders of imperial states in Central and Eastern Europe. While ethnographic maps of nationality would seem neutral in a social-scientific sense, maps commissioned by the Prussian-German, Austro-Hungarian, and Russian states did in fact inscribe biological notions of separate races and tribes (Ger. Stamm, Rus. *plemia*). The maps "measured" and categorized enclave populations, with an ambition to control or exclude populations within borders.

The effect of these maps for Ukrainians was manifest, since Ruthenian/Ukrainian nation-building intellectuals often internalized the labels given to them when faced with a limited number of cultural and political choices.[21] Leading up to World War I, German and Russian maps tended to label Ukrainians as "Ruthenians" (Ruthenen) or "Little Russians" (Malorusskie), often using mutually exclusive linguistic and religious or confessional units. An example from the collection is shown in two German-language maps by Professor Heinrich Kiepert (1818–99) and his son Richard Kiepert (1846–1915). Heinrich Kiepert served as the head of the Geographical Institute of Weimar from 1845 to 1852, worked in Berlin in the mid-1850s, and was professor of geography at the University of Berlin from 1859 to 1874. Kiepert published a vast assortment of German national and imperial school atlases; maps of the Ottoman Empire, Asia Minor, and Palestine; wall maps and globes; and maps of the ancient world. He served as the principal cartographer representing imperial Germany at the 1878 Congress of Berlin, where he had a hand in redrawing the Balkan nations and states in southeastern Europe. He appeared again with his maps at the Berlin Congress of 1884–85, which was attended by every European country except Switzerland. Part of the "scramble for Africa," cartographers carved up the map of Africa without extending an invitation to African populations. Their units of measurement were race and ethnicity/nationality, which often reduced peoples to spoken language, and offered a vast simplification of multidimensional tribal, cultural, religious and political lines.[22] Kiepert's 1857 maps *Das Europäische // Russland* (**figure 73, #280**) and *Ungarn // mit seinen Nebenländern // und // Galizien. Südliches // Dalmatien // im Anschluss an den untern Rand der Hauptkarte* (**figure 74, #281**) appeared in the *Hand-Atlas der Erde und des Himmels*, printed in Weimar at the Geographical Institute. The maps offered mid-nineteenth-century ethnographic classification schemes of Habsburg and Russian imperial lands, and were based on what were assumed to be monolingual enclaves, thus ignoring other variables such as in- and out-migration, the natural mutability of language and dialect, intermarriage, and use of language for official or religious/confessional purposes.

In the Russian Empire, where German geography had a long and enduring influence, such mapping styles were applied for census-taking purposes, including the important first All-Russian Census of 1897.[23] Finally, drawing from the work of Kiepert and the influential Russian-German ethnographic cartographers Petr Keppen and Aleksandr Rittikh, the 1910 map *Etnografichna karta Slavianshchyny po Niderle* (**figure 75, #840**) was printed in Ukrainian as an adaptation of Lubomir Niderle's ethnographic maps of Central and Eastern Europe. The map uses categories of nationality to depict all of Slavdom, and includes a green area designated as a homeland for the Ukrainians ("Ukraïntsi"). Printed just before the start of World War I, a 1912 map by the firm of I. D. Sytin in Moscow, *Etnograficheskaia karta Rossiiskoi Imperii* (**figure 76, #739**), sketches all the ethnographic collectives in the Russian Empire. By the end of the century and well into the Soviet period, the German-language scientific trends of mapping by race and nationality/ethnicity, which generally considered cultural identity in terms of a single spoken language or "mother tongue" (*rodnoi iazyk*), became the primary way for Russian sovereigns to organize "ethnic" territory for their own political purposes.

Krawciw's interest in the statist thrust of European cartography was an integral part of his experience, his interpretation of Ukrainian history, and his larger geopolitical and cultural agendas. Krawciw's collecting of maps of Ukraine was certainly one way for him to argue for the country's national distinctiveness. But he was not a simple romantic. This becomes apparent in his gathering of ethnographic and development maps—"scientific" evidence that recognized the existence of Ukrainian populations and promoted a sense of Ukrainian homeland identity. A refugee from Europe in the North American diaspora, he was aware not only of the plight of Ukrainians and their language under repressive regimes in the nineteenth and twentieth centuries, but also of the importance of maps that deploy the category of language as a symbol of identity and territory. By 1914, with categories such as race, class, language, and religion/confession in the political air as ordering principles, Ukraine in its geostrategic position as a Russian and European borderland had become even more an object of imperial and national agendas.

Conclusion

In the radically changing European political scene of the nineteenth century, which saw the partitions and annexations of former states, the aftermath of the French Revolution, and the Napoleonic wars, military and state cartography reflected Great Power attempts to categorize and claim borderlands for the sake of nation and state building in Central and Eastern Europe. With the settlements at the Congress of Vienna, Ukraine became incorporated into the neighboring Habsburg and Russian Empires. Comprehensive general European atlases of the ancient, medieval, and modern worlds, products of the Enlightenment, became part of romantic national movements that sought to define a continuous, distinctive, and historical identity for national collectives on the map of civilization. After 1848, Ukraine and Ruthenians/Ukrainians scattered across imperial borderlands were mapped—and often mapped themselves—according to Habsburg and Russian categories of nationality (Nationalitäten, nationalités, *narodnosti*). This novel trend generally reduced Ukrainian identity to spoken language. Politically motivated ethnographic maps, especially for census purposes, came to accompany military and historical maps and atlases in East-Central Europe. As Krawciw saw it, thanks in no small part to European Great Power policies and mapping projects, the Ukrainian Question had been transformed from the mapping of an early modern borderland to the definition of lands claimed by Europe's nations and states.

5 Twentieth-Century Geopolitics and Ukrainian Self-Determination

BOHDAN KRAWCIW WAS BORN in 1904 in Eastern Galicia, when the area was still part of the Habsburg (Austro-Hungarian) Empire. He had a prominent place among Ukrainian émigrés in West Germany and, after 1949, in North America. Krawciw's immediate experience was in the twentieth century, and he conceived of the course of Ukrainian history on the macro level of contemporary geopolitics during both hot and cold wars. Krawciw's organization of his collection was shaped by his hopes for Ukrainian independence, the struggle for political rights in the interwar period, and the events of World War II.

Krawciw's European and North American identities were not altogether in harmony; Ukrainian history for him was not limited to Austro-Hungarian, Polish, Russian, or Soviet conventions of historiography. Ukraine in an American context had significant resonance in the sense of Wilsonian "national self-determination" by language and territory, and Cold War liberationist agendas, but Ukraine in a European context could be imperial or simply nonnational in the sense that the historical boundaries of Ukraine-Rus´ were not confined to an "ethnic" nation-state. In his mind, Ukraine as a term came into common European parlance in 1648, and it designated all distinctive historical regions in the past, including the lands of Rus´.[1] Krawciw's geopolitical agenda for Ukraine as a nation and state was conditioned and circumscribed by competing great power and superpower geopolitical agendas in the Soviet Union, Europe, and North America. His vision of Ukraine's cultural and political role in world history was complex, and fundamentally revisionist in nature.

The twentieth-century maps Krawciw collected to tell the story of Ukraine's history read like a *Who's Who* of the Ukrainian intelligentsia abroad. Ukrainian scientific cartographers and other leaders of the postwar diaspora play a prominent role. Their work, taken together with his notebooks and an important essay on Beauplan, demonstrate his wide-ranging scholarship, his geographical and historical imagination, and his goal of Ukrainian independence. Krawciw's essay, "Guillaume Le Vasseur Sieur de Beauplan's 'Description of Ukraine' and His Military Maps of Ukraine," was published in 1959 as part of an English-language reproduction of Beauplan's work (**figure 77, #51**). The atlas was printed by the Organization for the Defense of Four Freedoms of Ukraine, which was headed by Krawciw and four others on the editorial board, J. F. Vovchuk, J. T. Petryshyn, O. Sokolyshyn, and V. Davydenko. Krawciw writes in a revealing passage, which should be quoted at length:

> Wider knowledge of Ukraine as a territory subject to the personal authority of Polish kings and ethnically and culturally separate was brought into world geography and history by the geographer Guillaume le Vasseur de Beauplan with his *Description d'Ukraine* and his general and special maps of Ukraine. What really established this name in the world was the uprising of Bohdan Khmelnytsky against Poland in 1648, which was crowned by the creation of the Ukrainian Kozak State and brought Ukraine onto the scene of world history. From the time of Khmelnytsky's revolt the name Ukraine entered into general use in the West. In all descriptions of Ukraine printed in those days, as in almost all maps of contempo-

> rary cartographers the Rus′ of that time—Russia Rubra, Volhyn together with Kiev land, Podilya—was already specified as Ukraine.[2]

In this way, by valorizing Beauplan, Krawciw followed in the path of Mykhailo Hrushevs′kyi and other important late-nineteenth-century intellectual founders of Ukrainian historiography, envisioning Ukraine separate from the grand narratives of the Habsburg and Russian Empires, and the boundaries of former Poland-Lithuania.

Wartime Drives for Ukrainian Independence, 1914–21

Krawciw's early years as a young man in East-Central Europe were spent in a climate of interwar activism to make political headway by using categories of class and nationality as modes of self- and collective identification, and the post-1914 maps in the collection reflect his activist participation among the Ukrainian intelligentsia in Galicia. One such map is by Volodymyr Kushnir (1881–1933; also spelled Kuschnir or Kušnir), a self-described neo-pan-Slav publicist born in Galicia whose transnational life of travel and migration was the shared experience of many politically and culturally active Ukrainian intellectuals. Kushnir came to Vienna in the early twentieth century and died at a relatively early age in interwar Prague. When in Vienna, Kushnir worked on the *Ruthenische Revue*, cofounded and edited *Ukrainische Rundschau* (1906–10), and founded and headed the Ukrainian Press Bureau there in the center of the Habsburg monarchy. In 1912, Kushnir edited *Dilo* in Lviv; *Bukovyna* and *Narodnyi holos* (1913–14) in Chernivtsi; and *Ukrainische Korrespondenz* (1917–18), *Volia* (1919), and *Na perelomi* (1920) in Vienna. He was the first head of the Union of Ukrainian Journalists and Writers Abroad. From 1923 until his death, he lived in Prague and taught Ukrainian at the Higher Pedagogical Institute. He contributed articles to German- and Ukrainian-language publications, and shortly before his death published a volume on Polish repressions of Ukrainians in West Ukraine.[3] Krawciw took an interest in Kushnir's work, and added into his collection one of the maps from *Ukrainische Rundschau* (**figure 78, #834**), an updated wartime map from 1915.

Some of the more intriguing maps in the collection are revisionist maps by another influential wartime mapmaker, Count Pál Teleki de Szék (1879–1941), the prime minister of Hungary in 1920–21 and 1939–41. Born in Budapest, Teleki became an expert in geography and cartography and an interwar member of the Hungarian Academy of Sciences. An unapologetic Hungarian ethnonationalist during the interwar period, Teleki is indeed a controversial figure. Posthumous interpretations of his life and work are particularly revealing for the parallels to Ukrainian-Jewish and other intrastate and cross-border minority relations in East-Central Europe.[4] The c. 1919–21 map, *Ethnographical Map // of Hungary // based on // Density of Population // by // Count Paul Teleki // Professor of Geography... According to the census of 1910* (**figure 79, #600**), is one of two maps by Teleki in Krawciw's collection. Printed with text and tables in English, and used at the postwar conferences that resulted in the Treaty of Trianon (1920), the map was drawn from prewar census statistics. Teleki produced it in support of both Hungarian irridentism and independence, showing Hungarian national objections to the Treaty of Trianon and claims to territories inhabited by Hungarian-speaking populations that were beyond the borders created by the treaty.

A final example of the cartographic problems facing Ukrainian activists can be seen in maps by Georg von (Georges de) Gasenko, which veer into the gray area between historical knowledge and propaganda. Gasenko's 1920 map, *Svitova Mapa // z rozmishenniam Ukraïntsiv po svitu. // Z dodatkovymy tablytsiamy* (**figure 80, #818**), was printed in Vienna in three languages, Ukrainian, German, and French. Using the concept of nationality as an organizing principle, it shows the scattering of Ukrainians across the world and promotes the idea of a Ukrainian territorial homeland. A second map from 1920, *Ukraïna v chyslakh zladyv G. Gasenko. Po nainoviishym ofitsiial′nym ukraïns′kym, rosiis′kym, frantsuz′kym i ynshym dzherelam iak tezh po pratsiam E. Rekliu, Bonmariazh-a, Sen′obos-a, Hrushevs′koho, Rudnyts′koho, Volkova, Penka, Frekha, Shefera...* (**figure 81, #819**), drawn in the historical style of the works of major French, German, and Russian late-nineteenth- and early twentieth-century mapmakers, was printed in Ukrainian with the assistance of the influential cartographer Stepan Rudnyts′kyi. The map again calls for the establishment of distinct Ukrainian political borders in East-Central Europe.

Interwar Soviet Ukraine and World War II Propaganda Maps

Since Krawciw aimed to gather as much information as he could about Soviet Ukraine from its origin, the collection includes Soviet topographic, economic, ethnographic, political-administrative, and even geological maps. Krawciw also took an interest in English-language Great Power, and later superpower, representations of Ukraine between

Europe and the Soviet Union. Most such maps are copies, or excerpted from larger publications, often without a clear indication of authorship. The anonymous English-language c. 1920s–30s map, *Union of Soviet Socialist Republics in Europe* (**figure 82, #747**), was printed in Indianapolis by the George F. Cram Companies as part of the Modern Series in geography. It shows the continuous Russian imperial/Soviet presence in "European" territories, and place names and claims to territory in Ukraine. Modern ethnographic maps became important for nation- and state-building elites in modern Central and Eastern Europe after 1848. The ethnographic maps, which later became common in the Soviet Union, made complex multilingual and multiconfessional areas such as Galicia seem geographically legible from a statist vantage. Some of the ethnographic maps also raise questions similar to those that arose in the nineteenth century, about whether Ukrainian was a dialect or language. Others offer territorial visions of Ukraine and the spoken tongues of its inhabitants, often projected back in time based on census statistics. The unauthored English-language *Map of Dialects in Ukraine* (**figure 83, #848**), for instance, was drawn in retrospect to represent the plurality of dialects and linguistic politics in Ukraine before Stalin's rise to power and his reversal of "affirmative action" policies, but was likely produced in the mid- to late 1960s as part of the *Encyclopedia of Ukraine* project.[5]

In addressing the geopolitical fate of Ukraine and its inhabitants during World War II, Krawciw's interests as a writer and journalist shaped his focus. He showed an interest not only in "high" cartographic productions by the likes of Beauplan and other state-affiliated individuals, but also impressions of Ukraine in the popular press and imaginations. He collected a number of propaganda maps from popular magazines. One curious map in the collection, *Hitler's 'Fortress' Germany* (**figure 84, #593**), is by the British short story and dime-novel writer Robert N. Cool and dates to 1940. Cool wrote pulp adventure stories for magazines in the United Kingdom, usually in a romantic tone. They included *A Sailor's Chance* (1936), *Guns on Romance Range* (1938), and *Six-Gun Love Talk* (1938). Another example of British wartime geopolitical imagining is the unauthored *Russia in Europe and its Border States* (**figure 85, #748**), which appeared in English in 1942. Published in London during wartime by the London Geographic Institute, the map was part of a larger series, called "Philips' Authentic Imperial Maps for Tourists & Travelers." The maps foreshadow the geopolitical pigeonholing of Soviet Ukraine in the postwar years.

Contested Places: The Postwar Mapping of Ukraine as a Captive Nation

The postwar cartography of Ukrainian lands was a cooperative but not unproblematic effort in the many Ukrainian diasporas. Until the establishment of centers for Ukrainian studies in both Western Europe and North America, Ukrainian cartographers and collectors of maps lacked a genuine state or institutional support that could function, at least in part, separately from commonplace Cold War geopolitical conceptions of "Russia and Eastern Europe" or "the Soviet bloc." Krawciw's postwar collecting of maps, and the idealization of them as a means to attain political recognition, was tied to the self- and collective identification, and ultimate political, intellectual, and social fate, of Ukrainian diaspora communities. Among those whose works he gathered were Volodymyr Kubiiovych, Mykola Kulyts´kyi, Mykola Diachyshyn, William Karpa, Mykhailo (Michael) Marunchak, Volodymyr Sichyns´kyi, Ivan Zilyns´kyi, Tymish Olesiiuk, Petro Kurinnyi, Iurii Polians´kyi, and Roman Drazniowsky. They played an active part in mapping Ukraine as a politically independent territorial entity.

After Stepan Rudnyts´kyi (1877–1937), Volodymyr Kubiiovych was probably the most influential Ukrainian geographer and cartographer in the twentieth century until his death in Paris in 1985. Born in Nowy Sącz in 1900 in Western Galicia, then part of the Habsburg Empire, Kubiiovych was a demographer and ethnographic geographer who lectured at Jagiellonian University in Kraków from 1928 to 1939. In 1931, he was elected full member of the Shevchenko Scientific Society, and headed its geographical commission. In 1940, during World War II, he was appointed Professor at the Ukrainian Free University in German-occupied Prague, and headed the newly formed Ukrainian Central Committee (UCC) in Kraków and Lviv. With Nazi institutional sponsorship, Kubiiovych in 1941 produced a bilingual German- and Ukrainian-language atlas of Ukraine, with titles in French, based on an earlier 1937 Ukrainian-language work, *Atlias Ukraïny i sumezhnykh kraïv* (An Atlas of Ukraine and Adjacent Countries), and the 1938 *Heohrafiia Ukraïny i sumezhnykh zemel´* (A Geography of Ukraine and Adjacent Lands). The 1937 Ukrainian-language work was published under the auspices of the Shevchenko Scientific Society. The 1941 bilingual work remained unpublished until 1993, when it was printed in Munich, Paris, and Wiesbaden through a partial grant from the Kubiiovych fund of the Canadian Institute of Ukrainian Studies.

After 1945, Kubiiovych emigrated to Germany

and France. He became the general secretary of the Shevchenko Scientific Society from 1947 to 1963, and was elected president of its European branch in 1952. In all, he published over eighty works on anthropogeography, demography, and statistics of Ukraine, particularly of Galicia and the Carpathian region. Kubiiovych also initiated the important *Encyclopedia of Ukraine* project in 1949, a massive project for which he served as chief editor.[6] Kubiiovych enlisted the scholarly and technical assistance of other Ukrainian specialists in geography and cartography, such as Mykola Kulyts´kyi (1903–70). Like Krawciw, Kulyts´kyi was also from a small town in Galicia (Hlyniany, Peremyshliany district). Like Kubiiovych, he was at Jagiellonian University in Kraków in the 1930s and 1940s. With Kubiiovych, Kulyts´kyi directed the cartography and statistics office of the Ukrainian Central Committee in Kraków and Lviv. Kulyts´kyi became a postwar refugee in Berlin, receiving his doctorate from the Ukrainian Free University in Munich. He later became a full member of the Shevchenko Scientific Society, and emigrated to Chicago in the early 1950s. The c. 1950 map of Kubiiovych and Kulyts´kyi, *Etnohrafichna Karta Ukraïny* [*Ethnographical Map of Ukraine*] (**figure 86, #832**), was published in Ukrainian and English. Another variant by the same two scholars was the extraordinarily large 1952 *Ethnographic Map of the Southwestern Ukraine (Halyčyna-Galicia)* (**figure 87, #830**). Their 1968 *Ethnographical Map of Ukraine, 1930* (**figure 88, #829**) was published as part of the multivolume series *Encyclopedia of Ukraine*. As a Ukrainian *encyclopédiste*, Kubiiovych and his compatriot Kulyts´kyi followed the precedents of earlier maps made by Stepan Rudnyts´kyi in gathering in the lands of historical Ukraine and representing them for the sake of independence in a world-historical context.

The eclectic postwar maps in the collection put Krawciw's international bent on display. Through his journalistic and literary activism on behalf of Ukrainian nation and state building, Krawciw became generally aware of maps of Ukraine produced not only in the Soviet Ukraine, but also in European and North American cities. The Ukrainian-language general map by Mykola Diachyshyn (Diaczyszyn) *Zahal´na // Karta // Ukraïny. // Zladiv M. Diachyshyn // zakhodom i nakladom chas. "Svoboda," organu Ukraïns´koho Nar. Soiuza v Amerytsï* (**figure 89, #813**) was published in 1950 in New York by Rand McNally & Co., and reprinted for the newspaper *Svoboda*. William Karpa's 1955 work, *The Map of // Ukraine // Sponsored by the League of Americans of Ukrainian Descent, State Organization of Ukrainian Congress Committee of America* (**figure 90, #824**), was printed by the Ukrainian diaspora in Chicago. O. Viryn's (Weeryn) 1957 *Karta ekonomichnykh raioniv Ukraïny* (**figure 91, #845**) was produced in Kyiv for use by the Central Committee of the Communist Party. The anonymous Soviet Ukrainian-language map, *Skhematychnyi plan Kyieva // naibil´sh tsikavi vyznachni mistsia i pam'iatnyky* (**figure 92, #675**), is a map of Kyiv in 1958 with familiar monuments and cultural landmarks. The unauthored 1959 map in English, *Map of Location of the Slave and Concentration Camps in the Soviet 'Union'* (**figure 93, #752**), produced after Khrushchev's secret speech and in the early years of the Thaw, indicates Soviet prison sites. The anonymous 1960 English-language propaganda map, *There Can Be No Peace in Europe until Ukraine Is Reunited and Forty-Five Million Ukrainians Are Free Again* (**figure 94, #868**), offers a geopolitical image of Ukraine as a captive nation under Soviet communism.

Judging from the collection, Krawciw was especially interested in maps of the Nazi and Soviet prison camp systems. He acquired an interesting map from 1963, *Map of Nazi Concentration Camps in Ukraine* (**figure 95, #839**), by the Ukrainian-Canadian historian Mykhailo (Michael) Marunchak. A fellow refugee who also came from Galicia originally (b. 1914, Dalesheva, Horodenka district), Marunchak taught many years at the University of Manitoba. He himself was interned during the war by the Nazis, and in his postwar years as an émigré in Germany (1947–48) was in charge of the League of Ukrainian Political Prisoners. Starting in 1961, Marunchak was a full member of the Shevchenko Scientific Society, and after 1983 president of the Ukrainian Academy of Arts and Sciences. He wrote about the concentration camp system in Ukraine and the politics of extermination, offering maps to illustrate his points. Marunchak's magnum opus, the two-volume *Istoriia ukraïntsiv Kanady* (1968, 1974), which appeared in English in 1970 (2nd ed., 1982) as *The Ukrainian Canadians: A History*, was a work Krawciw knew quite well.

Conclusion

Bohdan Krawciw treated maps as a matter of political recognition, not merely academic representation. With broad strokes, he sought a "self-determined" cultural and political identity for Ukrainians, and a territorial placement of Ukraine as a sovereign and independent homeland. He treated maps such as Beauplan's as sources of information about Ukraine, and also within the context of a "gathering of Ukrainian lands" between multiple polities from the Renaissance through the twentieth century. The writing of *Monumenta Cartographica Ucrainae*, a project he envisioned in his many notebooks but never completed, was a truly colossal ambition.

To be sure, Ukrainian claims in borderlands were not unproblematic, especially as the twentieth century wore on, nor was "Ukraine" by definition easily inclusive of all its historical populations. Krawciw's political, literary, and aesthetic configurations of Ukraine often collided with the historical memories of other politically motivated groups. Struggling to define what Ukraine was, Krawciw attempted to situate imperial and national naming and claiming of Ukrainian lands by European (including Russian) cartographers into a general, scholarly context over *la longue durée*. The mere collection of maps, he thought, would resonate among Ukrainian communities, and those researching them, in pluralistic settings of North America. Like his forerunners in Galicia and contemporaries in the Ukrainian political emigration, he knew that the history of cartography was a messy geopolitical affair, not innocuous antiquarianism. Through his nearly nine hundred maps and numerous notebooks, Bohdan Krawciw tried to understand, and perhaps even emplot, Ukraine's fragmented geopolitical existence as a stateless nation, and a potentially future state, under Western eyes.

Notes

Chapter One

1. For an introduction to the terms "Rus´" and "Ukraine," see Kubijovyč, *Ukraine*, 3–12.
2. On the Ptolemaic revival, see especially Relaño, "Cartography and Discoveries." For an analysis of the general patterns in Renaissance mapping, see Woodward, *Maps as Prints*; Varago, *Reframing the Renaissance*; Buisseret, *Mapmaker's Quest*; Randies, *Geography, Cartography and Nautical Science*.
3. Wolff, *Inventing Eastern Europe*; and Kordan, *Black Sea, Golden Steppes*.
4. For evaluations of these shifting cartographic paradigms, see Edson, *Mapping Time and Space*; and Hanawalt and Kobialka, *Medieval Practices of Space*. Recent scholarship on *mappae mundi* and medieval mapping practices is examined in Buisseret, *Mapmaker's Quest*, 1–9. See also Heffernan, *Meaning of Europe*; Lewis and Wigen, *Myth of Continents*.
5. For an overview of the political purposes of Renaissance and early modern atlases in connection with emerging concepts of state sovereignty, see Akerman, "Structuring of Political Territory."
6. Bagrow, *History of Cartography*, 77–86.
7. The map in the collection appears in a 1730 version, published in Amsterdam, *Orbis antiqui Tavlae Geographicae secundum Cl. Ptolemaevm...in vsvm geographiae veteris stvdiosorvm, Amstelaedami, Apud R. & J. Wetstenios & Guil. Smith. MDCCXXX*.
8. The dimensions and implications of the problem, of course, are more than merely geographical or cartographical. By original accounts, the Sarmatians were Persian-speaking tribes who lived near the northern regions of the Black Sea from roughly the second to the fourth century BC, until they were forced out by Huns. They created their own intertribal associations and attacked Roman possessions along the Danube. Ptolemy's influence was inescapable, but problems arise when using Ptolemy's flawed maps in attempting to discern an original tribal homeland somewhere in Sarmatia. In the late seventeenth and early eighteenth century, for instance, when the Roman Catholic and Uniate churches were on the rise and Orthodoxy in decline in Right-Bank Ukraine, Polonized nobles subscribed to a Sarmatian ideology (*Sarmatyzm*). They claimed to be the sole descendants of the ancient Sarmatians, and used this to justify *a priori* their status as the ruling political nation of the Commonwealth and protectors of Christianity against the Ottomans. In contested imperial borderlands after the French Revolution, Russian scholars advanced the Sarmatian theory, the notion that the Slavic homeland was on the Don River, and the idea that the Slavs were descended from the Scythians and Sarmatians. Stimulated by sporadic revivals of Ptolemy's maps and by references to the Primary Chronicle, the fate of a Slavic homeland and national and imperial claims—by Russians, Poles, Czechs, and Ukrainians, especially—became popular in the late nineteenth and twentieth centuries, the high epoch of imperialism and ethnolinguistic nationalism. On the origins of Rus´, and the multiple claims to homelands and contentious lines of descent, see Pritsak and Reshetar, *Ukraine and the Dialectics of Nation-Building*; and Ševčenko, *Ukraine between East and West*.
9. The map is in *Geographia Vniversalis, Vetus et Nova... succedunt tabulae Ptolemaicae, opera Sebastiani Munsteri nouo paratae modo...Basileae, 1545*, 48. On Wapowski and his influence, see Buczek, *History of Polish Cartography*, 32–40.
10. Background information on the making of the map can be found in Kordt, *Materialy po istorii russkoi kartografii*, 2:12–13; Olszewicz, *Kartografja polska*, 156; and Buczek, *Dzieje kartografi polskiej*, 25–26. See also Vavrychyn et al.,

Ukraïna na starodavnykh kartakh, 28–29.

11. On Münster's work, see Bagrow, *History of Cartography*, 150–52; and Hantzsch, *Sebastian Münster*.
12. Broecke et al., *Abraham Ortelius and the First Atlas*; and Karrow Jr. et al., *Abraham Ortelius, 1527–1598*.
13. For Mercator's atlas itself, see Watelet and Akerman, *Gerardi Mercatoris Atlas Europae*. On the technical aspects of Mercator's ambitious work, see Keuning, "History of an Atlas." For a popular, and very readable, history, see A. Taylor, *World of Gerard Mercator*.
14. Bagrow, *History of Cartography*, 133, 179–80.
15. See especially Akerman, *Cartographies of Travel and Navigation*; and the superb study of European humanist travel descriptions by Poe, *"A People Born to Slavery."*
16. On Herberstein's conventions and preconceived notions, see Poe, *"A People Born to Slavery,"* 117–44; Grudzińska-Gross, "Tangled Tradition."; Leitsch, "Herberstein's Impact." For an English translation of the Latin original, see Herberstein, *Description of Moscow and Muscovy, 1557*.
17. Poe, *"A People Born to Slavery,"* 5.
18. For an inventory of the earliest cartography of Muscovy, see Poe, *Foreign Descriptions of Muscovy*. Pre-Muscovite maps and Russian/Soviet scholarship on them are discussed at length in Postnikov, *Russia in Maps*, 7–35. See also Kleimola and Lenhoff, *Culture and Identity in Muscovy*.
19. On Hirschvogel's cartography and the Herberstein description, see Bagrow, *History of Cartography*, 156, 172.
20. Ibid., 156–57.
21. On Lazius' description, see Papp-Váry, *Magyarország története térképeken*, 84–87; and Szathmáry, *Descriptio Hungariae*.
22. Vavrychyn et al., *Ukraïna na starodavnykh kartakh*, 96–97.
23. Keuning, "XVIth Century Cartography in the Netherlands." On de Jode, see Ortroy, *L'oeuvre cartographique*.
24. The map appeared in *Specvlvm Orbis Terrae. Antverpiae, Sumptibus Viduae et Hered[u?] Gerardi de Iudaeis /1593/*. See Lelewel, *Épilogue de la géographie*, 202; Olszewicz, *Kartografja polska*, 163; Buczek, *Dzieje kartografi*, 34–35, 46.
25. Grodecki's name, originally spelled in Polish as Grodziecki after the village of Grodziec in Cieszyn Silesia, where he was born, was Latinized as Grodecius or Godreccius. On his background and work, see Buczek, *History of Polish Cartography*, 41–49.
26. Rastawiecki, *Mappografia dawnej Polski*, 14–15; Olszewicz, *Kartografja polska*, 159–60; Buczek, *Dzieje kartografi polskiej*, 36.
27. Mikoś, "Monarchs and Magnates"; and Mikoś, "Polish Kings and Cartography."
28. Buczek, *History of Polish Cartography*, 58–63.
29. Rastawiecki, *Mappografia dawnej Polski*, 116–20; Kordt, *Materialy po istorii russkoi kartografii*, 3:3–7; Olszewicz, *Kartografja polska*, 165–66; Buczek, *Dzieje kartografi polskiej*, 47–51.
30. Keuning, "Hessel Gerritsz."
31. Keuning, "Novus Atlas of Johannes Janssonius."
32. On the professional rivalry, see Bagrow, *History of Cartography*, 180–81.
33. See the important work by Kivelson, *Cartographies of Tsardom*; Kivelson, "Cartography, Autocracy and State Powerlessness"; Kivelson, "'Souls of the Righteous.'" For a recent overview of Soviet and Russian scholarship on the history of cartography, including the important work of V. S. Kusov, B. A. Rybakov, L. A. Gol′denburg, and D. M. Lebedev, see Postnikov, *Russia in Maps*, 7–31.
34. Cosgrove, "Mapping New Worlds"; Cosgrove, *Palladian Landscape*; and Tooley, "Maps in Italian Atlases."
35. The map in the collection is a reprint from a book by Swedish explorer and map collector A. E. Nordenskjöld, *Voyage of the Vega*, 516–17.
36. Long thought lost for good, the map was discovered by the Polish historian of cartography Krystyna Szykuła and debuted at the 13th International Conference on the History of Cartography in the Netherlands in 1989. Samuel H. Baron, the leading scholar on the map in its historical context, has examined the many interpretations of the content, sources, and motives behind Jenkinson's 1562 map. See Baron, "B. A. Rybakov on Jenkinson's Map"; Baron, "Lost Jenkinson Map of Russia." For earlier views, see Keuning, "Jenkinson's Map of Russia"; Berry and Crummey, *Rude and Barbarous Kingdom*; and F. Wilson, *Muscovy*. For an interesting Russian perspective, see Postnikov, *Karty zemel′ rossiiskikh*, 15–16.
37. Kordt, *Materialy po istorii russkoi kartografii*, 1:10–11; Tooley, *Maps and Map-Makers*, 51; Lebedev, *Ocherki po istorii geografii v Rossii XV i XVI vekov*, 46, 58, 60, 217, 296; Bagrow, *History of Cartography*, 249, 499.
38. Keuning, "Isaac Massa, 1586–1643." For an English translation of the original, see Massa, *Short History*.
39. On this trend of template reporting and stereotyping through European cultural lenses, see Edward Keenan's classic "Muscovite Political Folkways"; and Poe, "Truth about Muscovy."
40. Hondius, *Nova Europae Descriptio*; Keuning, "Jodocus Hondius, Jr."; and Keuning, "History of an Atlas."
41. Buisseret, *Mapmaker's Quest*.

Chapter Two

1. David Buisseret, *Monarchs, Ministers and Maps*; Kain and Baigent, *Cadastral Map*.

2. Biggs, "Putting the State on the Map"; Brotton, *Trading Territories.*
3. Other Central European centers for map-, atlas-, and globe-making that deserve mention here are Vienna, Nuremberg, and Augsburg.
4. The existing literature on Beauplan and his maps is quite rich. Especially important is the analysis by Pernal and Essar in Beauplan, *Description of Ukraine*; see also Pernal and Essar, "1652 Beauplan Maps"; Buczek, *History of Polish Cartography*, 64–74; Mikoś, "Polish Kings and Cartography"; and Mikoś, "Monarchs and Magnates." For commentary in Russian and Ukrainian, see Kordt, *Materialy do istoriï kartohrafiï Ukraïny*; Vavrychyn et al., *Boplan i Ukraïna*; Kordt, *Kartohrafiia ta istoriia Ukraïny*; and Sossa, *Istoriia kartohrafuvannia terytoriï Ukraïny vid naidavnishykh chasiv.*
5. For a short historiographical overview of the Khmel′nyts′kyi rebellion and the Hetmanate's diplomatic relations from 1648 to 1659, see Kubijovyč, *Encyclopedia of Ukraine*, s.vv. "Cossack-Polish War" and "Khmelnytsky, Bohdan."
6. Plokhy, *Tsars and Cossacks.*
7. On the terminology, see Kubijovyč, *Ukraine*, 5–7, 11–12.
8. Kordan, *Land of the Cossacks.*
9. On Dutch cartography and early modern colonialism, see Schmidt, "Mapping an Empire"; Egmond, "Secrets of a Long Life"; Zandvliet, *Mapping for Money*; Schilder, *Art of Map-Making*; Schilder, *Monumenta Cartographica Neederlandica.*
10. Bagrow, *History of Cartography*, 132.
11. Welu, "Map in Vermeer's 'Art of Painting.'"
12. Bagrow, *History of Cartography*, 181–83.
13. Kordt, *Materialy po istorii russkoi kartografii*, 1:13, 2:21.
14. Cross, "Dutch Cartographers."
15. The map is taken from volume 44 of *La Galerie Agréable du Monde... Divisée en LXVI. tomes...a Leide, Par Pierre van der Aa...1729.*
16. Keuning, "Cornelius Danckerts."
17. Almagià et al., *Vincenzo Coronelli*; and Rhodes, "Some Notes on Vincenzo Coronelli."
18. On cartography and Great Power Sweden, see Ellenius, *Baroque Dreams*; Lönborg, *Sveriges karta*. For Dahlberg's maps and text in Swedish, see *Svecia antiqua et hodierna*. Many of his maps are stored in the collection of Baron Nils Adolf Erik Norderskjöld in the library of the University of Helsinki, Finland.
19. Rastawiecki, *Mappografia dawnej Polski*, 28. The map in Krawciw's collection is a reproduction in Swedish by Adolf Hillman of Samuel von Pufendorf's *Sju Böcker om Konung Carl X Gustaf's Bragder*, vol. 1 (Stockholm, 1915).
20. For French trends in cartography, see Pedley, *Commerce of Cartography*; and Pedley, "Map Trade in Paris, 1650–1825." On the technical aspects and policies of French state cartography, see Konvitz, *Cartography in France*; and Pelletier, *Cartes des Cassini.*
21. Rastawiecki, *Mappografia dawnej Polski*, 33; Kordt, *Materialy do istoriï kartohrafiï Ukraïny*, 6–7. A 1675 version of the map (#456) is also in the collection.
22. Smith and Kain, *English Maps*; Edney, *Mapping an Empire*; Cumming, *British Maps of Colonial America.*
23. E. Taylor, "English Atlas."
24. The 1943 book reprint in Krawciw's collection includes text describing various lands, four maps, and a preface by the historian Oskar Halecki.

Chapter Three

1. On this standard trope in the maps of Ukraine in the University of Alberta Map Collection, see Kordan, *Land of the Cossacks.*
2. On the geographical imaginings of Voltaire, Rousseau, and other *philosophes* gazing eastward, see Wolff, *Inventing Eastern Europe.*
3. See the seminal article on cartographic positivism and its technical uses for European states by Edney, "Cartography without 'Progress'"; for imperial mappings, see Livingstone and Withers, *Geography and Enlightenment*. On maps and encyclopedias, see Sullivan, "Circumscribing Knowledge"; and Godlewska and Smith, *Geography and Empire.*
4. Appleby, "Mapping Russia"; on the history of French cartography, see Pelletier, *À la découverte de la terre*; and Pelletier, *Cartographie de la France.*
5. On Senex and the mapping of Scotland, see Withers, "Social Nature of Map Making."
6. The authorship of the two was usually entwined. Gilles generally signed maps as M. Robert or "Gilles Robert," while Didier used "de Vaugondy." On the family, see Pedley, *Bel et utile*; Pedley, "Subscription List."
7. Bellin was a frequent contributor to the *Encyclopédie.*
8. Vann, "Mapping under the Austrian Habsburgs"; Kretschmer et al., *Österreichische Kartographie*; Kretschmer, "Rolle Österreichs"; and Dörflinger, *Österreichische Karten.*
9. Grenacher, "Guide to the Cartographic History."
10. Hollstein, *Dutch and Flemish Etchings.*
11. On the Black Sea and cartography as it pertained to Ukraine, see Kordan, *Black Sea, Golden Steppes*; for an introduction to the history of Black Sea traffic and the history of naming and claiming, see Ascherson, *Black Sea.*
12. Keuning, "Nicolaas Witsen as a Cartographer."
13. On atlases and atlasmaking in context, including the Châtelain atlas, see Black, "Historical Atlases"; and Black, *Maps*

and History.

14. Châtelain's maps of so-called Tartary in 1732 and the Russian Empire in 1735 are also in the collection. The works show Russian imperial claims to lands across the Eurasian continent, and show the fabled lands of Tartary across Central Asia.
15. Heinz, "Programme for Map Publishing"; Sandler, *Johannes Baptista Homann.*
16. Ritter, "Seutter, Probst and Lotter."
17. On these eighteenth-century Russian imperial mapping practices within the framework of Russian/Soviet historiography relative to European Great Power politics, and to the post-Petrine borrowing of geographical science and technology from the West, see Postnikov, *Razvitie kartografii*; and Postnikov, "Outline of the History of Russian Cartography"; Shaw, "Geographical Practice"; Gnucheva, *Geograficheskii departament*; Lebedev, *Ocherki po istorii geografii v Rossii XVIII v.*; Lebedev, *Geografiia v Rossii petrovskogo vremeni*; Lebedev and Esakov, *Russkie geograficheskie otkrytiia i issledovaniia*; Appleby, "Mapping Russia"; Fel′, "Role of Petrine Surveyors"; and Vucinich, *Science in Russian Culture.*
18. Kordt, *Materialy do istoriï kartografiï Ukraïny*, 16; Gnucheva, *Geograficheskii departament*, 237.
19. Gnucheva, *Geograficheskii departament*, 242; Fel′, *Kartografiia Rossii*, 184–90.

Chapter Four

1. Magocsi, *Historical Atlas of Central Europe*; Kohut, *Russian Centralism and Ukrainian Autonomy*; Le Donne, "Territorial Reform"; Le Donne, *Russian Empire*; Martin, "Empire's New Frontiers."
2. Ritter, "Seutter, Probst and Lotter."
3. Wolff, "Inventing Galicia."
4. Łukowski, *Partitions of Poland*; Łukowski, *Liberty's Folly.*
5. Mikoś, "Polish Kings and Cartography."
6. Schlögl, "Cartography in the Service of Reform Policy"; Galkovich, "Russian Manuscript Maps."
7. On the importance of visualizing Russian imperial power, see Wortman, *Scenarios of Power*; and Whittaker, *Russia Engages the World.*
8. Successes and failures of such firms had many variations; see Harley, "Bankruptcy of Thomas Jeffreys."
9. Grimsted, *Trophies of War and Empire.*
10. There are other examples of this phenomenon; see my "Cartography and the Collected Nation."
11. Pedley, *Map Trade in the Late Eighteenth Century.*
12. Jäger, *Prussia-Karten*; Meurer, *Mappae Germaniae.*
13. Another map in the collection, *L'Empire Russe en Europe et en Asié // avec les Acquisitions graduelles et caracterisées* (#306), appeared in the same 1805 atlas. There, Le Sage included tables and texts for Russian history and territorial acquisitions: "Tableau Geographique de la Russie, Explications, etc"; "Tableau Statistique et Historique des Provinces Russes"; "Elemens Geographiques de L'Empire Russe, Bornes, Étendue, Montagnes, Fleuves, etc."; "Notice Historique sur les Turces"; "Famille Ottomane…Sultans en Asie…Empereurs a Constantinople."
14. Kaser, *Messrs. Carey & Lea*; Clarkin, *Mathew Carey.*
15. Magocsi, *Ukraine*, 313; Bassin, *Imperial Visions.*
16. The atlas, printed by Carey, is a manifest borrowing from Gros and Le Sage. With its elaborate title, it followed the Le Sage model: *A complete genealogical, historical, chronological, and geographical atlas; being a general guide to ancient and modern history: exhibiting an accurate account of the origin, descent, and marriages of the principal royal families from the earliest authentic records; together with the various possessions, wars, celebrated battles, and remarkable events, to the year 1821. According to the plan of Le Sage, greatly improved. The whole forming a complete system of history and geography. By M. Lavoisne. Philadelphia, M. Carey and sons, 1821.*
17. Smith and Kain, *English Maps*; Kain, *Cadastral Map*; on Cary, see Fordham, *John Cary.*
18. Herbert, "Royal Geographical Society's Membership"; Herbert, "'London Atlas.'"
19. Five major cities in Ukraine (Odesa, Kyiv, Kharkiv, Dnipropetrovsk, and Mykolaïv) tripled or quadrupled in size from 1863 to 1897, and doubled again before 1914; see Magocsi, *Ukraine*, 324. For an overview of nineteenth-century urbanization in Ukrainian cities, see Herlihy, "Ukrainian Cities"; Hamm, *City in Russian History*; and Balan, "Urbanization and the Ukrainian Economy."
20. Other city maps in the collection include St. Petersburg, Berlin, Rome, Moscow, Poltava, Danzig/Gdańsk, and Constantinople. For an excellent overview of urban cartography, its history and evolving practices, see Konvitz, *Urban Millennium.*
21. Seegel, "Beauplan's Prism."
22. Zögner, *Antike Welten, neue Regionen*. On the problematics of modern boundary settlements, see Hartshorne, "Franco-German Boundary of 1871."
23. Andriewsky, "Russian-Ukrainian Discourse."

Chapter Five

1. In describing the history of Rus′, Krawciw tended to project "state" and "empire" back in time. He writes, "The name

Rus´ (in Latin *Russia, Ruthenia*) which during the 10th and 12th centuries referred to the Kievan State—at first only to the Kiev land and later to the whole territory from the Syan and Buh rivers in the West to the Volga in the East—toward the end of this state's existence and particularly after the rise of the Halych-Volhynian State (1199) and the destruction of the Kievan Empire by the Mongols (1240), was attached to the western lands of Ukraine—Halych land and Volhyn" (Krawciw, "Guillaume le Vasseur," ix).

2. Ibid., ix–x.
3. On Kushnir, see Kubijovyč, *Encyclopedia of Ukraine*, vol. 2, s.v. "Kushnir, Volodymyr."
4. In the spring of 2004, a major dispute arose in the Hungarian media over the building of a statue in Budapest to honor him. Supporters of the statue claimed that he was a hero of Hungary who signed agreements with Nazi Germany in order to preserve the civic Hungarian nation, made genuine diplomatic attempts to avoid conflict with neighboring countries such as Yugoslavia from 1939–41, and made efforts to prepare for a government-in-exile, like the Polish and Czechoslovak ones, as Hitler increased military and political pressure on Hungary. Detractors pointed to his manifest anti-Semitism, since he issued twelve anti-Jewish laws, starting with the Numerus Clausus of 1920, and the second and third Anti-Jewish Laws of 1939 and 1940. He wrote the preamble to the 1939 law, personally signed fifty-two anti-Semitic decrees during his political career, proposed a "treaty of everlasting friendship" with Hitler in October 1940, and evidently favored the deportation of the Jews from Hungary by November 1940. On April 3, 1941, three days before the joint Nazi-Hungarian invasion of Yugoslavia, he committed suicide by poison.
5. Martin, *Affirmative Action Empire*; Hirsch, *Empire of Nations*; Luckyj, *Literary Politics in the Soviet Ukraine*.
6. Subtelny, *Ukraine*, 457–58, 470–72; Black, *Maps and History*, 127–28, 149–72; Snyder, *Reconstruction of Nations*; and Snyder, "Memory of Sovereignty."

Works Cited

Akerman, James R., ed. *Cartographies of Travel and Navigation*. Chicago: University of Chicago Press, 2006.

———, ed. "The Structuring of Political Territory in Early Printed Atlases." *Imago Mundi* 47 (1995): 138–54.

Almagià, Roberto, Rodolfo Gallo, and Giorgio Trentin, eds. *Vincenzo Coronelli nel terzo centenario dalla nascità*. Venice: Comune di Venezia, 1950.

Andriewsky, Olga. "The Russian-Ukrainian Discourse and the Failure of the 'Little Russian Solution,' 1782–1917." In *Culture, Nation, and Identity: The Ukrainian-Russian Encounter, 1600–1945*, ed. Andreas Kappeler, Zenon E. Kohut, Frank Sysyn, and Mark von Hagen, 182–214. Edmonton: Canadian Institute of Ukrainian Studies, 2003.

Appleby, John H. "Mapping Russia: Farquharson, Delisle and the Royal Society." *Notes and Records of the Royal Society of London* 55, no. 2 (May 2001): 191–204.

Ascherson, Neal. *Black Sea*. New York: Hill and Wang, 1995.

Bagrow, Leo. *History of Cartography*. Cambridge, Mass.: Harvard University Press, 1964.

Balan, Boris P. "Urbanization and the Ukrainian Economy in the Mid-19th Century." In *Ukrainian Economic History*, ed. I. S. Koropeckyj, 277–309. Cambridge, Mass.: Harvard Ukrainian Research Institute, 1991.

Baron, Samuel H. "B. A. Rybakov on Jenkinson's Map of Russia." In *New Perspectives on Muscovite History*, ed. Lindsey Hughes, 3–13. London: St. Martin's Press, 1993.

———. "The Lost Jenkinson Map of Russia (1562): Recovered, Redated and Retitled," *Terrae Incognitae* 25 (1993): 53–65.

Bassin, Mark. *Imperial Visions, Nationalist Imagination and Geographical Expansion to the Russian Far East, 1840–1865*. Cambridge: Cambridge University Press, 1999.

Beauplan, Guillaume le Vasseur, Sieur de. *A Description of Ukraine*. Introduction, translation, and notes by Andrew B. Pernal and Dennis F. Essar. Cambridge, Mass.: Harvard Ukrainian Research Institute, 1993.

Berry, Lloyd E., and Robert O. Crummey, eds. *Rude and Barbarous Kingdom: Russia in the Accounts of Sixteenth-Century English Voyagers*. Madison: University of Wisconsin Press, 1968.

Biggs, Michael. "Putting the State on the Map: Cartography, Territory, and European State Formation." *Comparative Studies in Society and History* 41, no. 2 (April 1999): 374–405.

Black, Jeremy. "Historical Atlases." *The Historical Journal* 37, no. 3 (Sep. 1994): 643–67.

———. *Maps and History: Constructing Images of the Past*. New Haven: Yale University Press, 1997.

Broecke, Marcel van den, et al., eds. *Abraham Ortelius and the First Atlas: Essays Commemorating the Quadricentennial of His Death, 1598–1998*. Utrecht: HES, 1998.

Brotton, Jerry. *Trading Territories: Mapping the Early Modern World*. Ithaca: Cornell University Press, 1998.

Buczek, Karol. *Dzieje kartografi polskiej od XV do XVIII wieku; zarys analityczno-syntetyczny*. Wrocław: Zakład Narodowy im. Ossolińskich, 1963.

———. *The History of Polish Cartography: From the 15th to the 18th Century*. Trans. Andrzej Potocki. Amsterdam: Meridian Publishing Co., 1982.

Buisseret, David. *The Mapmaker's Quest: Depicting New Worlds in Renaissance Europe*. Oxford: Oxford University Press, 2000.

———, ed. *Monarchs, Ministers and Maps: The Emergence of Cartography as a Tool of Government in Early Modern Europe*. Chicago: University of Chicago Press, 1992.

Clarkin, William. *Mathew Carey: A Bibliography of His Publications, 1785–1824*. New York: Garland, 1984.

Cosgrove, Denis E. "Mapping New Worlds: Culture and Cartography in Sixteenth-Century Venice." *Imago Mundi* 44 (1992): 65–89.

———. *The Palladian Landscape: Geographical Change and Its Cultural Representation in Sixteenth-Century Italy*. Leicester: Leicester University Press, 1993.

Cross, W. Redmond. "Dutch Cartographers of the Seventeenth Century." *Geographical Review* 6, no. 1 (1918): 66–70.

Cumming, William P. *British Maps of Colonial America*. Chicago: University of Chicago Press, 1974.

Dörflinger, Johannes. *Österreichische Karten des 18 Jahrhunderts: Österreichische Karten des frühen 19 Jahrhunderts*. Vienna: Verlag der Österreichischen Akademie der Wissenschaften, 1984, 1988.

Edney, Matthew. "Cartography without 'Progress': Reinterpreting the Nature and Historical Development of Mapmaking." *Cartographica* 30, no. 2–3 (1993): 54–68.

———. *Mapping an Empire: The Geographical Construction of British India, 1765–1843*. Chicago: University of Chicago Press, 1997.

Edson, Evelyn. *Mapping Time and Space: How Medieval Mapmakers Viewed Their World*. London: British Library, 1999.

Egmond, Marco van. "The Secrets of a Long Life: The Dutch Firm of Covens & Mortier (1685–1866) and Their Copper Plates." *Imago Mundi* 54 (2002): 67–86.

Ellenius, Allan, ed. *Baroque Dreams: Art and Vision in Sweden in the Era of Greatness*. Uppsala: Uppsala University Library, 2003.

Fel′, S. E. *Kartografiia Rossii XVIII veka*. Moscow: Izd-vo geodezicheskoi lit-ry, 1960.

———. "The Role of Petrine Surveyors in the Development of Russian Cartography during the 18th Century." In *Essays on the History of Russian Cartography, 16th to 19th Centuries*, ed. and trans. James R. Gibson, 27–42. Cartographica Monograph no. 13. Toronto: B. V. Gutsell, Dept. of Geography, York University, 1975.

Fordham, Sir Herbert George. *John Cary, Engraver, Map, Chart and Print-Seller and Globemaker 1754 to 1835*. Cambridge: The University Press, 1925.

Galkovich, Boris G. "Russian Manuscript Maps of the Dnieper Area, 1730–1760." *Imago Mundi* 33 (1981): 72–81.

Gnucheva, V. F. *Geograficheskii departament Akademii nauk XVIII veka*. Moscow: Izd-vo Akademii nauk SSSR, 1946.

Godlewska, Anne, and Neil Smith, eds. *Geography and Empire*. Oxford: Blackwell, 1994.

Grenacher, Franz. "Guide to the Cartographic History of the Imperial City of Augsburg." *Imago Mundi* 22 (1968): 85–106.

Grimsted, Patricia Kennedy. *Trophies of War and Empire: The Archival Heritage of Ukraine, World War II, and the International Politics of Restitution*. Cambridge, Mass.: Harvard Ukrainian Research Institute, 2001.

Grudzińska-Gross, Irena. "The Tangled Tradition: Custine, Herberstein, Karamzin, and the Critique of Russia." *Slavic Review* 50, no. 4 (Winter 1991): 989–98.

Hamm, Michael, ed. *The City in Russian History*. Lexington, Ky.: University Press of Kentucky, 1976.

Hanawalt, Barbara A., and Michal Kobialka, eds. *Medieval Practices of Space*. Minneapolis: University of Minnesota Press, 2000.

Hantzsch, Viktor. *Sebastian Münster: Leben, Werk, wissenschaftliche Bedeutung*. Nieuwkoop: B. de Graaf, 1965.

Harley, J. B. "The Bankruptcy of Thomas Jeffreys: An Episode in the Economic History of Eighteenth Century Map-Making." *Imago Mundi* 20 (1966): 27–48.

Hartshorne, Richard. "The Franco-German Boundary of 1871." *World Politics* 2, no. 2 (Jan. 1950): 209–50.

Heffernan, Michael. *The Meaning of Europe: Geography and Geopolitics*. New York: Arnold, 1998.

Heinz, Markus. "A Programme for Map Publishing: The Homann Firm in the Eighteenth Century." *Imago Mundi* 49 (1997): 104–15.

Herberstein, Sigismund von. *Description of Moscow and Muscovy, 1557*. Ed. Bertold Picard. London: J. M. Dent, 1969.

Herbert, Francis. "The 'London Atlas of Universal Geography' from John Arrowsmith to Edward Stanford: Origin, Development, and Dissolution of a British World Atlas from the 1830s to the 1930s." *Imago Mundi* 41 (1989): 98–123.

———. "The Royal Geographical Society's Membership, the Map Trade, and Geographical Publishing in Britain 1830 to ca. 1930: An Introductory Essay with Listing of Some 250 Fellows in Related Professions." *Imago Mundi* 35 (1983): 67–95.

Herlihy, Patricia. "Ukrainian Cities in the 19th Century." In *Rethinking Ukrainian History*, ed. Ivan L. Rudnytsky, 135–55. Edmonton: Canadian Institute of Ukrainian Studies, 1981.

Hirsch, Francine. *Empire of Nations: Ethnographic Knowledge and the Making of the Soviet Union*. Ithaca: Cornell University Press, 2005.

Hollstein, F. W. H. *Dutch and Flemish Etchings, Engravings, and Woodcuts, ca. 1450–1700*. Vols. 38–41. Amsterdam: M. Hertzberger, 1949.

Hondius, Jodocus. *Nova Europae Descriptio*. Reprint of 1606 edition with text. Edinburgh: John Bartholomew & Son, 1973.

Jäger, Eckhard. *Prussia-Karten, 1542–1810: Geschichte der kartographischen Darstellung Ostpreussens vom 16. bis zum 19. Jahrhundert*. Weissenhorn: A. H. Konrad, 1982.

Kain, Roger J. P., and Elizabeth Baigent. *The Cadastral Map in the Service of the State: A History of Property Mapping*. Chicago: University of Chicago Press, 1992.

Karrow, Robert W., Jr., et al., eds. *Abraham Ortelius, 1527–1598: cartographe et humaniste*. Turnhout: Brepols, 1998.

Kaser, David. *Messrs. Carey & Lea of Philadelphia: A Study in the History of the Booktrade*. Philadelphia: University of Pennsylvania Press, 1957.

Keenan, Edward L. "Muscovite Political Folkways." *Russian Review* 45, no. 2 (1986): 115–81.

Keuning, Johannes. "XVIth Century Cartography in the Netherlands (Mainly in the Northern Provinces)." *Imago Mundi* 9 (1952): 35–60, 62–63.

———. "Cornelius Danckerts and His 'Nieuw Aerdsch Pleyn.'" *Imago Mundi* 12 (1955): 136–39.

———. "Hessel Gerritsz." *Imago Mundi* 6 (1949): 48–66.

———. "The History of an Atlas: Mercator – Hondius." *Imago Mundi* 4 (1947): 37–62.

———. "Isaac Massa, 1586–1643." *Imago Mundi* 10 (1953): 65–79.

———. "Jenkinson's Map of Russia." *Imago Mundi* 13 (1956): 172–75.

———. "Jodocus Hondius, Jr." *Imago Mundi* 5 (1948): 63–71.

———. "Nicolaas Witsen as a Cartographer." *Imago Mundi* 11 (1954); 95–110.

———. "The Novus Atlas of Johannes Janssonius." *Imago Mundi* 8 (1951): 71–98.

Kivelson, Valerie A. *Cartographies of Tsardom: The Land and Its Meanings in Seventeenth-Century Russia*. Ithaca: Cornell University Press, 2006.

———. "Cartography, Autocracy and State Powerlessness: The

Uses of Maps in Early Modern Russia." *Imago Mundi* 51 (1999): 83–105.

———. "'The Souls of the Righteous in a Bright Place': Landscape and Orthodoxy in Seventeenth-Century Russian Maps." *Russian Review* 58 (1999): 1–25.

Kleimola, Ann M., and G. D. Lenhoff, eds. *Culture and Identity in Muscovy, 1359–1584*. Moscow: ITZ-Garant, 1997.

Kohut, Zenon E. *Russian Centralism and Ukrainian Autonomy: Imperial Absorption of the Hetmanate, 1760s–1830s*. Cambridge, Mass.: Harvard Ukrainian Research Institute, 1988.

Konvitz, Josef W. *Cartography in France, 1660–1848: Science, Engineering, and Statecraft*. Chicago: University of Chicago Press, 1987.

———. *The Urban Millennium: The City-Building Process from the Early Middle Ages to the Present*. Carbondale: Southern Illinois University Press, 1985.

Kordan, Bohdan S. *Black Sea, Golden Steppes: Antiquarian Maps of the Black Sea Coast and the Steppes of Old Ukraine*. Saskatoon: Heritage Press, 2001.

———. *Land of the Cossacks: Antiquarian Maps of Ukraine: An Exhibition / Zemlia kozakiv: starovynni karty Ukraïny: vystavka*. Winnipeg, Manitoba: Ukrainian Cultural and Educational Centre, 1987.

Kordt, Veniamin. *Kartohrafiia ta istoriia Ukraïny: Zbirnyk naukovykh prats´*. Lviv: NAN, 2000.

———. *Materialy do istoriï kartografiï Ukraïny: Monumenta cartographica Ucrainae*. Kyiv, 1931.

———. *Materialy po istorii russkoi kartografii*. Kyiv, 1899–.

Krawciw, Bohdan. "Guillaume le Vasseur Sieur de Beauplan's 'Description of Ukraine' and His Military Maps of Ukraine." Preface to *A Description of Ukraine*, by Guillaume le Vasseur, Sieur de Beauplan, ix–xiii. New York: Organization for the Defense of Four Freedoms of Ukraine, 1959.

Kretschmer, Ingrid. "Die Rolle Österreichs bei der Kartographischen Erschliessung Mitteleuropas." *Mitteilungen der Österreichischen Geographischen Gesellschaft* 132 (1990): 172–91.

Kretschmer, Ingrid, Johannes Dörflinger, and Franz Wawrik, eds. *Österreichische Kartographie von den Anfängen im 15. Jahrhundert bis zum 21. Jahrhundert*. Vienna: Institut für Geographie und Regionalforschung der Universität Wien, Kartographie und Geoinformation, 2004.

Kubijovyč, Volodymyr, ed. *Ukraine: A Concise Encyclopaedia*. Vol. 1. 2nd ed. Toronto: University of Toronto Press, 1970.

———. *Encyclopedia of Ukraine*. 5 vols. Toronto, Buffalo: University of Toronto Press, 1984–93.

Lebedev, D. M. *Geografiia v Rossii petrovskogo vremeni*. Moscow: Izd-vo Akademii nauk SSSR, 1950.

———. *Ocherki po istorii geografii v Rossii XV i XVI vekov*. Moscow: Izd-vo Akademii nauk SSSR, 1956.

———. *Ocherki po istorii geografii v Rossii XVIII v. (1725–1800 gg.)* Moscow: Izd-vo Akademii nauk SSSR, 1957.

Lebedev, D. M., and V. A. Esakov, eds. *Russkie geograficheskie otkrytiia i issledovaniia*. Moscow: Mysl´, 1971.

Le Donne, John. *The Russian Empire and the World, 1700–1917: The Geopolitics of Expansion and Containment*. New York: Oxford University Press, 1997.

———. "The Territorial Reform of the Russian Empire, 1775–1796." *Cahiers du Monde Russe et Soviétique* 23, no. 2 (April–June 1982): 147–85.

Leitsch, Walter. "Herberstein's Impact on the Reports about Muscovy in the 16th and 17th Centuries: Some Observations on the Technique of Borrowing." *Forschungen zur osteuropäischen Geschichte* 24 (1978): 163–77.

Lelewel, Joachim. *Épilogue de la géographie du moyen âge*. Bruxelles: Ve et J. Pilliet, 1857.

Lewis, Martin, and Karen Wigen. *The Myth of Continents: A Critique of Metageography*. Berkeley: University of California Press, 1997.

Livingstone, David N., and Charles W. J. Withers, eds. *Geography and Enlightenment*. Chicago: University of Chicago Press, 1999.

Lönborg, Sven Erik. *Sveriges karta, tiden till omkring 1850, af Sven Lönborg. Utg. med understöd från Vilhelm Ekmans universitetsfond*. Uppsala: Almqvist & Wiksells boktryckeri, 1903.

Luckyj, George. *Literary Politics in the Soviet Ukraine, 1917–1934*. Durham: Duke University Press, 1990.

Łukowski, Jerzy. *Liberty's Folly: The Polish-Lithuanian Commonwealth in the Eighteenth Century, 1697–1795*. London: Longman, 1991.

———. *The Partitions of Poland, 1772, 1793, 1795*. London and New York: Longman, 1999.

Magocsi, Paul Robert. *Historical Atlas of Central Europe*. Seattle: University of Washington Press, 2002.

———. *Ukraine: A Historical Atlas*. Toronto: University of Toronto Press, 1985.

Martin, Terry. *The Affirmative Action Empire: Nations and Nationalism in the Soviet Union, 1923–1939*. Ithaca: Cornell University Press, 2001.

———. "The Empire's New Frontiers: New Russia's Path from Frontier to *Okraina*, 1774–1920." *Russian History* 19, no. 1–4 (1992): 181–201.

Massa, Isaac. *A Short History of the Beginnings and Origins of these Present Wars in Moscow under the Reign of Various Sovereigns down to the Year 1610*. Trans. G. Edward Orchard. Toronto: University of Toronto Press, 1982.

Meurer, Peter H. *Mappae Germaniae: die schönsten und bedeutendsten Deutschlandkarten von 1482 bis 1803*. Bad Neustadt a.d. Saale: Dietrich Pfaehler, 1984.

Mikoś, Michael. "Monarchs and Magnates: Maps of Poland in the Sixteenth and Eighteenth Centuries." In *Monarchs, Ministers and Maps: The Emergence of Cartography as a Tool of Government in Early Modern Europe*, ed. David Buisseret, 168–82. Chicago: University of Chicago Press, 1992.

———. "The Polish Kings and Cartography." *Imago Mundi* 41 (1989): 76–86.

Nordenskjöld, A. E. *The Voyage of the Vega*. New York, 1882.

Olszewicz, Bolesław. *Kartografja polska XV i XVI wieku: Przegląd chronologiczno-bibljograficzny* [*La cartographie polonaise du XV-e et XVI-e siècles: Revue chronologique et bibliographique*]. Lviv: Książnica-Atlas, 1930.

Ortroy, Fernand van. *L'oeuvre cartographique de Gérard et de Corneille de Jode*. Amsterdam: Meridian Pub. Co., 1963.

Papp-Váry, Árpád. *Magyarország története térképeken*. Budapest: Kossuth Kiadó: Cartographia, 2002.

Pedley, Mary Sponberg. *Bel et utile: Work of the Robert de Vaugondy Family of Mapmakers*. Tring, Herts, England: Map Collector Publications, 1992.

———. *The Commerce of Cartography: Making and Marketing Maps in Eighteenth-Century France and England*. Chicago: University of Chicago Press, 2005.

———, ed. *The Map Trade in the Late Eighteenth Century: Letters to the London Map Sellers Jefferys and Faden*. Oxford: Voltaire Foundation, 2000.

———. "The Map Trade in Paris, 1650–1825." *Imago Mundi* 33 (1981): 33–45.

———. "The Subscription List of the 1757 Atlas Universel: A Study in Cartographic Dissemination." *Imago Mundi* 31 (1979): 66–77.

Pelletier, Monique. *À la découverte de la terre: dix siècles de cartographie: trésors du Départment des Cartes et Plans (mai–juillet 1979). Paris: Bibliothèque Nationale, 1979.*

———. *Les cartes des Cassini: la science au service de l'état et de les régions*. Paris: CTHS, 2002.

———. *Cartographie de la France et du monde de la renaissance au siècle des lumières*. Paris: Bibliothèque Nationale de France, 2001.

Pernal, Andrew B., and Dennis F. Essar. "The 1652 Beauplan Maps of the Ukraine." *Harvard Ukrainian Studies* 9 (1985): 61–78.

Plokhy, Serhii. *Tsars and Cossacks: A Study in Iconography*. Cambridge, Mass.: Harvard Ukrainian Research Institute, 2002.

Poe, Marshall. *Foreign Descriptions of Muscovy: An Annotated Bibliography of Primary and Secondary Sources*. Columbus, Ohio: Slavica, 1995.

———. *"A People Born to Slavery": Russia in Early Modern European Ethnography, 1476–1748*. Ithaca: Cornell University Press, 2000.

———. "The Truth about Muscovy." *Kritika: Explorations in Russian and Eurasian History* 3, no. 3 (2002): 473–86.

Postnikov, Aleksei V. *Karty zemel´ rossiiskikh: Ocherki istorii geograficheskogo izucheniia i kartografirovaniia nashego otechestva*. Moscow: Nash dom, 1996.

———. "Outline of the History of Russian Cartography." In *Regions: A Prism to View the Slavic-Eurasian World: Towards a Discipline of "Regionology,"* ed. Kimitaka Matsuzato, 1–50. Sapporo, Japan: Slavic Research Center, Hokkaido University, 2000.

———. *Razvitie kartografii i voprosy ispol´zovaniia starykh kart*. Moscow: Nauka, 1985.

———. *Russia in Maps: A History of the Geographical Study and Cartography of the Country*. Moscow: Nash dom – L'Age d'Homme, 1996.

Pritsak, Omeljan, and John S. Reshetar, Jr. "The Ukraine and the Dialectics of Nation-Building." *Slavic Review* 22 (1963):224–55.

Randies, W. G. L. *Geography, Cartography and Nautical Science in the Renaissance: The Impact of the Great Discoveries*. Aldershot: Ashgate/Variorum, 2000.

Rastawiecki, Edward. *Mappografia dawnej Polski*. Warsaw, 1846.

Relaño, Francesco. "Cartography and Discoveries: The Re-Definition of the Ptolemaic Model in the First Quarter of the Sixteenth Century." In *La cartografia europea tra primo Rinascimento e fine dell'illuminismo*, eds. Diogo Ramada Curto, Angelo Cattaneo, and André Ferrand Almeida, 49–61. Florence: Leo S. Olschki, 2003.

Rhodes, Dennis E. "Some Notes on Vincenzo Coronelli and His Publishers." *Imago Mundi* 39 (1987): 77–79.

Ritter, Michael. "Seutter, Probst and Lotter: An Eighteenth-Century Map Publishing House in Germany." *Imago Mundi* 53 (2001): 130–35.

Sandler, Christian. *Johannes Baptista Homann, die Homännischen Erben, Matthäus Seutter und ihre Landkarten: Beiträge zur Geschichte der Kartographie*. Amsterdam: Meridian Pub. Co., 1979.

Schilder, Günter. *The Art of Map-Making: Dutch Cartography from the Middle Ages to the Industrial Revolution*. Chicago: The Newberry Library, 1985.

———. *Monumenta Cartographica Neederlandica*. Alphen aan den Rijn: "Canaletto," 1986.

Schlögl, Daniel. "Cartography in the Service of Reform Policy in Late Absolutist Bavaria, c.1750–1777." *Imago Mundi* 48 (1997): 116–28.

Schmidt, Benjamin. "Mapping an Empire: Cartographic and Colonial Rivalry in Seventeenth-Century Dutch and English North America." *The William and Mary Quarterly* 54, no. 3 (July 1997): 549–78.

Seegel, Steven J. "Beauplan's Prism: Represented Contact Zones and Nineteenth-Century Mapping Practices in Ukraine." In *Rebounding Identities*, ed. Blair Ruble and Dominique Arel, 151–80. Baltimore and Washington, D.C.: The Johns Hopkins University Press and Kennan Institute for Advanced Russian Studies, 2006.

———. "Cartography and the Collected Nation in Joachim Lelewel's Geographical Imagination: A Revised Approach to Intelligentsia." In *Words, Deeds and Values: The Intelligentsias in Russia and Poland during the Nineteenth and Twentieth Centuries*, ed. Fiona Björling and Alexander Pereswetoff-Morath, 23–31. Slavica Lundensia 22. Lund: Lund University, 2005.

Ševčenko, Ihor. *Ukraine between East and West: Essays on Cultural History to the Early Eighteenth Century.* Edmonton: Canadian Institute of Ukrainian Studies, 1996.

Shaw, Denis J. B. "Geographical Practice and Its Significance in Peter the Great's Russia." *Journal of Historical Geography* 22, no. 2 (1976): 160–76.

Smith, Catherine Delano, and Roger J. P. Kain. *English Maps: A History*. Toronto: University of Toronto Press, 1999.

Snyder, Timothy. "Memory of Sovereignty and Sovereignty over Memory: Poland, Lithuania and Ukraine, 1939–1999." In *Memory and Power in Post-war Europe: Studies in the Presence of the Past*, ed. Jan-Werner Müller, 39–58. Cambridge: Cambridge University Press, 2002.

———. *The Reconstruction of Nations: Poland, Ukraine, Lithuania, Belarus, 1569–1999*. New Haven: Yale University Press, 2003.

Sossa, Rostyslav I. *Istoriia kartohrafuvannia terytoriï Ukraïny vid naidavnishykh chasiv do 1920 r.: korotkyi narys*. Kyiv: Naukova dumka, 2000.

Subtelny, Orest. *Ukraine: A History*. 2nd ed. Toronto: Canadian Institute of Ukrainian Studies, 1994.

Sullivan, Lawrence E. "Circumscribing Knowledge: Encyclopedias in Historical Perspective." *The Journal of Religion* 70, no. 3 (Jul. 1990): 315–39.

Svecia antiqua et hodierna. Ed. Erik Dahlberg, text by Erik Vernberg. Stockholm: Wahlström & Widstrand, 1983.

Szathmáry, Tibor. *Descriptio Hungariae: Magyarország és erdély nyomtatott térkepéi 1477–1600*. Cremona, 1987.

Taylor, Andrew. *The World of Gerard Mercator: The Mapmaker Who Revolutionized Geography*. New York: Walker & Co., 2004.

Taylor, E. G. R. "The English Atlas of Moses Pitt, 1680–1683." *The Geographical Journal* 95, no. 4 (Apr. 1940): 292–99.

Tooley, Ronald Vere. "Maps in Italian Atlases of the Sixteenth Century, Being a Comparative List of the Italian Maps Issued by Lafreri, Forlani, Duchetti, Bertelli and Others, Found in Atlases." *Imago Mundi* 3 (1939): 12–47.

———. *Maps and Map-Makers*. London: Batsford, 1952.

Vann, James. "Mapping under the Austrian Habsburgs." In *Monarchs, Ministers and Maps: The Emergence of Cartography as a Tool of Government in Early Modern Europe*, ed. David Buisseret, 153–67. Chicago: University of Chicago Press, 1992.

Varago, Claire, ed. *Reframing the Renaissance: Visual Culture in Europe and Latin America, 1450–1650*. New Haven: Yale University Press, 1995.

Vavrychyn, Mariia, Iaroslav A. Dashkevych, and Uliana Kryshtalovych, eds. *Ukraïna na starodavnykh kartakh: kinets´ XV – persha polovyna XVII st.* Kyiv: DNVP "Kartohrafiia," 2004.

Vavrychyn, Mariia, et al., eds. *Boplan i Ukraïna: Zbirnyk naukovykh prats´*. Lviv: NAN, 1998.

Vucinich, Alexander. *Science in Russian Culture: A History to 1860*. Stanford: Stanford University Press, 1963.

Watelet, Marcel, and James R. Akerman, eds. *Gerardi Mercatoris Atlas Europae*. Pleasant Hill, Ore: Walking Tree Press, 1997.

Welu, James A. "The Map in Vermeer's 'Art of Painting.'" *Imago Mundi* 30 (1978): 9–30.

Whittaker, Cynthia Hyla, ed. *Russia Engages the World, 1453–1825*. Cambridge, Mass.: Harvard University Press, 2003.

Wilson, Francesca. *Muscovy: Russia through Foreign Eyes, 1553–1900*. New York and Washington: Praeger Publishers, 1971.

Withers, Charles W. J. "The Social Nature of Map Making in the Scottish Enlightenment, c.1682–c.1832." *Imago Mundi* 54 (2002): 46–66.

Wolff, Larry. *Inventing Eastern Europe: The Map of Civilization on the Mind of the Enlightenment*. Stanford: Stanford University Press, 1994.

———. "Inventing Galicia: Messianic Josephinism and the Recasting of Partitioned Poland." *Slavic Review* 63, no. 4 (Winter 2004): 818–40.

Woodward, David. *Maps as Prints in the Italian Renaissance: Makers, Distributors and Consumers*. London: British Library, 1996.

Wortman, Richard. *Scenarios of Power: Myth and Ceremony in Russian Monarchy*. Princeton: Princeton University Press, 1995–2000.

Zandvliet, Kees. *Mapping for Money: Maps, Plans, and Topographic Paintings and Their Role in Dutch Overseas Expansion during the 16th and 17th Centuries*. Amsterdam: Batavian Lion International, 1998.

Zögner, Lothar. *Antike Welten, neue Regionen: Heinrich Kiepert, 1818–1899*. Berlin: Kiepert KG, 1999.

Figures

List of Figures and Corresponding Map Numbers

Sarmatici oceani pars
Sali
Hyper
borei
Terra incognita
Fons occidentalis Rha fluuij
Basilicę id est Regij Sarmatę
Modocę
Chersinus fluuius
Pagiritę
Zacatę
S A R
Riphei montes
Tanais fluuius
Borusci
Periérbidi gens magna
Alexandri arę
Nasci
Sarmatiæ
Europææ
Cęsaris arę
pars
Ophlones
Sturni
Turuntus fluuius
Tanaitę
Exapolis
Tanais fluuius
Alaunij
Alaunus mons
Osyli
Tanais
Iaxa
Nauaris
matę
Siraceni
Poritus fluuius
Carcea vicus
T I A
Alopetia quę & Tanais
Pamardis
Marubius fluuius
Hygris
Pataruę
Pessij
Amadoci montes
Lycus flu.
Saltus dei lucus
Mæotis palus
Theophanius fluuius
Azara
Themæotę
Agarus fluuius
Gerrus fluuius
Rhombites paruus flu.
Azaraba
Tyrambę
Azabetis mitra vel tenia
Acra
Cnemia
Tyrambe
Atticitus fluuius
Agarum prom.
Auchis
Zenononis chersonesus
Arichi
Noua mœnia
Byces palus
Achilleum apud os
Cimmerium prom.
Gerusa
Psatis fluuius
Mapeta
Heraclium
Parthenium
Myrmecium pro.
Taurica cherso
Bosph
Cimmericus
Apaturgus
Cona
pseni
Vardanus fluuius
Suraba
orani
Synda opp.
Phanagoria
Bata opp.
Scopelus
Corax mons
Pantica pea
Corocondame
Achęi
Psychrus flu.
Tazos
Me
Burca flu.
tibi
Abunis
Eupatoria
nesus
Tyrictata
Nymphęum
Theodosia
Cercitę
Amsalis
Heniochi
Cucunda
Batrache
Naana
Ago
Dandaca
Symbolorum portus
Ctenus portus
Lagyra
Charax
Bosphorus
Hermonassa
Synda portus
Bata portus
Achea vicus
Oenanthia
Suanocolchi
Corax fluuius
Colchi
Istrianus flu.
Cercetidus sinus
Toretice extrema
Fortia mœnia
Dioscuria quę & Sebastopolis
Hippus fluuius
Ponti Euxini pars
a
b
c
60
65
70
A
B
C
D
60
58
56
54
52
50
48

TAB. II. ASIÆ
SARMATIAM
ASIATICAM
repræsentans
Medius Meridianus 75. reliqui ad eum inclinati sunt iuxta rationes parallelorum 50.58.
Vlterior terra incognita
Hyperborei montes
Fons orientalis Rha fluuij
Sar ma tę
Rha fluuius occidentalis
Hippophagi Sarmatę
Rha fluuius orientalis
M A
Asæi
Nesiotis regio
Phthirophagi
Materi
Canodip sas regio
Rhymmici montes
Rha fluuius
Mithridatis regio
A S I A
Sapotrenę
Scythię intra
Melanchlęni
Coraxi
Scymnitę
Amazones
Imaum pars
Orinei
Surani
Alexandri columnę
Vali
Orgasi
Socani
Rhymnus fluuius
Rha fluuius
T I C A
Serbi
Ceraunij montes
Vdę
Vdon fluuius
Vardanus fluuius
Seraca
Diduri
Pylę Sarmaticę
Olondę
Alonta fluuius
Tusci
Ebrapa
Saua rei
Caucasus mons
Isondę
HYRCANI siue Caspij maris pars
Alma
Caucasus mons
Sarmaticę pylę
Gerri
Gerrus flu.
Teleba
Soana fluuius
Lubum
Mosega
Albanię pylę
Thabilaca
Thilbis
Iberię pars
Albanię
Chobata
pars

Figure 1

POLONIA ET VN
PRVSSIA
Thorn
Dobrin
Visthna
DVCATVS MASOVIAE
Narew fl.
Vladis lauia
Plotzkc
Gnezna
Glopo lacus
Warta fl.
Bug fl.
Turec
Crobia
Stradia
Lasko
Tarczin
Schedletz
Ianow
Vepres fl.
Radon
Schydlowetz
POL ONIA
Mõs S.Crucis
Pynzow
Solecz
Chelm
Turobin
Grodeck
SLESIA
Vratislauia
Odera fl.
Cracouia
Bochnia
Preslou
San fl.
Porilitz
RVS SIA
Bugk fl.
Oderburg
Fridec
Cantt
Fons Vistulæ
Bietz
Premisla
Leopol
PODO
Moravia
Vaga fl.
Proprut fl.
Rohatin
Posonia
Rosenberg
Casfouia
Nester fl.
Halitz
Culicow
Nitri
Lyfca
Lokow fl.
Pokutz
Bacda
Agria
Tetzin
Samos fl.
Strigonia
Buda
Pest
Debrecyn
Septem Castra
Regno
Cecnlia
Alba regal.
VNGARIA
TRANS SILVA NIA
Balaton lacus
Colloda
Varadinu
Segedyn
Moros fl.
Cibin
Drawa fl.
Bacia
Chanadinum
TRAN
Saua fl.
Dytel
Petz
Danubius fl.
Orsaw
Bodon
BOSNA
Saba
Zarnow
Galombetz
RASCIA
Drina fl.
SERVIA
Nys
DALMATIA
Suere lagno
Campus Cassouius
Scopy
Mare Adriaticu
Munster Basle 1545

ARIA XX NOVA TABVLA
Smolensky
Orsha
Mosthia
Vilna
Chronon fl.
Chrones lacus
nouagrodes
Nemen fl.
Tayona noua
Neper fl.
Homya
Popouabora
Starodub
Rylse
LITVANIAE ducatus
Copil
Bobruysko
MOSCOVIA
Pinsko
Grodeck
Lubyecz
RVSSIA ALBA
Putuila
Pripectus fl.
Modzyr
Visigrod
Volhinia
Owurz
Neper ul
Borysthenes fl.
Canio w
Czerkasi
Zaslau
Czudnow
Bog fl.
Sob fl.
Codimia
Suynigrod
Amadoca palus
Row
Braczlaw
Nigra sylua
TAR TARIA minor
Camenetz
Stoka
Orybow
Telikol
Tyras & Nester fl.
Precop
Bicis palus
Sotzana Metropol
Lacus Obidouo
Oczakow
Ostiam
Byalogrod
Sinus Carcinitus
Margosesi Metropol
BESSARABIA
TARTARIA Precopien
Kylia
Kerkel
VALACHIA MAGNA
Barlach
Moncop
Baba
Rebnyc
Galacz
Pontus Euxinus
Proud.
Varna
Ister fl.
BVLGARIA
Ternow
Philippol
Sophia
Cispol
Thracia Romania
Stagnaw
Con stātinopel
Scutari
Nicomedia
Carpy
NATVLIA
Bithyniæ pars

Figure 2

GRONLANDIAE PARS
OCEANVS HYPERBOREVS
DEVCALIDONIVS OCEANVS
MARE SEP
EVROPAE
OCEANVS GERMANICVS
OCEANVS BRITANNICVS
CANTABRICVS OCEANVS
OCCIDENTALIS OCEANVS
SVECIA
GOTIA
SICILIA
BARBARIA
AFRICAE
FESSA
Germanica
Gallica
Italica

OCEANVS SCHITICVS
CONDORA
TARTA
R
I A.
PERMIA
TVMEN
ASI
AE
PARS.
TVRCHESTAN
MARE DE BACHV olim CASPIVM
CATTA CHI
GEORGIANI
MARE MAGIORE
BOLLI
AMASIA
NATOLIA
AIDINELLI
CARMANIA
TVRCOMANI
MEDIA
MESOPOTAMIA
PERSIA
ARABIA
IVDEA
AEGYPTVS
BARCHA
CANDIA
Cum priuilegio

Figure 3

SCYTHIAE, SIVE SARMA
Rypaei montes, ad Tanais usque
Ex conatibus geographicis Abrahami Ortelij.
Hamaxobij, qui et Hamaxobitae.
Chuni.
Bodini.
Amadoci.
Basilides.
Nomades.
Callipidae.
Georgi.
Nauari.
ANTII, ANTAE, et ANTES.
Carpiani.
Basternae
Vrgi
Iazyges.
Tagri.
Eneocadlae.
Tyrangitae.
SCYTHIA PARVA.
SENDICA.
Sagaricus sinus.
HYLAEVM Mare.
Carcinites, qui et Tamyraces Sinus.
TAVRICA, QVAE ET SCYTHICA CHERSONESVS.
OCCIDENS.
GETAE.
DACIAE PARS.
PONTVS EVXINVS. Sed, Dictus ab a
Salmydessus sinus
THRACIAE PARS.
PROPONTIS.
Heneti.

ΠΟΝΤΟΣ ΕΥΞΕΙΝΟΣ.
PONTVS EVXINVS.
Aequor Iaſonio pulſa: tum remige primum.
Cum Priuilegio Imp. Reg. et Belgico, Ad decennium. 1590.
ÆVROPAEÆ PARS.
Tanaitae.
Sargatii.
Gynaecocratumeni.
Exobigitæ.
Roxolani.
Cares.
Laxij.
MAEOTIS PALVS: Quam Scythię gentes circundant vndique ripis, Et matrem Ponti perhibent Meotidis undam.
Arichi.
Corapseni.
ASIA:
SINDICA.
Aſpungitani
Achaei.
Zicchi.
Cercetij.
Mardi.
Sanegæ.
CAVCASEVM MARE.
Metibi.
Epageritae.
Heniochi.
Suani.
PHASIANVM MARE.
Heptacometæ.
s Axenus ille fuit.
Byzeres.
Moſynoeci.
Macrones.
Cilsij.
THIANNICA Regio
AR: MENI: AE PARS.
CAPPADOCIA.
PONTVS CAPPADOCICVS.
COLCHIS.
ORIENS.
CIMMERIVS.
BOSPORVS

Figure 4

EUR OPÆ PARS
BOHEMIA
VARSOVIA
VOLINIA
AVSTRIA
PODOLIA
GALLIÆ PARS
HVNGARIA
TRANSILVANIA
MOLDAVIA
SERVIA
BVLGARIA
CORSICA
SARDINIA
ITALIA
GOLFO DE VENETIA
SICILIA
GRECIA
CANDIA CRETA
LYCIVM MARE
MARE MEDITERRANEU
LYBICVM MARE
BARCHA
ÆGYPTVS
Tropicus Cancri
AFRICÆ
NVBIÆ
REGNI:
PA
PARS
TVRCICVM IMPERIVM

TARTARIÆ PARS.
ASTRACAN NOVA.
MARE DE BACHV ol. CASPIVM MARE.
ET HYRCANVM.
ARMENIA.
MEDIA.
PERSIÆ
PARS.
ASYRIA.
MESOPOTAMIA.
CALDEA.
ARABIA DESERTA.
AYAMAN.
ARABIA.
FELIX.
MARE ELCATIF ol. SIN. PERSICVS
ORMVS.
MARE RVBRVM olim SINVS ARABICVS
MARE DE MECCA
MARE ARABICVM
ET
INDICVM.
ADEN.
ZIBIT.
IRMIN.
HERIT.
THEAMA.
ALIBINALI.
GVBELAMAN.
Joannes Lhuilier sculp. t'Amsterdam

Figure 5

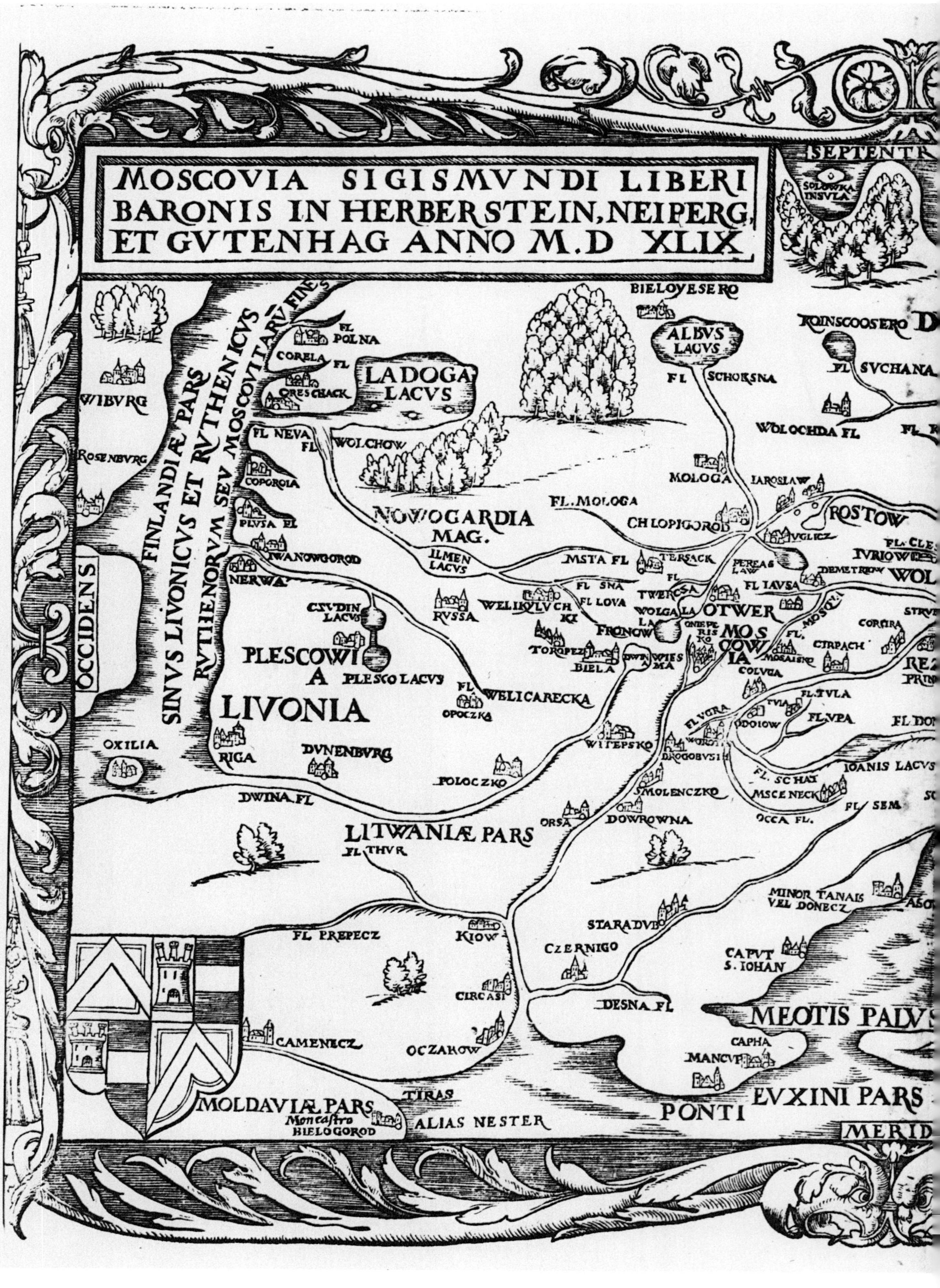
MOSCOVIA SIGISMVNDI LIBERI
BARONIS IN HERBERSTEIN, NEIPERG,
ET GVTENHAG ANNO M.D XLIX
SEPTENTR
SOLOWKA INSVLA
OCCIDENS
MERID
FINLANDIÆ PARS
SINVS LIVONICVS ET RVTHENICVS
RVTHENORVM SEV MOSCOVITARV FINES
WIBVRG
ROSENBVRG
LADOGA LACVS
CORELA
ORESCHACK
FL NEVA
WOLCHOW
COPOROIA
PLVSA FL
IWANOWGOROD
NERWA
BIELOYESERO
ALBVS LACVS
SCHORSNA
WOLOCHDA FL
SVCHANA
MOLOGA
IAROSLAW
FL. MOLOGA
ROSTOW
CHLOPIGOROD
NOWOGARDIA MAG.
ILMEN LACVS
MSTA FL
TERSACK
CSVDIN LACVS
PLESCOWIA
PLESCO LACVS
LIVONIA
RVSSA
WELIKIVLVCH KI
FL SNA
FL LOVA
TWERCSA
WOLGA
OTWER
FRONOW
MOS COWIA
TOROPEZ
BIELA
WELICARECKA
OPOCZKA
COLVGA
FL TVLA
FL VPA
OXILIA
RIGA
DVNENBVRG
WITEPSKO
DROGOBVSI
POLOCZKO
SMOLENCZKO
FL. SCHAT
MSCENECK
IOANIS LACVS
DWINA FL
ORSA
DOWROWNA
OCCA FL.
LITWANIÆ PARS
FL THVR
MINOR TANAIS VEL DONECZ
STARADVB
FL PREPECZ
KIOW
CZERNIGO
CAPVT S. IOHAN
CIRCASI
DESNA FL
MEOTIS PALVS
CAMENECZ
OCZAKOW
CAPHA
MANCVP
MOLDAVIÆ PARS
Moncastro
BIELOGOROD
TIRAS
ALIAS NESTER
PONTI
EVXINI PARS

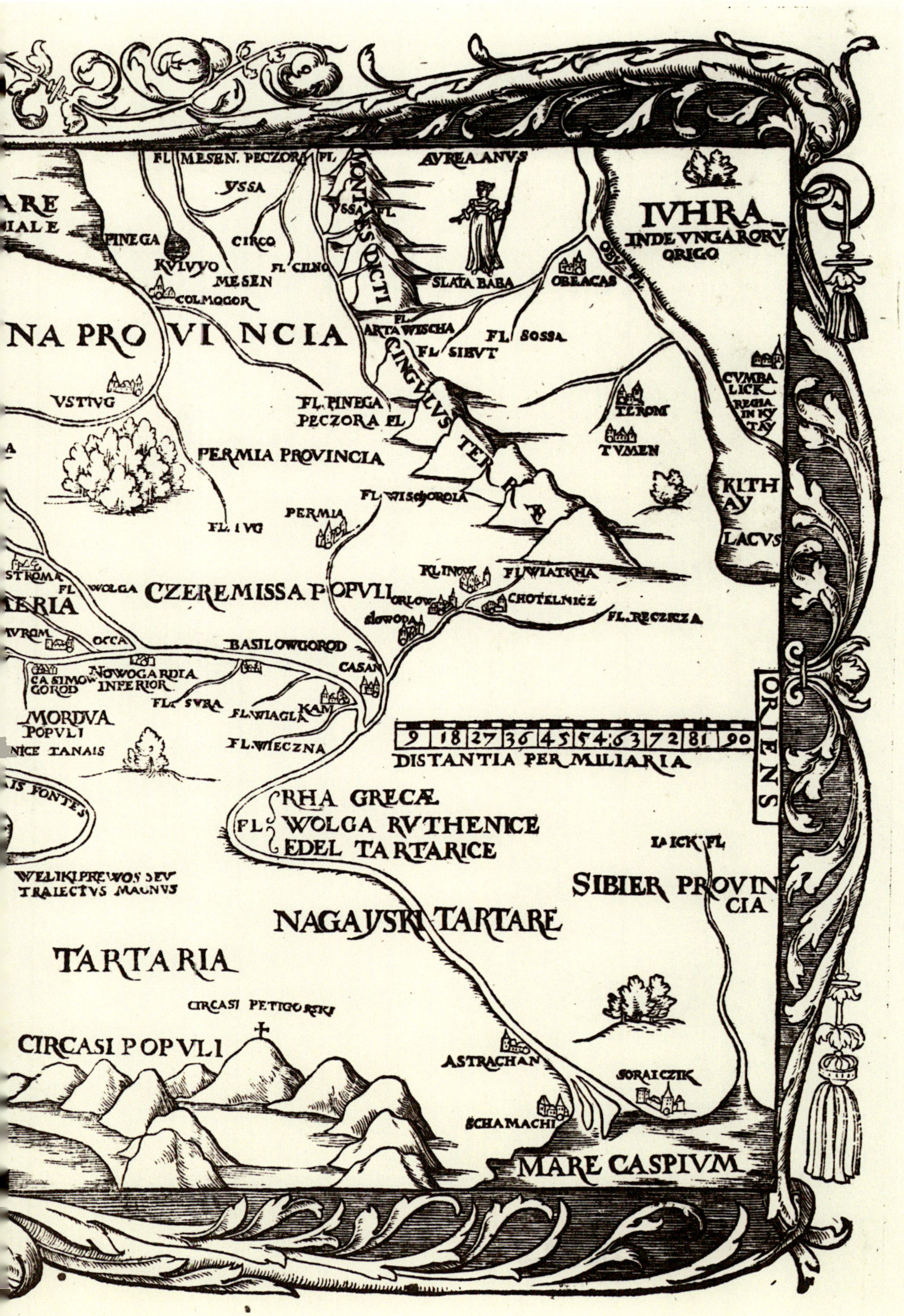

FL. MESEN. PECZORA FL.
YSSA
AVREA ANVS
MONTES DICTI
IVHRA
INDE VNGARORV ORIGO
PINEGA
CIRCO
KVLVYO
MESEN
FL. CILNO
SLATA BABA
OBEACAS
COLMOGOR
NA PRO VI NCIA
ARTA WISCHA
FL. SOSSA
FL. SIBVT
CINGVLVS TERRÆ
VSTIVG
FL. PINEGA
PECZORA FL.
CVMBA LICK
TEROM
TVMEN
PERMIA PROVINCIA
KITHAY
LACVS
FL. WISCHOROIA
PERMIA
FL. IVG
CZEREMISSA POPVLI
FL. WOLGA
KLINOW
FL. WIATKHA
ORLOW
CHOTELNICZ
FL. RECZICZA
OCCA
BASILOWGOROD
CASAN
NOWOGARDIA INFERIOR
FL. SVRA
FL. WIAGLA
FL. WIECZNA
MORDVA POPVLI
TANAIS
ORIENS
9 18 27 36 45 54 63 72 81 90
DISTANTIA PER MILIARIA
RHA GRECÆ
FL. WOLGA RVTHENICE
EDEL TARTARICE
IAICK FL.
WELIKIPREWOS SEV TRAIECTVS MAGNVS
SIBIER PROVINCIA
NAGAYSKI TARTARE
TARTARIA
CIRCASI PETIGORSKI
CIRCASI POPVLI
ASTRACHAN
SORAICZIK
SCHAMACHI
MARE CASPIVM

Figure 6

HVNGARIAE DESCRIPTIO, V
Cum Priuilegio.
POLONIAE PARS.
MORAVIAE PARS.
Marcomannoru campi.
AV STRIAE PARS.
OCCIDENS.
VERTHES mons hungarice BATON germanice den Schiltberg. quod Henricum III imp. istic castrametatum referunt, contra Belam Hungarie Rege, fame pressum relictis scutis castra deseruisse.
PANONIA SVPERIOR VEL PRIMA CONSVLARIS.
PANONIA INFERIOR ud SECV DA PROCONS
VALERIA, uel INTERAMNIA PANONIA
REGNI SLAVONIÆ PARS, HVGARIÆ VECT GALIS, olim SAVIA PANONIA
BOSNIA
CROATIA olim LIBVRNIA
Centum colles
Balaton lacus
Windischland.
Steirmarck

OLFGANGO LAZIO AVCT.
RVTHENIAE PARS.
MVLDAVIAE PARS.
Harayta mons
SICVLI.
TRANSILVANIA
HERDEL, olim
Dacia mediterr.
Clausenburg
Weissenburg
HACZAG VALLIS.
TRANSASPINA
HVNG
HAVASALFEVLDE
olim
DACIA ALPESTRIS
WALACHIAE PARS
MOESIA INFERIOR
BVLGARIA
RASCIANI.
VASCAPV
ORIENS.
Danubius fluuius
Tibiscus flu.
Marusius fl.
Saw fluuius
Temeswar.
Debrezen.
Hungaricæ uoces quomodo legendæ sunt.
CH. litteræ in vocibus Hungaricis CZ. S, littera pro SCH. Z, littera pro S simplici in vocibus Hungaricis. W, litteram in fine pro V. Y.
Scala Milliarium Hungaricorũ.
Nissa olim Nessus Moesiæ
Taurunum, Griechschweissenburg. Nandor alba.
Zingidunum. Zenderin.
Pontis Traiani vestigia.
Vlpia Traiana Zarmisegetusa.
40.
41.
42.
47
48
49
46

Figure 7

POLONIÆ
AMPLISSIMI
REGNI . . TYPVS
GEOGRAPHICVS
BALTICVM MARE
Scala miliarium Germanicorum
Dantziger mer
SEPTETRIO
OCCIDENS
ORIENS
MERIDIES
POMERANIA
POLONIA
MAIOR
MARCA ANTIQVA
SILESIÆ PARS
MORAVIÆ PARS
SIGVSMVN . III . D . G . REX Pol. Mold.
XXXVII . prox. hæres REGIS Suæd.

MAGNVS
DVCATVS
LITVANIAE
LITVANIA
DVCATVS
VOLHINIA
MASOVIA
RVSSIA
MOLDAVIÆ PARS
Vilna
Kowno
Grodno
Brizestre
Cobrin
Pinsko
Lublin
Leopolis
Sendomiria
Jaroslaw
Luczko
Belz

Figure 8

Scala miliarium Germanicorum
5 10 15 20 25 30
Venedicus sinus.
BALTICVM MARE.
SILESIAE PARS.
OCCIDENS.
Cum privilegio Imp. Regiæ etc. decennali.
POLONIAE, LITVANIÆQ. DESCRIPTIO.
Auctore Wenceslao Godreccio; et correctore Andrea Pograbio Pilsnensi.

TENTRIO.
LIVON: IAE PARS.
Bersany
MAGNVS
DVCATVS
LITVANIAE.
VOLHINIA.
PODOLIA, pecore opima.
Codima solitudo uastissima.
MOLDAVIAE PARS.
ORIENS.
PONTI EVXINI PARS.
DIES.
Smolensko
Vilna
Troki
Kowno
Grodno
Borisowo
Mogilow
Orscha
Dobrawna
Dobranczko
Nowogrodeck
Brzescye
Cobrin
Pinske
Petrilow
Cernobel
Kijow
Kanyow
Czyrkasy
Zwinigrod
Ostrog
Rowna
Lucko
Olesko
Dunayow
Trebowla
Bar
Braclaw
Sroka
Vinnicza
Culicow
Chotym
Snatyn
Colomia
Margoseth
Seret
Oczakow
Byalogorod
Maiak
Tubarcza
Obidowo lacus
Borysthenes
Nijeper flu.
Bog flu.
Nijester flu. Ptol. Tyras

Figure 9

MAGNI DVCATVS LITHVANIÆ,
et Regionum Adiacentium exacta Descriptio
Signorum declaratio.
Urbs
Civitas
Civitas cum Palatinatu
Domicilium Ducum
Oppidum
PARS MINORIS POLONIÆ.
RVSSIA RVBRA.
WOLYNIA CITERIOR.
PODOLIA.
Pokutiæ pars.
Paludes Poleliæ
Campi deserti citra Boristenem.
Campi deserti ultra Boristenem.
quæ tum VKRAINA ab aliis vocitatur.
Rzeczycensis terra.
SEVERIENSIS DVC:
Ioannes Blaeu Excudebat.

Occidens
MARE BALTICVM.
Vulgo
DE OOST ZEE.
Curonius Lacus
Osiliæ pars.
CVR: LANDIÆ
DVCA: TVS
SINVS MAGNVS
LIVONIÆ.
SAMOGITIA
SEMIGALLIA
Riga
Vilna
Spectator
Habes hic Magni Ducatus Lithvaniæ, ut anti: quis terminis continebatur, descriptionem accu: ratam, minus verò regionum adjacentium. Quævis igitur loca in longitudine circino expressa quanti: tati non respondere, noli mirari; iter enim obli: quum id facit, et inæqualitas milliarium; quæ Li: thvanica Polonicis, Russica et Podolica Lith: vanicis amphora sunt.
Illustratio finium M.D. Lithvaniæ.
Septentrio
Polocsk
MOSCOVIÆ
CONFINIA
MOHILOVIENSIS DISTRICTVS.
MAGNI DVCIS
Smolensko
Territo:
PARS.
Oriens

Figure 10

TVRCICVM
IMPERIVM
Concordia res parvæ crescunt
Discordia maximæ dilabuntur.
MARE MAGIORE
Regni
Tuneta-
ni pars.
Occidens.
LIBYCVM MARE.
MARE ÆGYPTICVM.
BARCHA.
ÆGYPTVS
A F R I C Æ
Linea sub Tropico Cancri.
LYCIVM
MARE.
CANDIA
CRETA.
MOLDAVIA
PODOLIA
CYPRVS

TARTARIÆ PARS.
Magni Ducis
Moscoviæ confinia.
ASTRACAN NOVA.
MARE DE BACHU ol. CASPIUM MARE ET HYRCANUM.
EUXINUS
GIORGIA.
ARMENIA
TURCO MANIA
MESOPOTAMIA
DIARBECH.
ASYRIA
SERVAN.
GILAN.
MEDIA
CUSISTAN
CALDAR
FARCI
PERSIÆ PARS.
ARDEN.
ARABIA DESERTA.
CALDEA.
ORMUS.
MARE ELCATIF, ol. SINUS PERSICUS.
AYAMAN.
ELCATIF
ARABIA
MASCALAT.
FELIX.
AMAN SIRIFDIN.
ALIBINALI.
THEAMA
GUBEL AMAN.
IRMIN.
ZIBIT
HERIT.
ADEN.
MARE RUBRUM
SINUS ARABICUS
MARE DE MECCA
MARE ARABICUM et INDICUM
AMAMIR.
Oriens.
Ioannes Ianssonius excudit.
Milliaria Germanica communia.

Figure 11

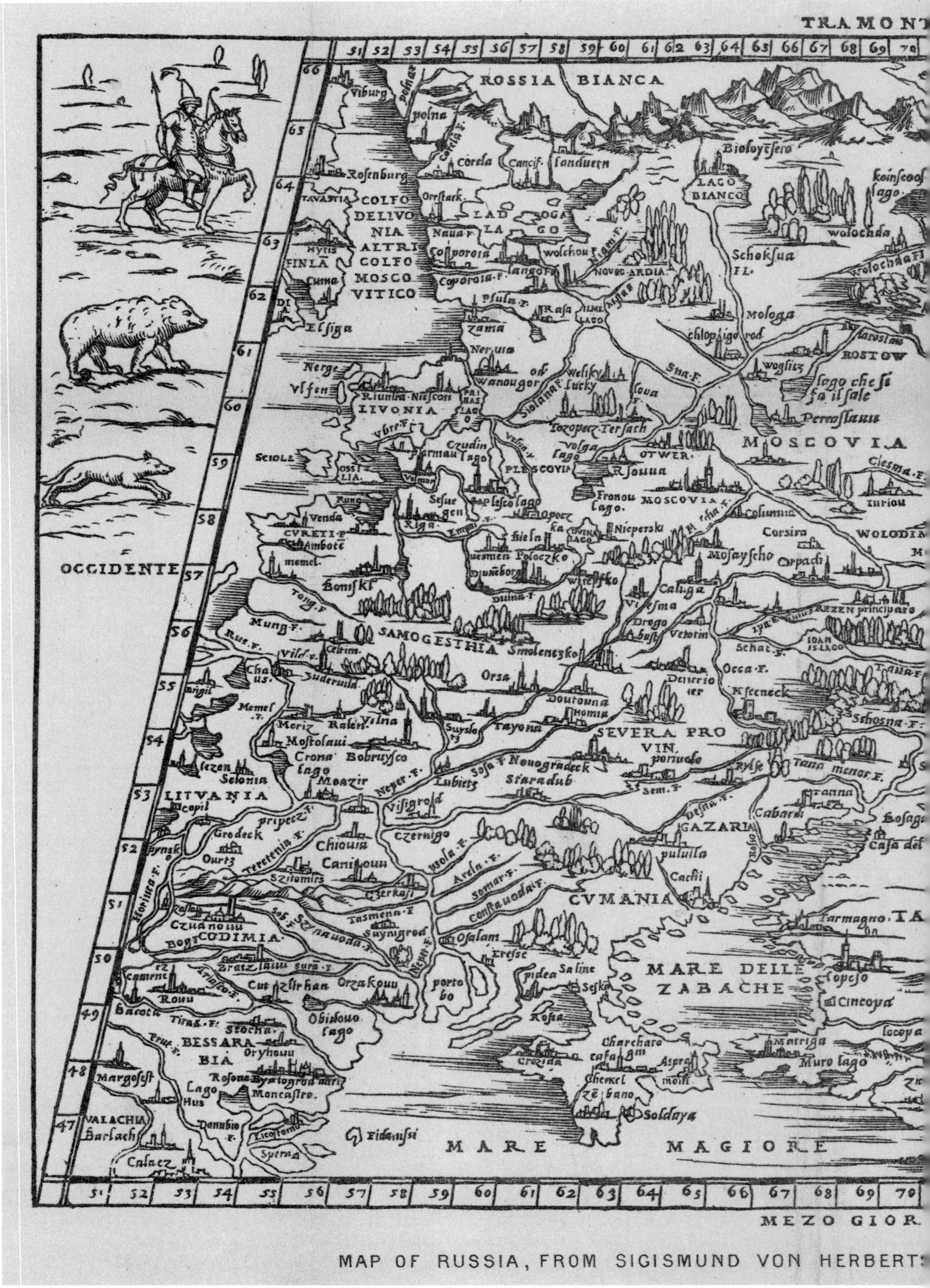

MAP OF RUSSIA, FROM SIGISMUND VON HERBERT

COMMENTARI DELLA MOSCOVIA. VENICE, 1550.

Figure 12

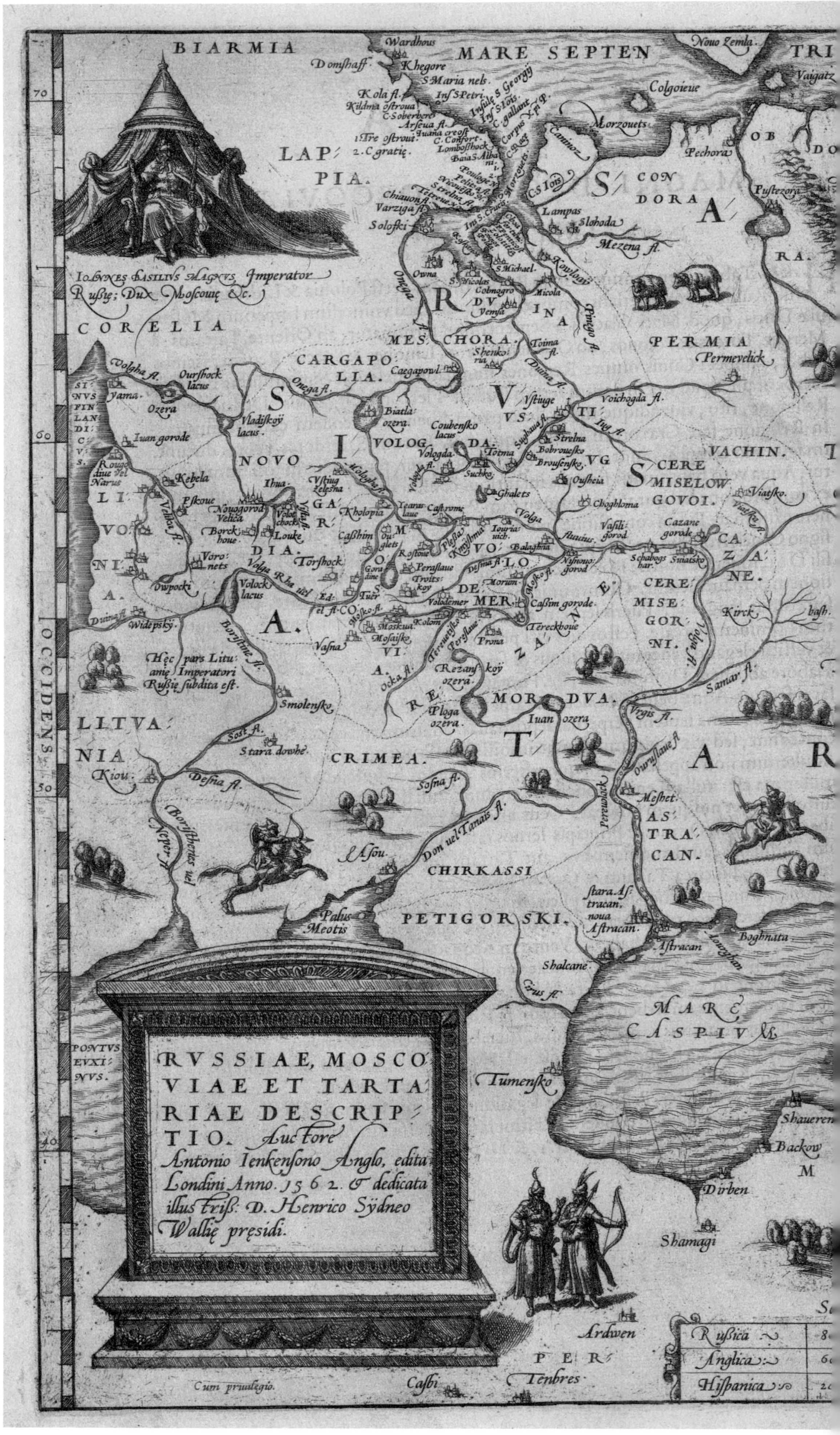
RVSSIAE, MOSCOVIAE ET TARTARIAE DESCRIPTIO. Auctore Antonio Ienkensono Anglo, edita Londini Anno. 1562. & dedicata illus triss: D. Henrico Sydneo Wallię presidi.
Iohannes Basilius Magnus Imperator Russię; Dux Moscouię &c.
BIARMIA
MARE SEPTEN
LAPPIA.
CORELIA
PERMIA
CARGAPOLIA.
LITVANIA
CRIMEA.
CHIRKASSI
PETIGORSKI.
MORDVA.
ASTRACAN.
MARE CASPIVM
PONTVS EVXINVS.
OCCIDENS
Hęc pars Lituanię Imperatori Russię subdita est.
Smolensko
Kiou.
Astracan
Shamagi
Tumensko
Rusšica
Anglica
Hispanica
Cum priuilegio.

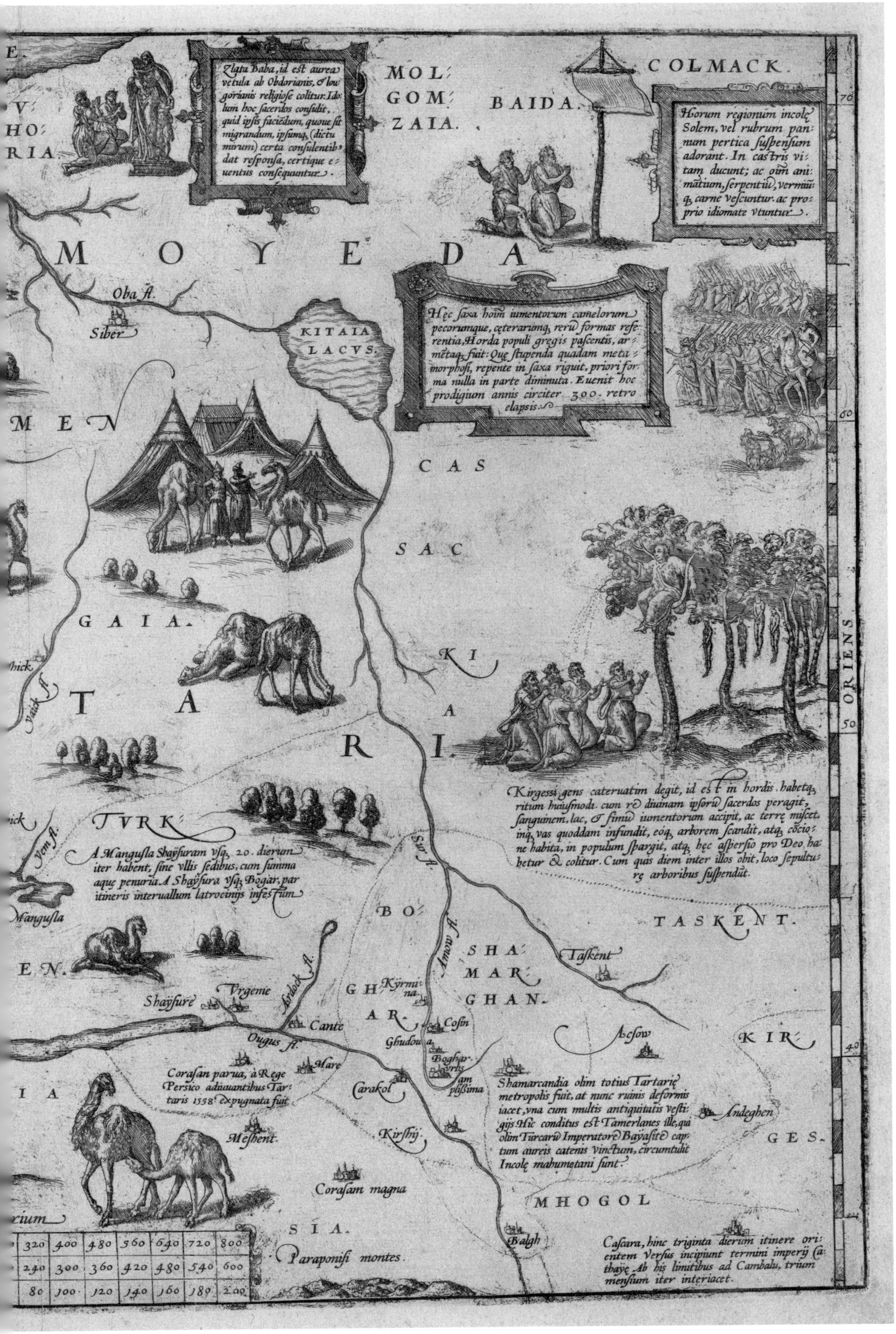

Zlata baba, id est aurea vetula ab Obdorianis, & low gorianis religiose colitur. Idolum hoc sacerdos consulit, quid ipsis faciendum, quoue sit migrandum, ipsumq, (dictu mirum) certa consulentib dat responsa, certique euentus consequuntur.
MOL GOM ZAIA.
BAIDA.
COLMACK.
Horum regionum incolę Solem, vel rubrum pannum pertica suspensum adorant. In castris vitam ducunt; ac oim animatium, serpentiū, vermiūq, carne vescuntur. ac proprio idiomate vtuntur.
M O Y E D A
Oba fl.
Siber
KITAIA LACVS.
Hęc saxa hoim iumentorum camelorum pecorumque, cęterarumq, rerū formas referentia, Horda populi gregis pascentis, armētaq, fuit: Quę stupenda quadam metamorphosi, repente in saxa riguit, priori forma nulla in parte diminuta. Euenit hoc prodigium annis circiter 300. retro elapsis.
M E N
C A S
S A C
G A I A.
K I
A
T A R I.
Kirgessi gens cateruatim degit, id est in hordis. habetq, ritum huiusmodi. cum rē diuinam ipsorū sacerdos peragit, sanguinem. lac, & fimū iumentorum accipit, ac terrę miscet, inq, vas quoddam infundit, eoq, arborem scandit, atq, e concione habita, in populum spargit, atq, hęc aspersio pro Deo habetur & colitur. Cum quis diem inter illos obit, loco sepulturę arboribus suspendūt.
TVRK
A Mangusla Shayfuram vsq, 20. dierum iter habent, sine vllis sedibus, cum summa aquę penuria. A Shayfura vsq, Bogar, par itineris interuallum latrocinijs infestum.
Mangusla
Yem fl.
Yaick fl.
B O
T A S K E N T.
S H A M A R G H A N.
Tasƙent
E N.
Vrgeme
Shayfure
Ardock fl.
G H A R
Kyrmina
Amow fl.
Sir fl.
Cante
Cosin
Ghudowa
Oxus fl.
Boghar
Acsow
K I R
Corasan parua, à Rege Persico adiuuantibus Tartaris 1558. expugnata fuit.
Mare
Samarcanda
Carakol
Shamarcandia olim totius Tartarię metropolis fuit, at nunc ruinis deformis iacet, vna cum multis antiquitatis vestigijs. Hic conditus est Tamerlanes ille, qui olim Turcarū Imperatorē Bayasitē captum aureis catenis vinctum, circumtulit Incolę mahumetani sunt.
Andeghen
I A
Meshent.
Kirshy.
G E S.
Corasam magna
M H O G O L
S I A.
Balgh
Paraponisi montes.
Cascara, hinc triginta dierum itinere orientem versus incipiunt termini imperij Cathayę. Ab his limitibus ad Cambalu, trium mensium iter interiacet.
ORIENS
70
60
50
40
320 400 480 560 640 720 800
240 300 360 420 480 540 600
80 100 120 140 160 189 200

Figure 13

MARE DELLE ZABACHE
olim
MÆOTIS PALVS
Taurica Chersonesus
CRIMEA SEU
DIKOIA
OGRAINA
RESANIA
PENSIS
DVCATVS
SEVERI
ENSIS
MERA
WOLOD
GRINA
Moscua
Ieroslaw
Wologda
Pereslaw
Smolensko
Occidens

MOSCOVIÆ
PARS AVSTRALIS.
Auctore
Isaaco Massa.
Czeremissi
Logowoy.
MORDO:
Czeremissi
Nagornoy.
WITI.
MORDUA.
Montes Boglowy
POLE
IA PRZECO:
Rosdori Donski.
Fossa Kamouz
TIGORI
a Regio
Oriens
Mill. Germanica quorum 15. uni grad. respond.
Mill. Russica. quorum 80. uni gradui respondent.

Figure 14

EUROPÆ PARS
MARE SEPTRIONALE
NOVA ZEMBLA
MOSCOVIA
CRIMEA
Precopen ses Tartari
PONTI EUXINI PARS
CHIRKASSI
Petigorski
GEORGURIA
ARMENIA
SERVAN
PERMIA
SAMOGEDA
JUGHORIA
PEROSITAS
NAGAIA Tartaro. Horda
ESACHADA
TUMEN
KALMUCKI TARTARI
MARE DE SALA vel de BACHU, Ruthenis Chualenske more ol: Caspium et Hircanũ
Ziebierairorum Horda
SAGATAI Tart.
OCRAGE
ELSABAR
Bactriana
TAGALISTAN
Iarchem
Care Desertum
Cara Kithay id est Nigra Cath
Tartari imago.
Samogedi imago.
Jodocus Hondius lectori salutem
In hac Tartariæ descriptione omnem adhibuimus diligentiam ut regiones nuper ab Anglis vel à Belgis alijsve detectæ suo quæq; loco haberentur, quamvis mihi ipsi non satisfecerim, quis enim vastum illud et incognitũ regnum vere describet? at ut possimus dum ut volumus non licet. usq; dum melius quoddam
Milliaria Germanica communia
15 30 45 60 75 90 105

TARTARIA
AMERICAE PARS
MARIS PACIFICI PARS
ANIAN FRETUM
Tabin promontoriu
Bargu Campestria
UNG qua GOG nonnullis dicitur
TENDUC Regnum in quo Christiani regnabant anno 1290.
Sumongul que MAGOG Mercatori
Tartar fluvius
CARLI.
Mons Alchai in quo Regum Tartarie sepulchra sunt.
Unchiam castru contra irruptiones Tartarorum
Xandu
C de Fortuna
BARSOL
CATAIO
Belgian desertum
Cavona
COLMAK
Cambalu R
Canglu
Camul
STINGUI
In hac regione mons est ex quo terra effoditur (terra) Abestus Plinio) tenuissimis fibris, instar graminis, que aptate in filum texuntur, pannusque ex eo in ignem infectus non comburitur.
Murus quadringentarum Leucarum inter montium crepidines à Rege Chinæ contra irruptiones Tartarorum extructus.
CIARTIAM
LOP DESERT.
Xuntien als Quinsay id est cæli civitas quod rex Chinæ hic sedem habeat.
CHINÆ REGNI PARS
Tauxem
COREA INS.
SANCII
Lacus Cincuthay
70. Parallelorum.

Figure 15

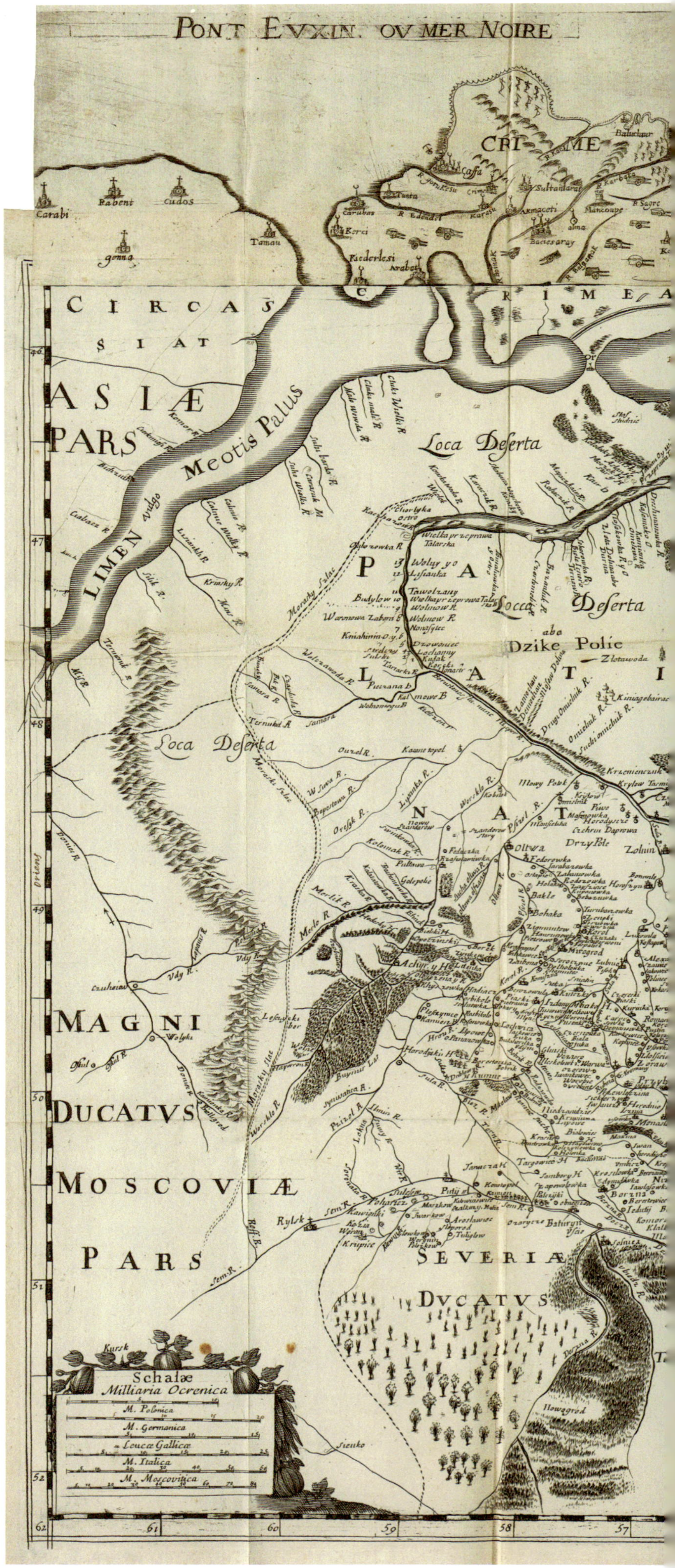
PONT EVXIN. OV MER NOIRE
CRIME
CIRCASSIAT
ASIÆ PARS
Meotis Palus
LIMEN vulgo
Loca Deserta
PALATI
Loca Deserta
Dzike Polie
NAT
MAGNI DUCATVS MOSCOVIÆ PARS
SEVERIÆ DVCATVS
Schalæ Milliaria Ocrenica
M. Polonica
M. Germanica
Leucæ Gallicæ
M. Italica
M. Moscovitica

Carte d'vkranie
Contenant plusieurs Prouinces comprises entre les Confins de Moscouie et les Limittes de Transiluanie
Dressez par G.L.V. sieur de Beauplan Ingenieur et Capitaine de l'Artillerie du serenissime Roy de Pologne.
A Rouen
Chez Jacques Cailloue dans la Cour du Pallais
CZARNE MORZE
Pontus Euxinus
BULGARIA
MOLDAVIÆ PARS
BUDZIAK
Loca Deserta
WALACHIA
PODOLIA
VOLYNIÆ PARS
RVSSIÆ PARS
KIIOVIENSIS
MAGNI DUCATUS LITHUANIÆ PARS
Meridies
Septentrio
Occidens

Figure 16

UKRAINÆ PARS,
QVÆ
KIOVIA PALATINATUS
Vulgo dicitur.
Per Guilhelmum le Vasseur de Beauplan
S.R.M. Poloniæ Architectum militarem etc.
Milliaria Polonica
M. Germanica
Ducatus
Czernichovien
Occidens
K I O V I
P A L A
N A T U S
B R A C L A V I
E N S I S
Kiiow
Biala Cerkiew
Bialo Grudki
Ostrze
Perealaw
Kaniow
Czyrkasy
Korsum
Boguslaw
Koniecpole
Woronowka
Czarny Szlak
Kuczmanski Szlak
Septen
Mer

SEVERIENSIS DUCATUS
PARS.
Signorum et Caracterum declaratio
Urbs munita
Oppidum
Sloboda nova
Colonia
Pagus
R. Flumen
D. Vallis
O. Insula
Ruinæ
Loca Paludosa
Fons
Molendinum
B. Transitus
Magni Ducatus Moscoviæ Confinia
MOS
CO
VIÆ
PARS.
Magni Ducatus Moscoviæ Confinia
Oriens
Konotop
Borzna
Putywl
Rylsk
Lochwica
Piaski
Hadiacz
Mirogrod
Lubnie
Lukomla
Korol
Horoszyn
Borumle
Pultawa
Oltwa
Drzy Pole
Dabrowa
Krylow
Krzemienczuk
deserti
Boristenem.
Campi deserti ultra
Boristenem.
Blaeu exc.

Figure 17

Typus Generalis
UKRAINÆ
sive
PALATINATUUM PODO:
LIÆ, KIOVIENSIS
et Braczlaviensis terras
nova delineatione
exhibens.
MAGNI DU
LITHUANIÆ
VOLYNIÆ
PARS
RUSSIÆ
PARS
PODOLIA
WALACHIA.
Loca Deserta
MOLDA:
VIA.
BUDZIAK
BULGARIA
BRACL
Occidens.
Signorum vel Characterum declaratio.
Civitates Polesiæ.
Olim Paludes

Terra Czer-
nichovien-
sis.
SEVERIÆ
DUCATUS.
MAGNI
DUCATUS
MOSCOVIÆ
PARS.
SCHALÆ.
Kursk
Czernihow
Biello grod
Loca
Deserta.
Loca Deserta
abo
Dzike Polie
Loca Deserta
CZARNE
MORZE
O. Tandra
Ocsakow
CRIME
LIMEN vulgo Meotis Palus.
ASIÆ
PARS
CIRCAS
SIA.
Oriens.
Meri 49 dies.

Figure 18

CELEBERRIMA INGENS ET IMPERIALIS CIVITAS MOSCUA.
A Kremelenagorod
B Kitay-gorod
C Tzar-gorod
D Skorodum
E Streletskasloboda
HABITUS RUSSORUM
TABULA
RUSSIÆ
Magno Domino, Tzari,
Nicolao Ioannis
Piscatore
1651.
Wirst seu
Milliaria Russica quorum 80 uni gradui respondent
Milliaria Germanica quorum 15 uni gradui respondent
OCEANUS GERMA
NORWEGIÆ PARS
DANIA
NOORT ZEE
OOST ZEE
MARE BALTICUM
SWECIÆ
FINLANDIA
Sinus Botnicus
Ladago Lacus
Onega Lacus
LIEFLANT
PRUSSIA
LITTAW
POLONIA
MOSCOVIA
SEVERA
PODOLIA
MOLDAVIA
WALACHIA
BULGARIA
THRACIA
CRIMEA
PALUS MEOT
PONTUS EUXINUS qua ZORNO
Sinus Carcinitus
NATOLIA nunc A

NOVA ZEMLA
MANSKOY MORE
Fretum Waygats
obdora
SIBIRIA
Iugoria
Tin goesi.
BAIDA
TARTA
R I Æ
Condora
Permia
Wiatka
CAZAN
Ceremissi Nagorni
MORDWA.
ASTRACAN.
PETIGORI
PARS IVRMEN
MARE CASPIVM, quod Russi vocant GVALENTSCHA MORE etiam vocatur MARE DE SALA vel de BACHU
GILAN REGIO.
PERSIÆ, CONFINIA.
ZIRPHANSCKA REGIO
ARMENIA
Iverla sive Grusiuæ Imperium
Visscher Excudebat
Rusche Narva sive Ivane-gorod
Archangel sive Archangelscka-goroda
Curia Imperatoris Russi appellata Kremelena-gorod
Hypocausta Russorum
Molendina Russorum

Figure 19

MARE
BALTICVM
Vulgo DE OOST ZEE.
Gottlandia
Oelandia
Bornholm
Oesel
Daghö
Wener Lacus
Palatinatus
WARSOVIA
Sandomi
Lublinen
Cracovi
Siradiensis
Pomerel
Wladislaw

REGNI
POLONIÆ
et Ducatus
LITHVANIÆ
VOLINIÆ, PODOLIÆ
VCRANIÆ
PRVSSIÆ, LIVONIÆ
et CVRLANDIÆ descriptio
emendata per F. de Wit.
Amstelodami.

Figure 20

Explication des Chiffres marquez sur ce Plan.
1. Quartier du Grand Général de la Couronne avec sa Cavalerie.
2. Quartier du Palatin de Belzo, Mestre de Camp du Royaume.
3. Celui du Palatin de Kiovie avec ses Esquadrons.
4. Celui du Palatin de la Russie avec sa Cavalerie.
5. Celui du premier Capitaine des Gardes avec ses Esquadrons et Bataillons.
6. Celui du Général de l'Artillerie ayant sous soi les Gardes du Corps avec 20. pièces de canon.
7. Celui du Général Major Korycki avec les Regimens d'Infanterie de l'Aile droite et 8. Canons.
8. Celui du Gén. Major Denhof avec les Regimens d'Infanterie de l'Aile gauche et 8. Canons.
9. Celui du Gén. de la Cavalerie de Lithuanie.
10. Celui du Duc de Radzivil, Gén. de Lithuanie.
11. Celui du Gén. d'Infanterie avec 12. Canons.
12. Celui du Prince de Moldavie, qui s'étoit rangé du côté des Polonois.
13. Celui du Gén. de l'Infanterie de ce Prince.
14. Troupes sous le commandement des Gén. Russczyc et Miaczynski, qui avoient reçu ordre de passer à la nage le Niester pour poursuivre l'Ennemi s'enfuyant vers Kaminiek.
15. Les Dragons commandez par le Palatin de Russie et le premier Capitaine des Gardes avec 4. Canons.
VIRTUS
TYRA FLUVIUS vulgo DNIESTER
Plan exact des Fortifications du Château de CHOCZIM sur le Niester
Général en Chef des Polonois ensuite Roi de Pologne, sous le nom de Jean III.
a Leide, Chez Pier

Suite de l'Explication.
16. Camp des Turcs sur des rochers escarpez muni d'un bon retranchement.
17. Château de Choczim situé sur un rocher de très-difficile accès.
18. Camp du Prince de Moldavie, où se fait le prémier assaut.
19. Retranchement au-delà du Niester pour défendre le Pont.
20 La ville de Choczim avec des fortifications imparfait qui defendoient le Camp du Prince de Valachie.
21. Lieux où les Gardes du Corps à pié donnerent le prémier assaut.
22. Postes emportez par l'Infanterie de l'Aile droite.
23. Retranchement forcez par l'Infanterie de l'Aile gauche.
24. Chemin, par où les Piquiers et les Chevaux legers allerent pour soutenir l'Infanterie qui commencoit à plier sous les efforts de la Cavallerie Turque.
25. Chemin de Zwaniecz et Kaminick, par où les Turcs se retiroient ayant abandonne leur Camp.
26. Fosse naturelle d'une extrême profondeur, qui fut remplie par les cadavres des Turcs et des chevaux.
27. Lieu, où un grand nombre de Turcs se précipiterent poussez par notre Cavallerie.
28. Porte du Camp du côté de Iassy.
29. Tente du Hussain Bassa, où l'on chanta le Te Deum apres la Victoire.
30. Pont rompu par le trop grand poids des fuyards.
VIRTUTI
NULLA EST
les Forts et retranchemens de l'Armée des Turcs défaite en 1673 par Jean Sobieski,
r les lieux par François Grata, Secretaire de Sa Majesté, et gravé par Rom. de Hooge.
Aa.
16

Figure 21

Exactissima
Asiæ
Delineatio.
in præcipuas regiones
Cæteraſq; Partes diviſa.
et denuo in lucem edita
Per
T. Danckerts
Amſtelodami cum Privilegio.
Mare Mediterraneum
Mare de Sala als Caspium
Arabia Deserta
Mare Arabicum
Persia.

Mare
Tartaricum.
Circulus Arcticus
Oceanus Septen-
trionalis.
Tartaria
Magna
Imperium
Mogolis
Cochinchina.
Formosa.
Tropicus Cancri
Golfo di
Bengala
Olim
Gangeticus
Sinus
Ceylon
Siam.
Manilla.
Mindanao
Celebes
Terra dos Papos.
Moluccæ.
Sive
Oceanus
Orientalis.
Hollandia Nova.

Figure 22

MARE BALTICUM VULGO DE OOST ZEE
SUECIÆ PARS
OELANDIA
Bornholm I.
CURLANDIÆ
SAMOGITIÆ
POLONIA
POMERANIÆ
HUNGARIÆ REGNI PARS
MORAVIÆ PARS
BOHEMIÆ PARS
GERMANIÆ PARS
Cracow
Belgrado
Temeswar
Wien
Oder Rx
Danube Rx

REGNI
POLONIÆ,
MAGNI DUCATUS
LITHUANIÆ,
CŒTERARUMQ; REGI POLONIÆ
SUBDITARUM REGIONUM
TABULA;
In Omnes suos Ducatus,
Palatinatus &c. divisa
et in lucem edita
a
CAROLO ALLARD,
Amstelo Batavo.
NOWOGROD WELIKI PRINCIPATUS.
NOVOGRODENSIS
SEVERIENSIS
DUCATUS
TARTARIÆ MINORIS PARS
PONTUS EUXINUS
HODIE
MARE NIGRUM
GERMANIS
DAS SCHWARTZE MEER.
SIC NOTANTUR

Figure 23

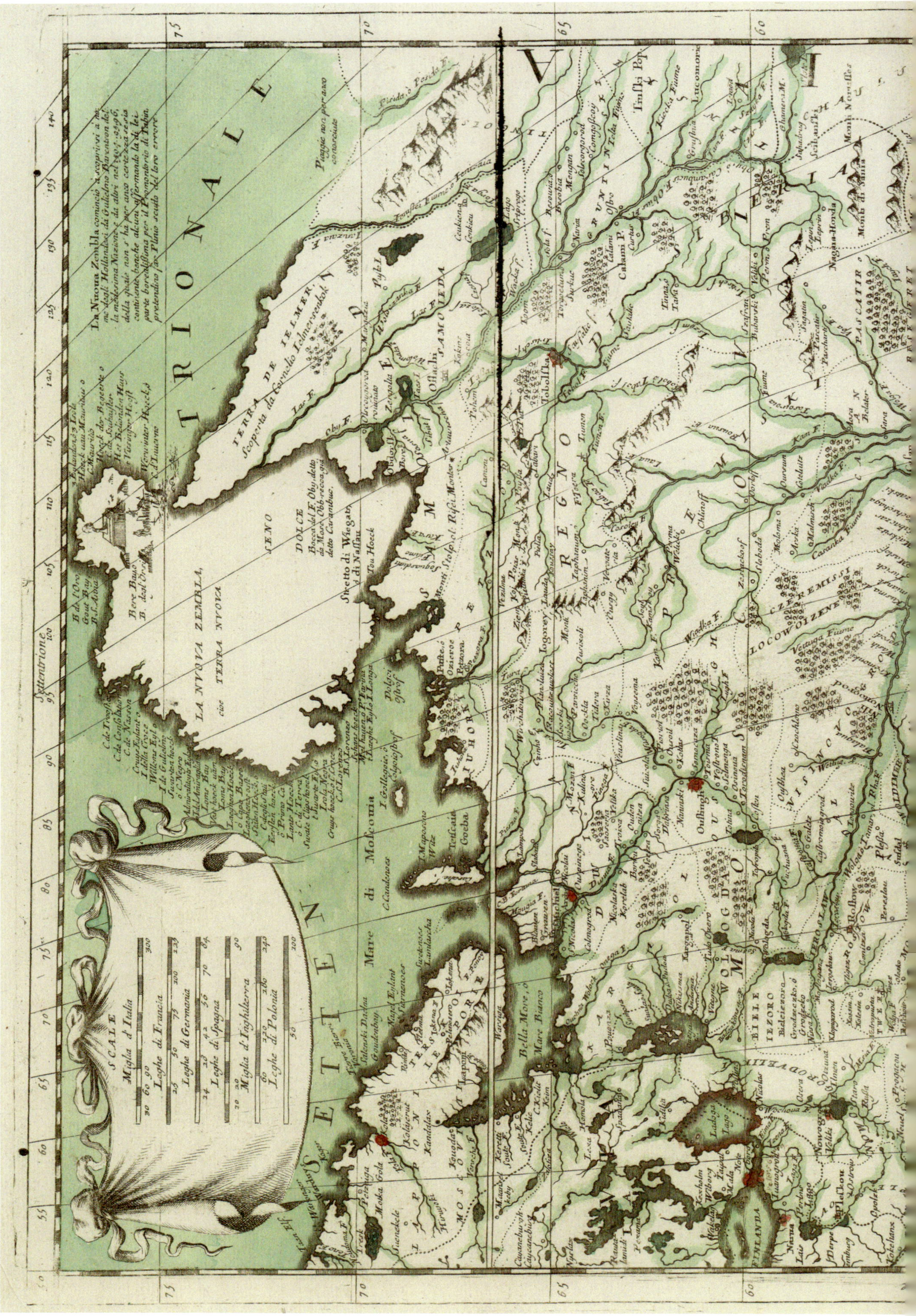

PARTE ORIENTALE
DELL' EUROPA
Descritta, e Dedicata
Dal P. Cosmografo Coronelli
All' Illustrissimo, et Eccellentissimo Signore
GIOVANNI DA MULA
Senatore amplissimo della Republica di Venetia.
Nella Serenissima &c.
MARE NEGRO, ò MARE MAGGIORE
MARE CASPIO

Figure 24

MARIS BALTICI PARS
Auctior et Correctior Tabula Chorographica
REGNI POLONIÆ,
Vicinarumque Regionum, ubi itinera, quæ Seren.
Rex Sueciæ
CAROLUS GUSTAVUS
una cum Exercitibus suis ab initio usque ad finem belli fecit,
nummeris denotantur, per supr. Castr. Metat. loc. E.I. Dahlberg.
Stationes et loca ubi Serenis. Rex ipsemet fuit Corona indicantur;
reliqua Vero itinera per Eiusdem Generales fuere confecta.
POMERANIA
BRANDENBURG
SILESIÆ PARS
REG. POLONIA
Milliaria Polonica communia
DANTZIG
STETIN
Colberg
Frankfort
Custrin
Crossen
Glogov
Breslau
Brieg
Oppeln
Görlitz
Kalisch

SAMOGETHIÆ PARS
Palatinatus Vilnensis
Palatinat. PARS Trokensis.
BORUSSIA
DUCALIS
MAGNI
PALATINATUS
NOWOGRODI:
DUCATUS
PODLASSIÆ.
LITHUANIÆ
PALATINATUS
BRIESTIENSIS
Palatinatus
Lublinien
Palati natus
Chelmensis
RUSSIA
PALATINATUS
BELSENSIS.
RUBRA
Palatinat.
Czerceviens
MAIOR
PALATI
NATUS SAN-
DOMI RI
MINOR
ENSIS
Warsovia
Grodno
Kowno
Vilna
Troki
Tykozin
Lublin
Lembery

Figure 25

LES ESTATS DELA COURONNE de POLOGNE, subdivisés en leurs PALATINATS, qui compren
LES DUCHÉS DE PRUSSE, MAZOVIE, SAMOGITIE, LES PROVINCES DE VOLHYNIE, ET DE P
ESTATS DE POLOGNE
Subdivisés suivant l'estendue des
PALATINATS
Par le S. SANSON
PRESENTÉ
A MONSEIGNEUR
LE DAUPHIN
Hubert Iaillot
MER BALTIQUE.
Golfe de Riga
Golfe de Dantzick
CONFINS DES ESTA
DE SUEDE
DUCHE DE CURLANDE
DUCHE de SAMOGITIE
GRAND DUCHE
Palatinat de Troki
Palatinat de Wiln
DUCHÉ DE PRUSSE
POMERANIE
BRANDEBOURG
GRANDE POLOGNE
Palatinat de Ploczko
DUCHE DE MAZOVIE
Palatinat de Sirad
Palatinat de Lublin
Palatinat de Chelm
POLESIE
HAUTE VOL
PETITE POLOGNE
Palatinat de Cracou
RUSSIE NOIRE
Palatinat de Lemberg
HAUTE PODO
ROYAUME DE BOHEME
D'AUSTRICHE
ROYAUME DE HONGRIE
FRONTIERES
TRANSILVANIE
ET CONFINS DES E
OCCIDENT
SEPTENTRION
MIDY

LE ROYAUME DE POLOGNE, LE GRAND DUCHE DE LITHUANIE,
E, ou est compris L'UKRAINE ou PAYS DES COSAQUES. Sur les memoires les plus Nouveaux 1692.
FRONTIERES DES ESTATS DU CZAR DE MOSCOVIE
Palatinat de Poloczk
Palatinat de Witepsk
Duché de Smolensko
LITHUANIE
Palatinat de Mseislaw
Duché de Novogorod Sewierski
Territoire de Rohaczow
Rzeczica
de Czernihow
VOLHYNIE
Palatinat de Kiow
UKRAINE ou PAYS DES COSAQUES
BASSE VOLHYNIE
Dzike Pole ou Campagnes Desertes et Inhabitées
Campagnes Desertes et Inhabitées
Palatinat de Braclaw
BASSE PODOLIE
TARTARES D'OCZAKOW
TARTARES BUDZIAK
DES TURQS
MER NOIRE
FRONTIERES DES PETITS TARTARES
ORIENT

Figure 26

THE ENGLISH ATLAS.

Volume I.

CONTAINING A

DESCRIPTION

OF THE

Places next the North-Pole;

AS ALSO OF

Muſcovy, Poland, Sweden, Denmark,

And their ſeveral Dependances.

WITH

A General Introduction to Geography, and a Large Index, containing the Longitudes and Latitudes of all the particular Places, thereby directing the Reader to find them readily in the ſeveral Maps.

OXFORD,

Printed at the *THEATER*, for *Joh. Janſonius a Waesberge*, and *Steven Swart*, Bookſellers in *Amſterdam*. MDCLXXX.

Figure 27

A New Map of EUROPE according to the Present General Divisions and Names of its Countries together with their Capital Cities Chief Rivers Mountains &c
Dedicated To His Highness William Duke of Gloucester.
Scale of English Miles
According to Vulgar Computation
According to Norwoods Computation
NORTH
The Arctick Circle.
First Meridian
ICELAND.
Hola.
Skalholt
Hecla Mount
Fero I.
Schetland I.
Orkney I.
Western I.
SCOTLAND
Edinburg.
Drontheim
NORWAY.
Dofrine Hills
DENMARK
BRITISH ISLES.
Dublin
IRELAND
Shannon R.
London
ENGLAND
Thames R.
Severn R.
Lands end
Lizard P.
British Channel.
Amsterdam
Brussels
NETHERLANDS
Weser R.
Elbe R.
GERMA
Paris.
Seyne R.
Loyre R.
FRANCE.
Garonne R.
Rhone R.
Rhine R.
Basil
SWITZERLAND
Alpes M.
Venice
Po R.
ITALY
Tiber R.
Bastia
Corsica I.
Sardinia I.
Cagliari
ATLANTICK or WESTERN OCEAN.
Cape Finister
Pyrene M.
Doure R.
Ebro R.
PORTUGAL.
SPAINE.
Madrid.
Tao R.
Lisbon
Guadiana R.
Guadalquivir R.
Minorca I.
Majorca I.
Yvica I.
E U R
Azores I.
C. Vincent
Cadiz
Straits of Gibraltar.
MEDITER
Madera I.
Fez
Algiers
Tunis
B A R B A
Atlas M.
Canaries I.
PART of AF
M.Burg. sculp.
West.

North Cape
OCEAN.
NOVA ZEMBLA.
Straits of Weygats
GREAT
THE WHITE SEA
Dwina R.
Oby R.
G. of BOTHNIA.
DEN.
RUSSIA
or
G. of FINLAND
Wolga R.
P E.
Moscow.
MOSCOVIE.
L. Dwina R.
POLAND.
Weissel R.
Don R.
Cracow.
Crapack M.
Nieper R.
LESSER or CRIM TARTARY.
Wolga R.
HUNGARIE.
TRAN SILVANIA Herman stat.
WALA CHIA.
Buda
Targovisco.
MOLDAVIA.
Danube R.
ZABACHE SEA.
Caffa.
Caucase M.
Straits of Caffa.
Straits of Constantinople.
TURKY in EUROPE.
Constantinople.
Balkan or Æmus M.
BLACK SEA.
GEORGIA
THE CASPIAN SEA.
SEA of MARMORA.
NATOLIA
Strait of the Dardanells
ARCHIPELAGO
Smirna.
TURKY in ASIA.
Rhodes I.
Aleppo
Tygris R.
MOREA
Misithra
Cyprus I.
Candia.
Candia I.
LEVANT SEA
SYRIA.
Euphrates R.
Nile R.
ASIA.
Barca
Cairo
ARABIA.
EGYPT.
RED SEA.
GOLF OF BALSERA.
PART of ASIA.
East.
University.
Bishops see.
Archbishops see.
Patriarchal see.

Figure 28

L'EUROPE
Dreßée sur les Observations de M.rs de l'Academie Royale des Sciences et quelques autres; & sur les memoires les plus recens.
Par G. DE L'ISLE Geographe.
A PARIS,
Chéz l'Auteur sur le Quai de l'Horloge a l'Angle d'or Avec Privilege du Roy pour 20. ans
1700.
AVERTISSEMENT.
Comme il y à plusieurs choses sur cette Carte et sur les autres que j'ay mises au jour qui sont differentes de ce qui se trouve sur les Cartes qui ont paru jusqu'icy, il est apropos d'avertir icy que ce la n'est point arrivé par inadvertence, et que je rends raison de ces changemens dans ma Nouvelle Introduction a la Geographie.
MER GLA
Cercle Polaire
ISLANDE
MER DU NORD
ISLES BRITANNIQUES
ECOSSE
IRLANDE
ANGLETERRE
LA MANCHE
OCEAN.
Premier Meridien
GOLFE DE GASCOGNE
FRANCE
ESPAGNE
CASTILLE
I. de Madere
ISLES CANARIES
R.E DE FEZ
R.E DE MAROC
ROYAUME D'ALGER
BARBARIE
Desert de Mesila
R.E DE TAFILET
AFRI

DUCHÉ DE PERMIE
La Grande Perme
DE SIBERIE
MOSCOVIE
ASIATIQUE
DUCHÉ DE WIATKA
DUCHÉ DE BULGAR
MOSCOVIE EUROPE
RUSSIE MOSCOVITE
ROYAUME D'ASTRACAN
NOGAIS D'ASIE
CIRCASSIE
CABARDINIE
POLOGNE
LITHUANIE
UKRAINE
PETITE TARTARIE
MER NOIRE
HONGRIE
TURQUIE EUROPE
BULGARIE
ROMANIE
GEORGIE
PERSE
ARMENIE
CURDISTAN
NATOLIE
TURQUIE ASIATIQUE
MEDITERRANÉE
Desert de Barca
GOLFE DE LA SIDRE
ECHELLE
H. van Loon Sculpe

Figure 29

Tome I. page 75.
L'EUROPE
Par De Fer
Liebaux Sculp.
OCEAN SEPTENTRIONAL
Premier Meridien
ISLES BRITANNIQUES
MER DE NORWEGE
MER DE MOSCOVIE
MER DE FRANCE
MER D'ESPAGNE
MER MEDITERRANEE
MER NOIRE
AFRIQUE
BARBARIE
ARABIE
PERSE
FRANCE
ESPAGNE
ALLEMAGNE
POLOGNE
MOSCOVIE
ANATOLIE ou ASIE MINEURE
PARIS
Madrid
Lisbonne
Toulouse
Bourdeaux
Lyon
Nantes
Rouen
Dijon
Valladolid
Toulon
Barcelone
Mayence
Cologne
Prague
Cracovie
Dantzick
Kiow
Smolensko
Moscow
Casan
Astrachan
Belgrade
Constantinople
Candie
I. Cypre
Alexandrie
Tripoli
Tunis
Alger
Sardaigne
Corse
Sicile
Malthe
Naples
Nort Cap
Shetland
Fero I.
Edimbourg
Stockholm
Riga
Revel
Abo
Wologda
Kargapol
Caffa
Derbent
Damas
Sinopoli
Trebisonde

Figure 30

A NEW MAP OF EUROPE
From the latest Observations
Inscrib'd to the Hon.ble Samuel Molyneux Esq. Secretary to his R. H. the Prince
by Jn.o Senex
ICELAND
THE NORTHERN OCEAN
THE WESTERN ISLES
NORTH SEA OR GERMAN OCEAN
ST. GEORGES CH.
BAY OF BISCAY
FRANCE
SPAIN
PORTUGAL
GRANADA
MINORCA I
MAJORCA I
YVICA I
THE MEDITER
PART OF AFRICA

PART OF ASIA
RUSSIA
MOSCOVY IN EUROPE
WHITE SEA
THE CASPIAN SEA
CIRCASSI TARTARY
ARMENIA
MAJOR
THE LESSER ASIA
SYRIA
EUXINE OR BLACK SEA
LITTLE TARTARY
CRIMEA
POLAND
HUNGARY
TURKY IN EUROPE
GREECE
MOREA
CANDIA
CYPRUS
THE LEVANT
SEA
AFRICA
BULGAR
Aleppo
Jerusalem
A Scale of Miles 60 to a Degree
Henry Wilson delin.
I. Senex sculp.

Figure 31

PARTIE ORIENTALE
DE LA RUSSIE POLONOISE
où se trouvent
LA BASSE VOLHYNIE et LA
BASSE PODOLIE.
Par le Sr. Robert de Vaugondy
Fils de Mr. Robert Geog. du Roi.
Avec Privilege 1749
Echelle
Lieues d'une heure
Septentrion
Midi
Occident
Orient
LITHUANIE
EMPIRE DE RUSSIE
HAUTE VOLHYNIE
BASSE VOLHYNIE
HAUTE PODOLIE
BASSE PODOLIE
Palatinat de Kiovie
Palatinat de Braclaw
MOLDAVIE
Campagnes Desertes
COSAQUES ZAPOROSKI
PETITE TATARIE
Des erts

Figure 32

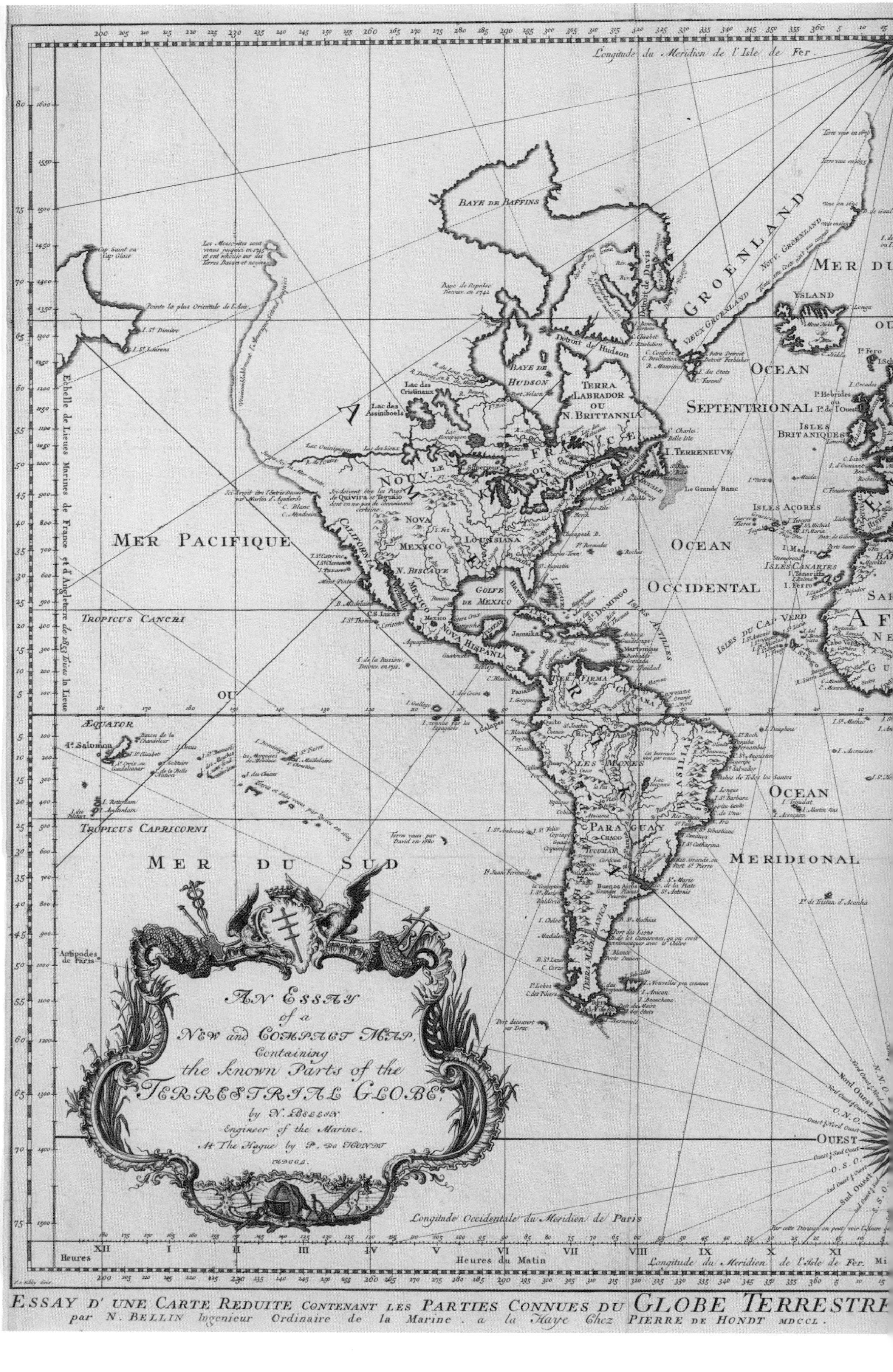
An Essay
of a
New and Compact Map,
Containing
the known Parts of the
Terrestrial Globe,
by N. Bellin
Engineer of the Marine.
At The Hague by P. de Hondt
MDCCL.
Longitude du Meridien de l'Isle de Fer.
Longitude Occidentale du Meridien de Paris
Echelle de Lieues Marines de France et d'Angleterre de 2853 Toises la Lieue
Mer Pacifique
Mer du Sud
Ocean Septentrional
Ocean Occidental
Ocean Meridional
Groenland
Baye de Baffins
Baye de Hudson
Terra Labrador ou N. Brittannia
Nova Mexico
Louisiana
Golfe de Mexico
Nova Hispania
Tropicus Cancri
Æquator
Tropicus Capricorni
Paraguay
Isles Antilles
Isles du Cap Verd
Isles Canaries
Isles Açores
Isles Britaniques
Ysland
Heures du Matin
ESSAY D'UNE CARTE REDUITE CONTENANT LES PARTIES CONNUES DU GLOBE TERRESTRE
par N. BELLIN Ingenieur Ordinaire de la Marine. a la Haye Chez PIERRE DE HONDT MDCCL.

MER GLACIALE
NOVA ZEMBLA
SAMOJEDA
JAKUTI
LAPPONIA
LES OSTIACS
TUNGUSI
JUKAGRI
KAMTSCHATKA
MOSCOVIE
POLONIA
HONGARIA
MER NOIRE
TURQUIE ASIATIQ
PERSIA
ARABIE
TARTARIE
PAYS DES ELUTS
PAYS DES KALKAS
TARTARIE CHINOISE
MONGALIA
TIBET
CHINA
ETATS DU MOGOL
BENGALE
GOLFE DE BENGALE
Ceylon
MER DES INDES
OCEAN
ORIENTAL
Formosa
ISLES PHILIPINES
MARIANES
Borneo
Nova Guinea
Nova Britannia
Terre de Diemen
HOLLANDIA NOVA
Terre de With
Terre d'Endract decouv. en 1628.
Terre de Nuits
Terre de Lewin
Nouvelle Zelande
Terre de la Compagnie
Terre vue par Don Juan de Gama
LES ISLES DU JAPON
NIPON
I^s SALOMON
Antipodes de Paris
ABISSINIE
CONGO
I. MADAGASCAR
Canal de Mozambique
Pays des Hottentots
DESERT
Echelle reduite de Lieues Marines de France et d'Angleterre de 20 au Degré
Longitude Orientale du Meridien de Paris
Heures du Soir
Heures
EST
E.N.E.
VAN EENE AAN-EENGESCHAKELDE KAART DER BEKENDE-DEELEN DES GANTSCHEN AARDBOLS
door N. BELLIN Ingenieur des Fransen-Zeevaards . in 's Hage by PIETER DE HONDT MDCCL .

Figure 33

Samt
CROATIEN, DALMATIEN, SCLAVONIEN,
bis an das
40
45
Brinn
Mont Crapack
Leopolstatt
Presburg
Bistrica
Eperies
Caschaw
Zemlin
Tokay
Donau
WIENN
Neuheusel
Neustatt
Rab
Erla
Chege
Zatmar
Novigrad
Gran
Pest
Gross Waradyn
Gratz
Stulweissenburg
BUDA OFEN
Zolnock
Pettau
Kanischa
Gyula
HUNGARIA
Copranitz
Colocza
Segedin
SCLAVONIA
Petsch
Chonad
Gradiskia
Temeswar
Sioeth
Posega
Brodt
Esseck
Peterwardein
CROATIA
Belgrad
Zeng
BOSNIA
Semendria
Zenobicz
Serato
Bagnialuch
Nona
Zara
Scardona
Sebenico
Traw
DALMATIA
Coronata
GOLFO
DI
VENETIA
Lesina
S. Andrea
Melissello
Cuza
Augusta
Ragusa
Stagno
Castel Novo
SERVI
Pristina
Pelagosa
I. de Tremithi
S. Domino
la Caprari
S. Nicola
Budoa
Antivari
Scutari
Ducagnit

N G A R N
räntzenden Ländern
NBÜRGEN, MOLDAU, WALLACHEU, BULGARIEN u. ROMANIEN,
tze Meer und Constantinopel . * *
PODOLIA
MOLDAVIA
BESSARABIA
WALACHIA
BULGARIA
ROMANIA
PONTUS EUXINUS hodie MARE NIGRUM
MARE DE MARMORA
CONSTANTINOPEL
Silistria
Sophia
Oczacow
Tandra
Ilanada
Varna
Sisopoli
Adrianopel
Philippopoli
Scutari

Figure 34

REGNORUM
HUNGARIÆ,
DALMATIÆ, CROATIÆ, SCLAVONIÆ, BOSNIÆ, SERVIÆ,
ET PRINCIPATUS
TRANSYLVANIÆ
Typus accuratior, et plane novus;
Editore PETRO SCHENCK caelatur.
Cum privileg. Regis Polon. et
Elect. Sax. et Ord. Holl. et Westfrisiæ.
Amstelaedami 1705.
MARE ADRIATICUM
OCCIDENS

SEPTENTRIO.
MERIDIES.
BUCHEREST
TERGOVISCO
TERNOVA
Campus Merlinius
Cassovius
Pristina
Procupie
Cratavo
Giustandil
Milliaria Hungarica.
Milliaria Germanica communia.
Milliaria Italica sive Geometrica.
OBIENS.

Figure 35

Borysthene R
le Bourg de Perewoloczna dans le quel sa Majeste Suedoise se retira et ou il passa le Boristhene apres la Bataille
Campement de l'Armée Suedoise pendent le Siege de la Place
Attaques des Suedois
POLTAWA
Redoute Moscovite pour Couvrir les Ponts
Contretranchemens des Suedois
Worskla R
Retranchement des Moscovites
Infanterie moscovite

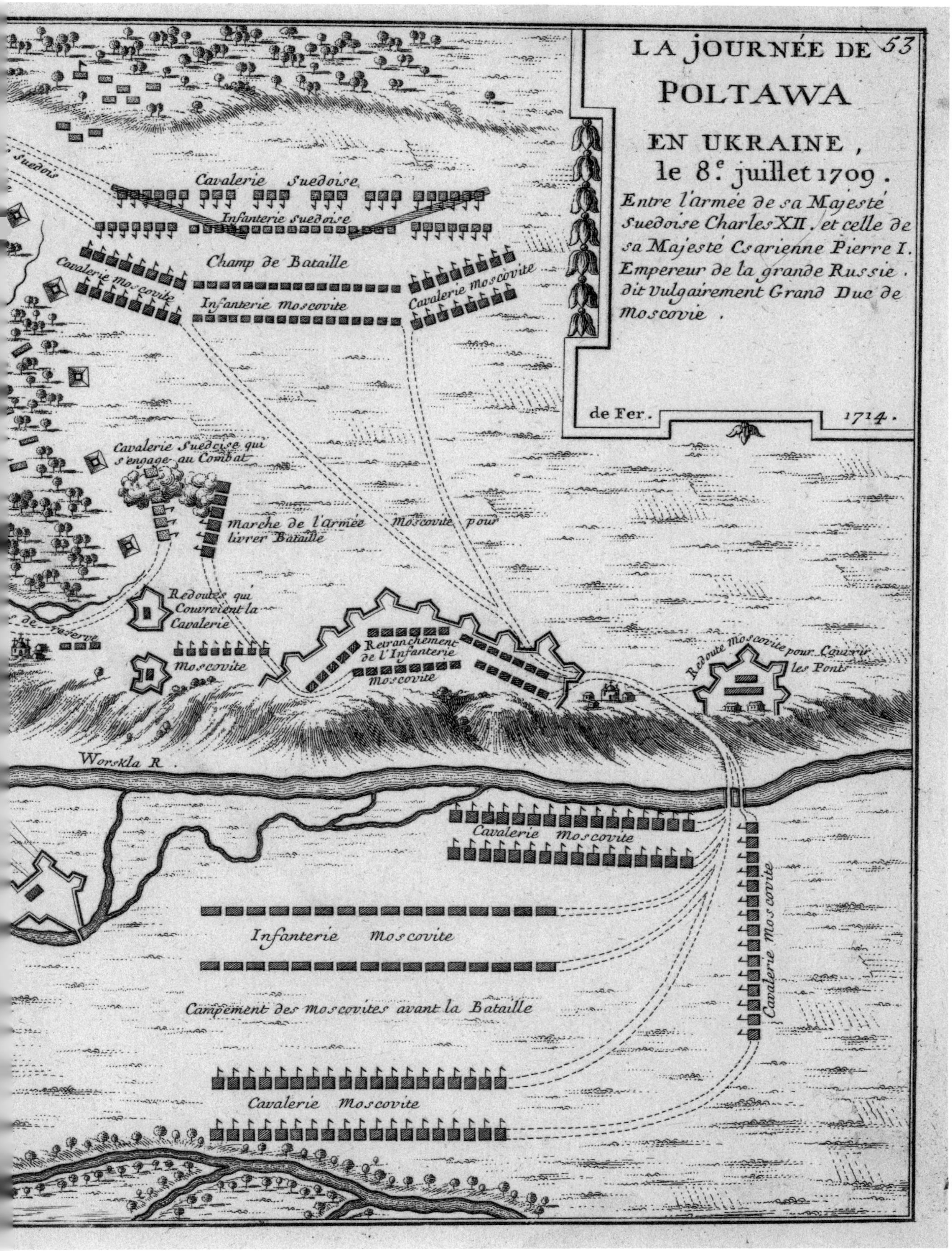
LA JOURNÉE DE
POLTAWA
EN UKRAINE,
le 8.e juillet 1709.
Entre l'armée de sa Majesté Suedoise Charles XII. et celle de sa Majesté Czarienne Pierre I. Empereur de la grande Russie. dit vulgairement Grand Duc de Moscovie.
de Fer.
1714.
53
Cavalerie Suedoise
Infanterie suedoise
Champ de Bataille
Cavalerie moscovite
Infanterie moscovite
Cavalerie moscovite
Cavalerie Suedoise qui s'engage au Combat
Marche de l'armée Moscovite pour livrer Bataille
Redoutes qui Couvroient la Cavalerie
Moscovite
Retranchement de l'Infanterie moscovite
Redoute Moscovite pour Couvrir les Ponts
Worskla R.
Cavalerie Moscovite
Infanterie Moscovite
Campement des moscovites avant la Bataille
Cavalerie Moscovite
Cavalerie Moscovite

Figure 36

Mer d'Azof
Mer Noire
Cosaques du Don
Pays des Cara-Nogais ou des Tartares Noirs
Petit Nagaia
Pays dependans du Kan des Tartares
Circassie ou Pays de Ladda
Cabardinie
Royaume de Mingrelie
Guria
Haute Georgie
Armenie
Erivan
Bayazit
Natolie
Frontieres des Etats du Grand Seigneur
Diarbeck
Curdistan
Royaume des Bochtans
Lac de Van
Ouroumi
Tabristan
Irac
Echelle
A Paris Chez le Sr. Delisle, Quay de l'Horloge.

Carte des Pays voisins de la Mer Caspiene dreßée pour l'usage du Roy Sur la Carte de cete Mer faite par l'ordre du Czar, Sur les Memoires manuscrits de Sofkam-Sabbas Prince de Georgie, Sur ceux de Mrs. Crusius, Zurabek, et Fabritius Ambaßadeurs a la Cour de Perse, et sur les éclairciβemens tirez d'un grand nombre de persones intelligentes du païs. Aßujetie aux Observations Astronomiques Par Guillaume Delisle Premier Geographe du Roy, de l'Academie Royale des Sciences.
Royaume d'Astracan
Grand Etats de Russie
Calmoucs
Nagaia
Mer Bleue
Golfe Vaseux d'Yemba
Tartarie
Isle de Koulal
Mer Bleue
Pays des Sablonneux Turcmens
Nation errante et Mahometane
Baye d'Alexandre
Ghiwie
Mer Caspiene
Royaume de Corgang ou de Carezem
Tartares Usbecs
Golfe de Balkan
Isle de la Nafte
Isle d'Agourgin
Golfe de Guilan
Golfe d'Astrabat
Zaweh
Jorjan
Chorasan
Masanderan ou Tabristan
Astrabat
Cumas
Roi de Perse
Kouhestan
Bouches du Volga
Marin sculpsit.

Figure 37

LAPPONIÆ
SUECIA
NORVEGIÆ PARS
Lapponia
Mouremanskoy Leporie
LAPPONIA
Treskoy
Leporie
SIN. BOTNICUS
FINLANDIA
RUS
Bella More hoc est ALBUM MARE
RUSSIA
KARGAPOLIA
Onega Lacus
Ladoga Lacus
SIN. FINNICUS
LIVONIÆ
WOLOGDA
Biela Ozera Duc
ROSTOVIA
Novogorod
DUCATUS
LITHUANIÆ
SMOLENSCO
MOSCOVIA
NOVOGARDIA SEVERIA DUCAT
RESCAN
Ocraina ubi Dikoia
TARTARIA MINOR
POLONIÆ
UCRANIA
MARE ZABACHIÆ
Ruthenis
LIMEN MORE
Mare Nigrum
Notar
Urbs Præcipua
Urbs
Urbs Parva
Burgus et Pagus
Arx

Nova Zemla
Niaren More
Petzorke More
Tartariæ Pars
Samoiedia
Regnum Sibiriæ
Ducatus Permia
Kasan Regnum
Bulgaria
Albi Kalmacki
Astracan Chanum
Mare Caspium
Imperii
Russici
Sive
Moscoviæ,
Status Generalis, in ejus
Regna, Ducatus, Provincias, Populos
Subjacentes Divisus, et ex Tabula
Spectat.mi Domini N: Witsen,
Urbis Amstelodamensis Consulis etc.
pro Maiori Parte Excerptus. Per F. de Witt
Mullaria Germanica Sive Geometrica 15 in uno Gradu
Milliaria Gallica Magna Sive Hora Itineris 20 in uno Gradu
Witsen, N. (1641-1717)

Figure 38

NOUVELLE CARTE GEOGRAPHIQUE DU GRAND ROYAUME DE
Par NICOLAS VISSCHER a Amstel
DUCHE DE LA GRANDE NOVOGOROD
SEIGNEURIE DE PLESKOV OU DE PSKOV
PROVINCE DE RZEVA
PRINCIPAUTE DE BIELA
GRAND DUCHE ET PALATINAT de SMOLENSCO
LITHUANIE
Palatinat de Vitepsk
Palatinat de Mcislave
RUSSIE POLONOISE
DUCHE DE SEVERIE
DUCHE et PALATINAT de CZERNIGOW
P. KIOVIE
UKRAINE PAYS DE COSAQUES
PARTIE DE POLOGNE
VOLHYNIA
DUCHE DE TVERE
COSAQUES ZAPOROSKI
Moskou
Deserts sans ea
Verstes ou Milles de Moscovie
Lieues communes de Pologne
Lieues communes d' Ukraine

OSCOVIE REPRESENTANT LA PARTIE MERIDIONALE &c.
Privilege des Etats Generaux. Nunc apud Petrum Schenk Junior.
CZARSTVO CAZANSKOIE
ou
ROYAUME DE CAZAN
DUCHÉ DE VOLODIMER
NISOVAIA DERIAVA
ou SEIGNEURIE DE
LA BAS NOVOGOROD
TARTARIE
LES MORDVA PEUPLES
CZEREMISSA NAGORNAIA
ou
DES MONTAGNES
DUCHÉ DE BULGAR
Campagnes Desertes
DE REZAN
MOSCOVITE
CALMOUCS
Orlokouri Horde
NAGAY
LE GRAND NAGAY
CZARSTVO ASTRAKANSKOIE
ou
ROYAUME D'ASTRACAN
Zolotaia horda
DIKIA POLE ou
CAMPAGNES DESERTES
COSAQUES DONSKI ou CZERCASSES
RASDOLIE DONSKI
ou
CAMPAGNES DU DON
GREBENSKAIA
COSAQUES
TARTARES NOGAIS
ou
TARTARES D'ASTRACAN
AZOFSKAIA STEP
ou
DESERT D'AZOF
les Autels d'Alexandre
CIRCASSIE
PAYS DES CIRCASSES
PETIGORIE
CABARDINIE
MEOTIDES
Bouches du Volga

Figure 39

NOUVELLE CARTE DE MOSCOVIE OÙ SONT REPRESENT
LE CHEMIN D'UN DE SES AMBASSA
OCEAN SEPTE
Mer du Nord
LAPPONS NORWEGIENS
LAPPONS MOSCOVITES
LAPPONS
LAPPONS SUECICI
NORWEGE
FINLANDE
Mer Blanche
Archangel
ETATS DE L'E
LIVONIE
Novogrot
Moscow
Smolensko
ROYAUME DE KASAN
LITHANIE
UCRAINE
Kiof
PETITE TARTARIE
ROYAUME D'ASTRACAN
LE PONT EUXIN ou MER NOIRE
GEORGIE
MER CASPIENE ou MER MORTE
NATOLIE
ARMENIE
TURQUIE ASIATIQUE
CURDISTAN
TABRISTAN
IRACAGEMI
SYRIE
Mer Mediterane
Remarque sur cette Carte
Table des principales Provinces de cette Car

DIFERENTS ETATS DE SA MAIESTÉ CZARIENNE EN EUROPE ET EN ASIE ET
PEKING VILLE CAPITALE DE L'EMPEREUR DE LA CHINE ET SON SEJOUR ORDINAIRE
Tome 4 N.29.
ONAL ou MER GLACIALE
OCEAN
ORIENTAL ou
Mer de Tartarie
E DE MOSCOVIE EN ASIE
SIBRIA
TUNGUSIA
ZUS TAINU
IASAKKI
IAKUTI
PIASIDA
DAURIA
LA GRANDE TARTARIE
PEUPLES DE BURATTI
PEUPLES GILIAKI
TARTARIE CHINOISE
QUES PEUPLES VAGABONDS
PROVINCES DES KIRGISES
MOGOLS NOIRS
ALBI KALMAKI
NIKANSKOI
COREA
LEAOTUNG
KATHAY CHINA
Desert de Xamo ou Desert sabloneux
PROVINCE DE LAGATAY
ZAGATAY KABUL
R. DE PEKING
XANSI
XANTUNG
HONAN
NANKING
ROYAUME DE TIBET
ETATS DU MOGOL
ETATS DE LA CHINE
Peking
Mergoen
Table alphabetique des Villes les plus considerables de cette Carte
Remarque sur les Calmouës ou Calmuchs

Figure 40

Accurate Vorstellung der Orientalisch-Kayserlichen Haupt-und Residenz-Stadt
Hellesponto oder dem freto der Dardanellen
EVROPÆ
PARS
Profundus saltus
Sepultura Iudæorum
ROMANIA
Portus Constantinopolis
STAMBUL oder CONSTANTINOPEL
Galata
Pera
BOSPHORUS THRACIUS
Scutari
Calcedonia
PROPONTIS
MARE di M
PERA

STANTINOPEL samt ihrer Gegend und zweyen berühmten Meer-Engen Bosphoro Thracio und
HANN BAPTIST HOMANN, DER RÖM. KAIS. MAJEST. Geographo in NÜRNBERG.
MARE DE NORT
FRETUM CONSTANTINOPOLIS
PONTUS EUXINUS
ASIÆ PARS
RMORA.
Castel Nova in Europa
Belgrado
Prospectivische Vorstellung der DARDANELLEN vor CONSTANTINOPEL
Hellespontus
Haupt und Residentz-Stadt STAMBUL oder
STANTINOPEL, von Mitternacht anzusehen.
GALATA

Figure 41

OCEANUS
Spitzberg
Beeren Eyland
MARE NORVEGUM
MARE MOSCOVITICUM
NOVA
LAPPONIA
SUECIA
FINNIA
SINUS BOTHNICUS
SINUS FINICUS
LIVONIA
LITHUANIA
Moscow
Smolensko
Archangel
IMPERII
OCCIDENS
PONTUS EUXINUS
MARE CASPIUM
NATOLIA
PERSIA
Generalis Totius
Imperii Moscovitici
Novissima Tabula
Magnam Orbis terrarum partem à Polo Arctico usque ad mare Iaponicum, et Chinæ Septentrionalis confinia exhibens
ex conatibus
Iohannis Baptistæ Homanni
Norimbergæ.

SEPTENTRIONALIS
Milliaria Germanica communia
Worst sive Milliaria Moscovitica
MARE TARTARICUM et GLACIALE
CIRCULUS ARCTICUS
OCEANUS
ORIENTA-
LIS
DAURIA
TARTARIA CHINENSIS
CHINA
TIBET
LASSA REGNUM
INDIA
KASHMIRE
Mergeen
COREA
ORIENS
MERIDIES

Figure 42

VKRANIA
quæ et
TERRA COSACCORVM
cum vicinis
WALACHIÆ, MOLDAVIÆ, MINORISq3
TARTARIÆ PROVINCIIS
exhibita
à IOH. BAPTISTA HOMANNO
Noribergæ.
VOLHYNIA SUPERIOR
VOLHYNIA
VOLHYNIA INFERIOR
RUSSIA RUBRA
POKUTIA
PODOLIA SUP
UKRAINA
PAL. BRACLAVIENSIS
PODOLIA INFERIOR
WALACHIA
TRANSYLVANIA
MOLDAVIA
TARTARIA BUDZIACENSIS
BULGARIÆ PARS
TRANSALPINA SANGIACATUS
NICOPOLKIENSIS
Occidens

Septentrio
R U S S I A
MOSKOU
DUCAT MOSKOU
DUCAT SMOLENSCO
DUCAT RESANIENS
M O S C O V I T I C A
TARTARIÆ
MOSCOVITICÆ PARS
MORDUANI
NOVOGRODENSIS
SEVERIENSIS DUCATUS
Deserta sine aqua
CAMPI DESERTI
Ubi nulla hominum habitatio, nec planta reperienda est.
COSACCI CIRCASSI
DESERTUM ASOVIENSE
KIOVIA
MOSCOVITI
TARTARIA NAGAJENSIS
SINUS ASOVIENSIS
MARE DELLE ZARACHE
PALUS MÆOTIS
TARTARIA MINOR
CIRCASSI TARTARI
KABARDINKI
EUXINUS hodie
MARE NIGRUM.
Milliaria Germanica.
Milliaria Polonica.
Meridies
Oriens

Figure 43

Spatiosissimum
Imperium
Russiæ Magnæ
juxta recentißimas Obſervationes
Mappa Geographica
accuratißime delineatum
opera et ſumtibus
Matthæi Seutteri
S. Cæſ. et Reg. Cathol. Majeſt. Geographi
et Chalcogr. Augustæ Vindel.
Persiæ Pars
Pontus Euxinus
Mare Caspium
Tartaria Minor
Occidens

SEPTENTRIO
GLACIALE
Circulus Arcticus
MARE
IAPONICUM
Penschinskoy More
Lamskoy More
Kamtsatzkoy
SINUS KAMTSCHATKE
PEKIN
CHEKIANG
MIAKO
SINICUM
I. Lequeo magna
I Formosa
Macao
MARE
Nota
Scala
LAOQUO
MERIDIES

Figure 44

A General MAP of
TURKY in EUROPE,
HUNGARY &c.
By H. Moll Geographer.
GERMANY
HUNGARY
SCLAVONIA
CROATIA
BOSNIA
DALMATIA
SERVIA
ALBANIA
EPIRUS
ITALY
GULF OF VENICE
NAPLES to the Emperor
SICILY to the Emperor
MEDITERRANEAN SEA
Miles of Great Britain 60 in a Deg.
Malta I. subject to a Grand Master and the Knights: Hospitallers, who are always at War with the Turks.
Degrees East from London

POLAND
PODOLIA
VOLINIA
UKRAIN
COSSACKS
Poeltava, noted for the defeat of the King of Sweden by the Moscovites in 1709.
Chotzyn, famous for a victory obtain'd over ye Turks by ye late K. of Poland in 1673.
Kamieniec
Braclaw
Bog R.
Konetzpol
Niester R.
Prut R.
Bender, Remarkable for ye Residence of ye K. of Sweden after his defeat by the Moscovites.
Zigirin
Black Forest
Siera
LITTLE TARTARY
Allies to ye Turk
Islam
Selenaia Dolina
Zernaja
Perekop
MEOTIS
BUDZIACK
TARTARY
Belgorod
Oczakow
Sereth
Soczowa
Iazy
Lapucizna
Targorod
MOLDAVIA
Tributary to the Turk
Famous for a Battel fought concluded between the Moscovites and Turks in 1711.
Borlat
Kilia Nova
Bojar
Nowajakalia
Kilia
I. Elenada
Danube River
I. Tandra
Arbatok
Koslof
Kirtz
Korzibas
Backzisarai
CRIM
Bolusna
Kafa
Str. of Caffa
Belbeg
BLACK SEA
WALACHIA
Turkish
Targovisco
Dembrowitz
Buckhorst
Danube R.
Silistria
Rotzig
BULGARIA
Tomi
Nicopoli
Cara Foui
Kutenha
Mankalia
Varna
Bulham
Bantzak
Kallat
Serais
Sophia
Adrinople
Philippipoli
Esbaba
Attabuli
EUROPE
ROMANIA
Constantinople
Scutari
Araclea
Propontus
Galipoli
Bursa
Strymon
Saloniki
Contessa
G. of Contessa
Gulf of Saloniki
Limni
Negropont
Athena
Lemnos
Casandria
Metelina
Audo
The Dardanello
TURKY
IN
ASIA
Smirna
Samos
ARCHIPELAGO
Morgo
Santurine
Stampalia
Longo
Rho.
Rhodes
Canea
Istantea
Candia
CANDIA
Spinalongo
BELGRADE
TEMESWAR
Morais
Temez R.
a. b. c. &c. shew where Considerable Battles have been Fought. a. Zenta, the Turks defeated here by P. Eugene 1697. b. here P. Eugene defeated ye Turks 1716. c. Salankemen, P. Lewis of Baden defeated ye Turks 1691. d. Zarno, the Elec. of Bavaria defeated ye Turks 1688. e. Semedri noted for ye defeat of Sigismund K. of Hungary by ye Turks 1409. f. Viddin, noted for ye defeat of ye Turks by Hunniades. 1443. g. Hunniades killed 34000 Turcks. h. the Field of ye Sepulchre, where ye Despot of Servia was defeated and Sul. Amurath Stabbed in 1390. i. Sultan Bajazet defeated Sigismund K. of Hungary 1396. k. Ladislaws K. of Hungary defeated by ye Turks in 1444. The Turks def. by P. Eugene before Belgrade in 1717.

Figure 45

EIN THEIL
VON
POHLEN
OBER WOLHINIEN
WOLHINIEN
NIEDER WOLHINIEN
OBER PODOLIEN
MOL
DAU
PODOLIEN
NIEDER PODOLIEN
DIE
BUDCZAKISCHE
TARTA-
-REY
DAS
SCHWARTZE
MEER
DIE WALACHEY
EIN THEIL
VON BULGARIEN
Kaminieck
Chozim
Soroka
Bendern
Bielgorod
Perewolozno
Saporoger Setsh
R: Donau

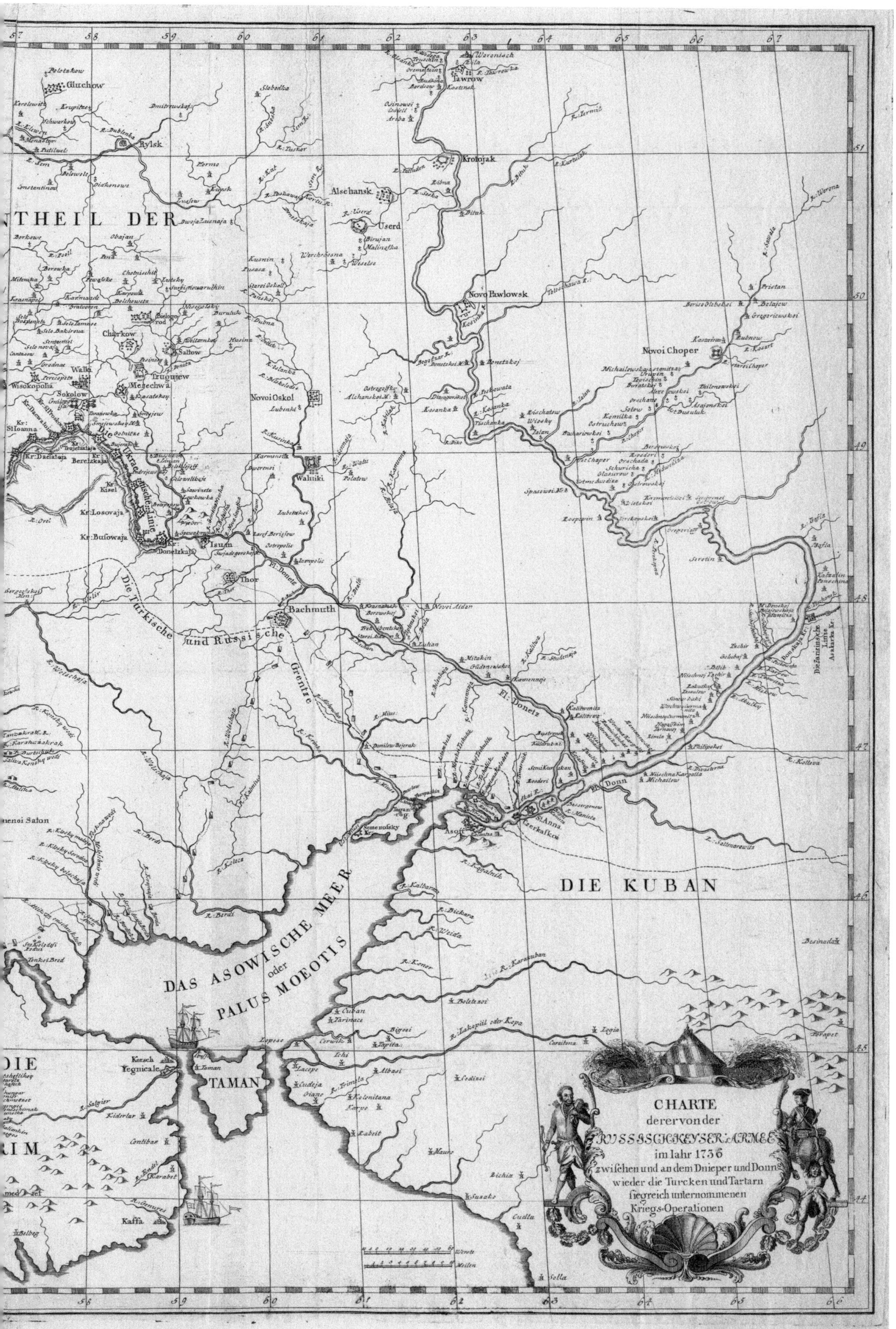
CHARTE
derer von der
RUSSISCH KEYSER. ARMEE
im Iahr 1736
zwischen und an dem Dnieper und Donn
wieder die Turcken und Tartarn
siegreich unternommenen
Kriegs-Operationen
DAS ASOWISCHE MEER
oder
PALUS MOEOTIS
DIE KUBAN
TAMAN
THEIL DER
Die Türkische und Russische Grentze
Gluchow
Rylsk
Krotojak
Alschansk
Userd
Novo Pawlowsk
Novoi Choper
Charkow
Novoi Oskol
Walniki
Isum
Thor
Bachmuth
Fl. Donetz
Donn
Asof
St. Anna
Tawrow
Kaffa
Werste
Meilen

Figure 46

MAPPA
GENERALIS
TOTIUS IMPERII
RUSSICI
SIEWE
ZEMLIA
NOWAIA
ARCHANGELO
GORODSKAIA
GUBER
NIIA
TOBOL'
SKAIA
CZERNOE
MORE
ASTRA
CHAN
SKAIA
GUBER
NIIA
CHWA
LYN
SKOE
MORE
ARAL'
SKOE
MORE
KRIEKIEKAZA
CZEI
ORDY
STEP
Scala milliarium Ruthenicorum

NOI OKIAN
TICHOE MORE
KAMCZACKOE MORE
TICHOE MORE
GOSUDARSTWA
CZAST
Ost Timaczkoi
Ost. Kyrlak
JAKUCK
Ost Tri Sestry
Ost Nadezda
Ost. Cytronnoi

Figure 47

Carte G
L'Empire D
17
Mer Septentrion
Novaia Zemlia
ou Nouvelle Terre
Mer Karskoe
Gouvernement d'Arkangel
Province de Vologda
Mer Noire
Mer Caspienne
Mer d'Aral
Samoyede
Tounguses
Frontieres de
Echelles
(Planche Premiere)

AMÉRIQUE SEPTENTRIONALE
PAYS DES TCHOUKTCHI
ISLES ALEOUTIENNES ET AUX RENARDS
PARTIE NORD DE LA MER PACIFIQUE
ou de
L'OCEAN ORIENTAL
MER D'OKOTSK appellée par les Toungouses Mer de Lama
ISLES KOURILES
KAMTCHATKA
Isle d'Amour ou Sahalien-Oula
OBSERVATION.

Figure 48

Mappa Geographica ex novissimis observationibus repraesentans Regnum Poloniae et Magnum Ducatum Lithuaniae, Cura et Sumptibus Tobiae Conradi Lotter, Geogr. Aug. Vindel. 1759.
MARE BALTICUM
DIE OOST SEE
SUECIAE
GOTLAND I.
OELAND I.
Bornholm I.
DANTZIG
KÖNIGSBERG
WARSAU
THORN
POSEN
CRACAU
LUBLIN
WIEN
Memel
Palatinatus Posnaniensis
Palatinat. Mazoviae
Palatinatus Siradiensis
Palatin. Culmens
SAMOGI
CURLAND
Sambia
GERMANIAE PARS
AUSTRIAE P.
TRANSYL
Clausenburg
OCCIDENS
Longitudines Geographicae posita Longit. Parisiorum XX Graduum.

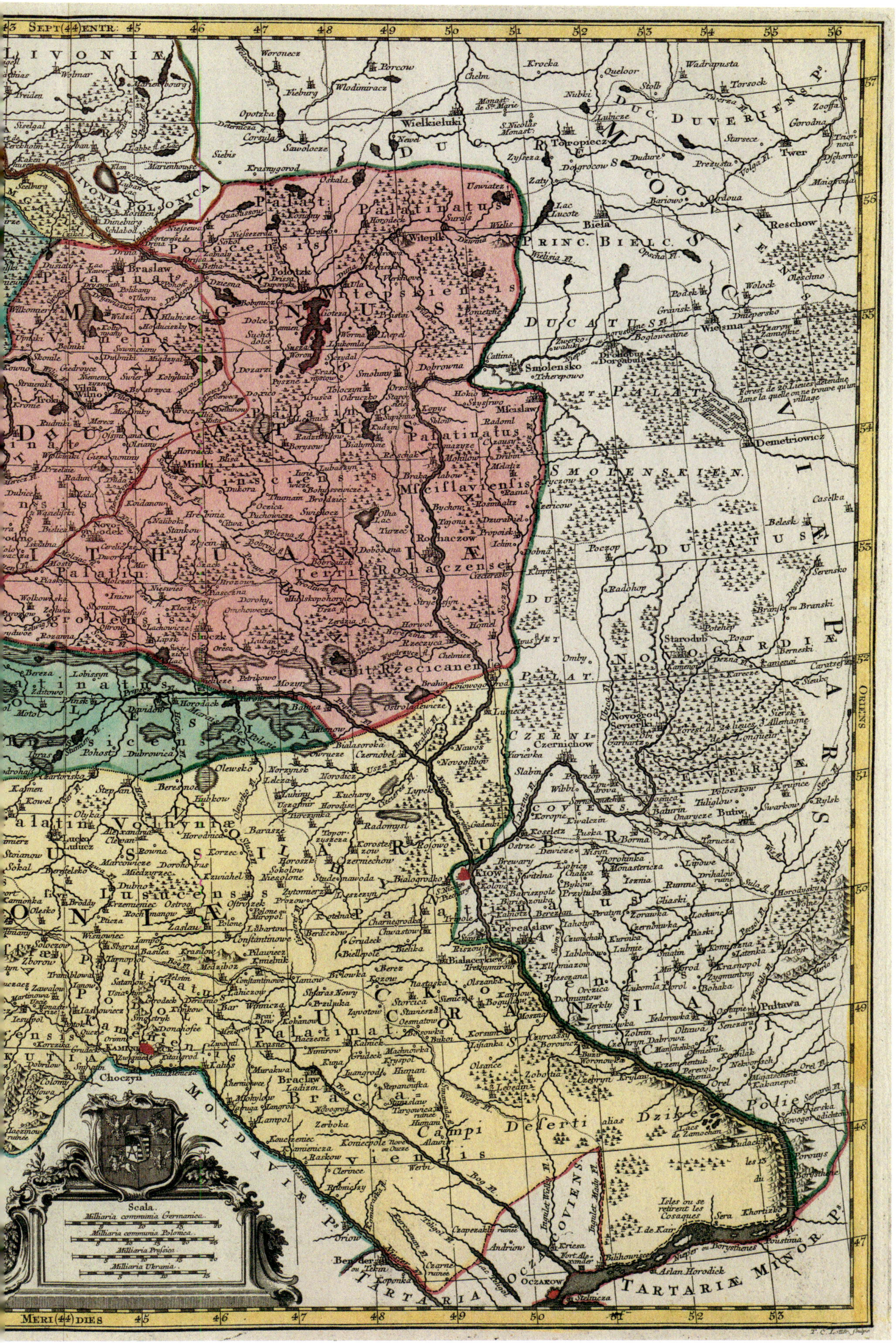

SEPT(44)ENTR:
LIVONIÆ
LIVONIA POLONICA
DUC. DUVERIENS P.
PRINC. BIELC.
DUCATUS SMOLENSKIEN.
Smolensko
Witepsk
Polotzk
Braslaw
Minsk
Mscislaw
Rohaczow
Palatinatus Rzecicanensis
Kiow
Czernichow
Pereaslaw
Pultawa
Choczyn
MOLDAVIÆ P.
TARTARIÆ MINOR P.
TARTARIA OCZ.
Oczakow
Campi Deserti
Scala.
Milliaria communia Germanica.
Milliaria communia Polonica.
Milliaria Prussica.
Milliaria Ukrania.
MERI(44)DIES

Figure 49

LE ROYAUME DE POLOGNE
Par le Sr. ROBERT DE VAUGONDY Géog. ord. du Roi de S. M. Pol. Duc de Lorr. et de Bar et de l'Academie royale de Nancy
A PARIS
Chez le Sr. Delamarche Géographe. Rue du Foin St. Jacques, au College de Me. Gervais
Avec Privilege 1767.
MER BALTIQUE
Golfe de Dantzik
Golfe de Riga
ECHELLE
Milles pas géométriques, de 60 au degré.
Lieues d'Allemagne et de Pologne, de 15 au degré.
Lieues communes de France, de 25 au degré.
Lieues d'une heure, de 20 au degré.
cette couleur indique ce qui est à l'Allemagne
à la Prusse.
à la Russie.
Septentrion
Occident

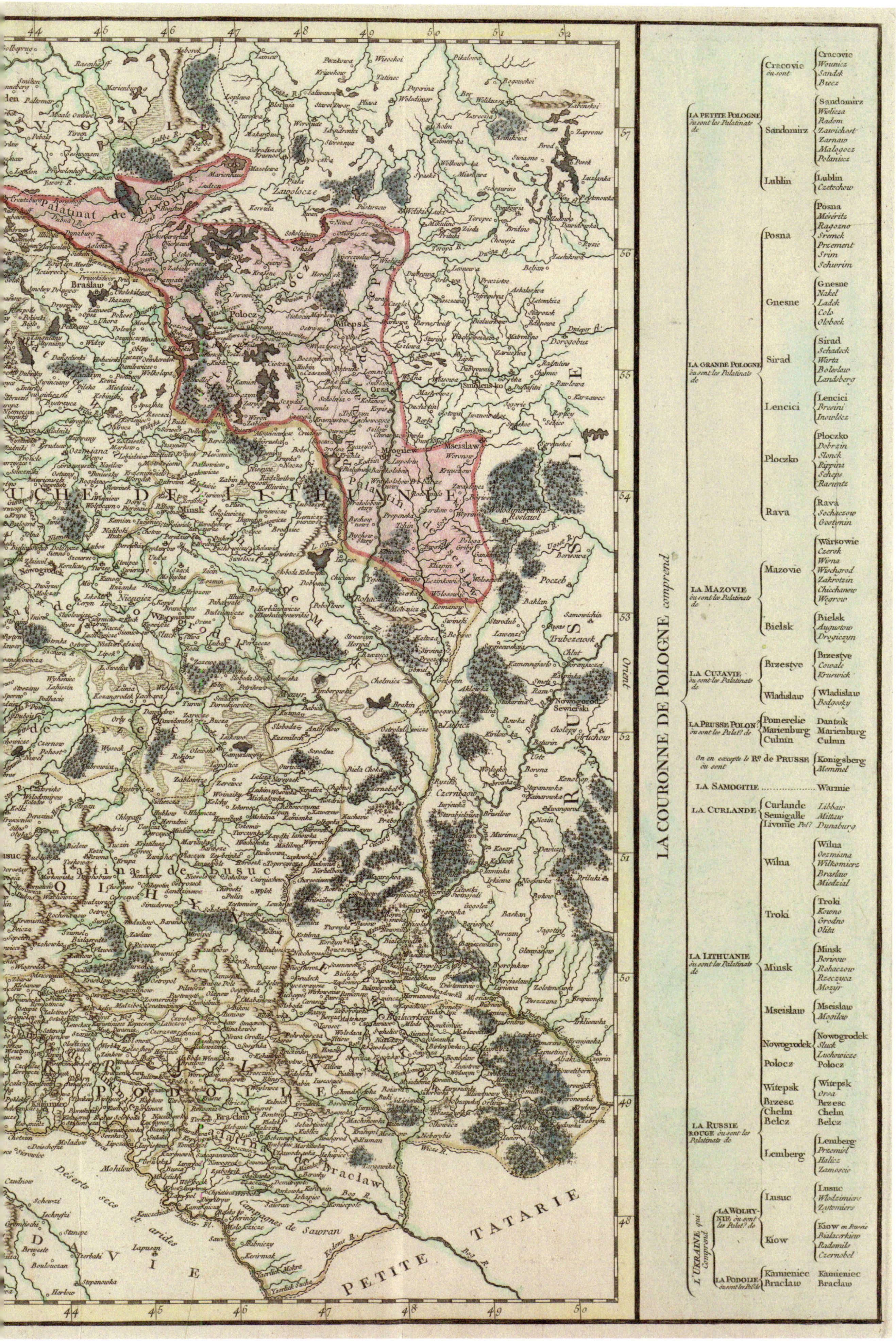
LA COURONNE DE POLOGNE comprend
LA PETITE POLOGNE ou sont les Palatinats de
Cracovie
Sandomirz
Lublin
LA GRANDE POLOGNE ou sont les Palatinats de
Posna
Gnesne
Sirad
Lencici
Ploczko
Rava
LA MAZOVIE ou sont les Palatinats de
Mazovie
Bielsk
LA CUJAVIE ou sont les Palatinats de
Brzestye
Wladislaw
LA PRUSSE POLON.
Pomerelie
Marienburg
Culmn
LA SAMOGITIE
Warmie
LA CURLANDE
Curlande
Semigalle
Livonie
LA LITHUANIE ou sont les Palatinats de
Wilna
Troki
Minsk
Mscislaw
Nowogrodek
Polocz
Witepsk
Brzesc
Chelm
Belcz
LA RUSSIE ROUGE ou sont les Palatinats de
Lemberg
L'UKRAINE qui comprend
LA WOLHYNIE
Lusuc
Kiow
LA PODOLIE
Kamieniec
Braclaw
Palatinat de Livonie
DUCHE DE LITHUANIE
PETITE TATARIE
Déserts secs et arides
Orient

Figure 50

SANDOMIRZKI WOIEWODZTWO
KRAKOWSKIE WOIEWODZTWO
POLONIAE
POWIAT SANDEKI
Erklærung der Zeichen und Buchstaben.
Stadt ... Miasto ... Urbs
Staedtchen ... Miasto Otwarte ... Oppidum
Dörfer ... Wies ... Pagus.
Abtey ... Opactwo ... Abatia.
Gold Gruben ... Kruszce Złota . fodinae auri.
Silber ... Srebra ... argenti.
Kupfer ... Miedzi ... cupri.
Saltz ... Żupy Solne ... Salinae.
Schwefel ... Gory Siarczyste ... Sulphuris.
Alt ... St. ... Starry ... Vetus.
Neu ... N. ... Nowy ... Novus.
Ober. Hoch ... G. ... Gorny albo Wysoki . supra.
Unter. Nieder ... D. ... Dolny albo Nisky ... infra.
Herzogthum ... X ... Xiestwo . Principatus vel Ducatus.
Geographische Meilen 15 auf einen Grad.
Roth Reußis und Pocutis Meilen 11¼ auf einen Gr.
Hungarische Meilen 19 auf einen Grad.
Longitudines a primo Meridiano qui a Parisiensi 20° distat versus Occidentem, numeratae sunt.

LUBOMERIÆ
ET
GALLICIÆ
REGNI
Tabula Geographica
Impensis Homannianorum
Hæredum
1775.
Cum Priv. Sac. Caes.
Majestatis.
TRANSILVANIA
PODOLSKI
WOIE WODZTWO
MOLDA
VIÆ PARS

Figure 51

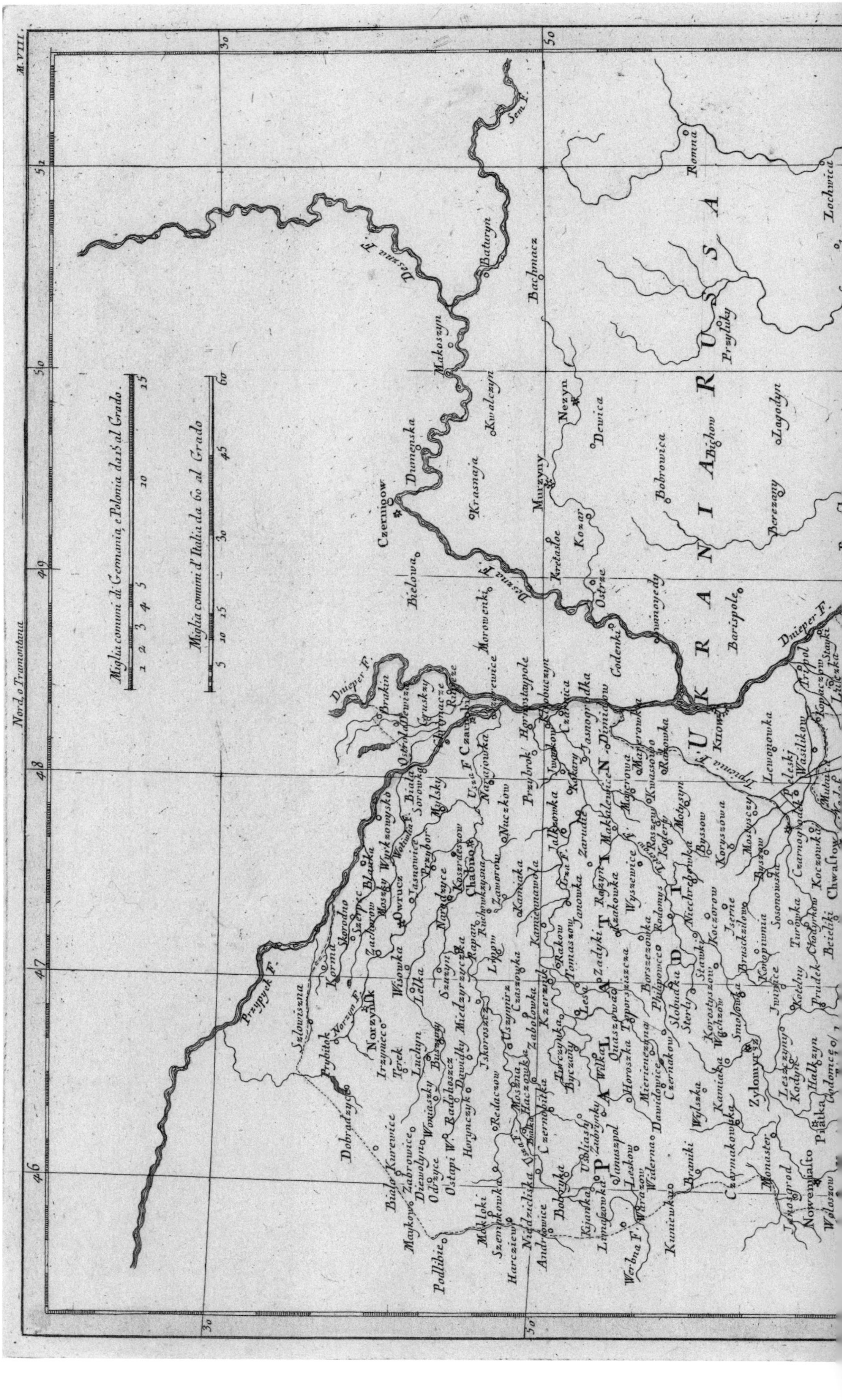
M. VIII.
Nord, o Tramontana
Miglia communi di Germania e Polonia da 15 al Grado
Miglia communi d'Italia da 60 al Grado
UKRANIA
RUSSA
Dnieper F.
Desna F.
Przypyek F.
Sem F.
Czernigow
Nezyn
Baturyn
Makoszyn
Romna
Lochwica
Przyluky
Baryspole
Kiow
Owrucz

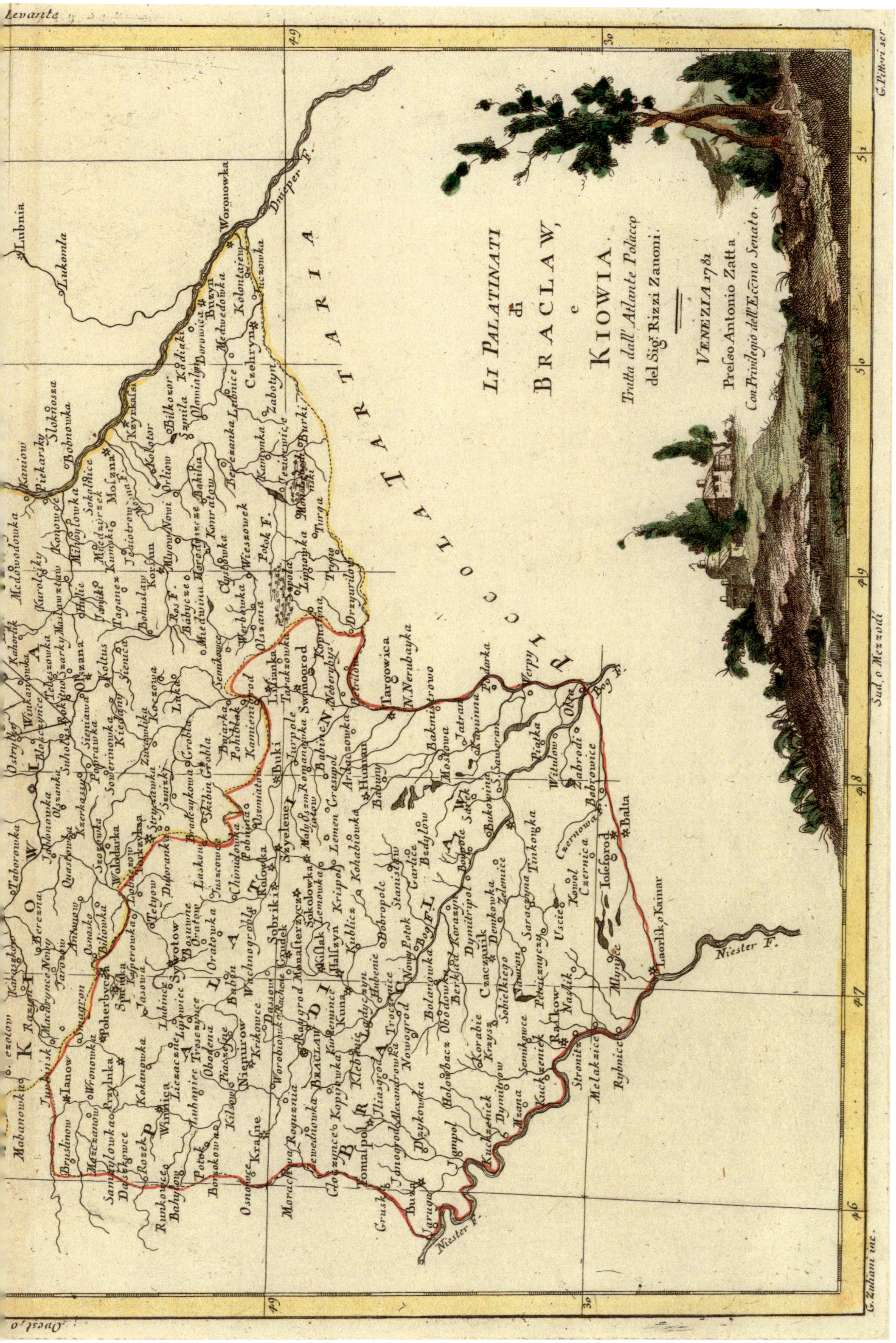
Li Palatinati
di
Braclaw,
e
Kiowia.
Tratta dall' Atlante Polacco
del Sig.r Rizzi Zanoni.
Venezia 1781
Presso Antonio Zatta
Con Privilegio dell'Eccmo Senato.
Levante
Sud o Mezzodi
Dnieper F.
Niester F.
Bog F.
Braclaw
G. Zuliani inc.
G. Pitteri scr.

Figure 52

LA PICCOLA
TARTARIA
divisa
NE' SUOI TERRITORI
Di nuova Projezione
VENEZIA 1783
Presso Antonio Zatta
Con Privilegio dell' Eccmo Senato
Ovest o Ponente

Tramontana
Mezzodi
G. Pitteri scr.
Est o Levante
Komarowka
Lewkowka
Lysagon
Gurgewka
Belidka
Nowoi Aidar
Izum
Kay Gorodok
Tor
Bachmut
St. Aidar
Lugan
Gondorowskoi
Glukoi Bojarak
Samara
Alexandrowskaja
Winok
Nikitinskaja
Ismail Mirza
Kazay Myrza
Kabil Teliak
Kirilowskaja
Alexicewskaja
Zachariewskaja
Kabil Nugay
Jamlaja
Juchan Kangli
Petrowskaja
Berdinskaja
Semenowska
Tangarod
Lutik
Wolga F.
AZOF
Sakarni Peski
I. Mandare
Aktar
Pa. Wisarionowa
Teniczi
P. Teodorowa
Azchuez
MAR DELLE ZABACHE ov. PALUDE MEOTIDE
Misir Isaka Burun
Arabat
Kusadge
Tatar Jermung
Topraghi
Kopil
Kuban F.
Kaffa
Degamen
Sudach
Taman
Chatukay
Anapa
Tsham
Sunelgik L.
Ghelingdig L.
Bowidial
Askafi
Biszjunta
Wardau
Mamac
Sutziak
Zaslunka
Medine Abaiku
Monte Caucaso
CIRCASSIA
TERRITORIO DE NOGAJESI TARTARI
RUSSIA
MAR NERO

Figure 53

XIX.
PARTIE OCCIDENTALE
de l'Empire
DE RUSSIE,
Extrait de l'Atlas Russien
et d'autres Cartes
A PARIS
1785.
MER GL
GOLFE DE LA MER GLACIALE
Cercle Polaire Arctique
GOLFE DE FINLANDE
ESTONIE
LIVONIE
CURLANDE

Figure 54

Longitude West from London
BOWLES'S
NEW ONE-SHEET MAP
of
EUROPE,
DIVIDED INTO ITS
EMPIRES, KINGDOMS, STATES,
and
SUBDIVISIONS;
laid down from Observations
of the most Celebrated
Geographers.
LONDON:
Printed for the Proprietors BOWLES and CARVER, No. 69 St. Paul's Church Yard.
NORT
Arctic Circle
ICELAND
Faro Islands
Shetland Isles
BRITISH ISLES
ATLANTIC
OR
WESTERN
OCEAN
GERMAN OCEAN
IRELAND
GREAT BRITAIN
ST. GEORGE'S CHANNEL
ENGLISH CHANNEL
BAY OF BISCAY
FRANCE
SPAIN
CORSICA
SARDINIA
MEDIT
MOROCCO
FEZ
ALGIERS
TUNIS
PART OF
BILEDULGERID
SCALES
British Statute Miles 69½ to a Degree.
Common French Leagues 25 to a Degree.
German Miles 15 to a Degree.
Sea Leagues 20 to a Degree.
Geographical Miles 60 to a Degree.
Russian Wersts 105 to a Degree.
North Latitude

Longitude East from London
FROZEN OCEAN
GULF OF OBI
PART OF SIBERIA
BALTIC SEA
WHITE SEA
LITHUANIA
POLAND
HUNGARY
LITTLE TARTARY
BLACK SEA
CASPIAN SEA
Desert of Astrakan
Plains of Mamai
GREAT NOGAYAS
SEA OF ASOF
KUBAN
MEDITERRANEAN SEA
EGYPT
ARABIA
SYRIA
AFRICA
A METHODICAL DIVISION OF EUROPE.
British Isles — Gr. Britain — England & Wales 52 Counties; Scotland 33 Counties. Ireland — 4 Provinces 32 Counties.
Denmark — N. Jutland, S. Jutland, or Sleswick. Isles of Zealand, Funen &c. in the Baltick. Norway, Danish Lapland & adjacent Isles. Iceland & Ferro Iss. in the Northern Ocean.
Sweden — Sweonia or Sweden proper, Gothland Swed. Lapland, Bothnia, Swed. Finland. Is. of Gothland, Oeland, Aland, Rugen &c.
Muscovy or Russia in Europe — Russ. Lapland, Archangel, Olonitz, Carelia, Kexholm, Ingria, Livonia, Bielezero, Usting, Solkam, Wiatka, Galitzk, Wologda, Novogorod, Moscow, Alatir, Kasan, Woronesk, Kiovia, Russ. Lithuania, Lit. Tartary, Crimea.
France — Formerly 37 Military Governments. Now 102 Departments. Island of Corsica.
Netherlands — United Prov. or Batavian Republic — Holland, Zealand, Friesland, Groningen, Overyssel, Gelderland & Utrecht. United to France — Brabant, Antwerp, Malines, Limbourg, Luxemburg, Namur, Hainault, Cambresis, Artois & Flanders.
Germany — Germany proper — Up. Saxony, Lo. Saxony, Westphalia, Up. Rhine, Lo. Rhine, Franconia, Austria, Bavaria, Swabia. States — Bohemia, Moravia, Silesia, Lusatia.
Switzerland — 13 Cantons
Poland Divided between Rußia, Prußia & Außtria — Gr. Poland, Lit. Poland, Galicia, Podolia, Volhinia, Polesia, Lithuania proper, Polish Livonia, Samogitia, Courland.
Prußia — Prussia proper, with many other States &c.
Hungary — Hungary proper, Transilvania, Sclavonia, Croatia, and part of Dalmatia.
Spain — 14 Districts, beside Gibraltar. Islands of Majorca, Minorca & Yvica.
Portugal — Kingdom of Algarve & 5 other Provinces
Italy — Savoy, Piedmont, Parma, united to France. Milan, Mantua, Modena, or Italian Republic. Venice, Genoa or Liguria, Lucca, Estates of the Church, Tuscany or Etruria, Naples &c. Islands of Sardinia, Sicily, Malta &c.
Turkey in Europe — Bessarabia, Moldavia, Wallachia, Bulgaria, Servia, Bosnia, Romania, Macedonia, Larissa, Livadia, Albania, Dalmatia, Morea, Repub. of Ragusa; Is. of Candia, Negropont &c.

Figure 55

The Third Part of
EUROPE,
Drawn & Engraved
from
D'ANVILLE'S
Four Sheet Map
for
J. Harrison, No. 115, Newgate Street, LONDON.
Feby 15, 1791
GREAT RUSSIA
BALTIC SEA
LITHUANIA
POLAND
HUNGARY
MOLDAVIA
WALACHIA
CRIMEA

MEDITERRANEAN SEA
CYPRUS
CRETE OR CANDIA
LIBYA
EGYPT
MOUTHS OF THE NILE
PALESTINE
SCALES
Gulf of Satalia
Gulf of Samsoun
Gulf of Arabes
SEA OF MARMARA
ARCHIPELAGO
RHODAS
GULF OF SALONIQUE
F OF VENICE
Barrai al Sham or Desart of Syria
Almotanah or Dead Sea

Figure 56

A MAP OF
THE KINGDOM OF POLAND
AND GRAND DUTCHY OF LITHUANIA
INCLUDING SAMOGITIA AND CURLAND,
Divided according to THEIR DISMEMBERMENTS
WITH THE KINGDOM OF PRUSSIA
By WILLM FADEN, Geographer to THE KING
M DCC XCIX.
Russian
Prussian
Imperial
BALTIC OR EAST SEA
GULF OF DANTZICK
WARSZAWA
WARSAW
Longitude East of London

GOVERNMENT OF NOWOGOROD
PALATINATE
1772
OF POLOCK
PALATINATE
OF WILNO
1793
LITHUANIA
PALAT. OF MSCISLAW
PALATINATE OF WITEPSK
PALATINATE OF NOWOGRODEK
1793
1795
1772
PALAT. OF WOLHYNIA
POLISH UKRAINE
PALAT. OF KIOW
1793
PAL. OF PODOLIA
PALAT. OF BRACLAW
1793
1772
MOLDAVIA
BOGDAN
GOV. OF CATHARINOSLAV
RUSSIAN UKRAINE
GOVERNMENT OF BIELOGOROD
GOVERNMENT OF SMOLENSKO

Figure 57

A NEW MAP OF
THE SEAT OF WAR,
Comprehending
GERMANY; POLAND,
WITH ITS DISMEMBERMENTS,
PRUSSIA; TURKEY in Europe,
ITALY &c.
From the Maps of Chauchard, Zannoni, &c.
ENGLAND
GERMAN SEA
THE CHANNEL
BATAVIAN REPUBLIC
REPUBLIC OF FRANCE
WESTPHALIA
LOWER SAXONY
HESSE
FRANCONIA
SWABIA
BAVARIA
SWITZERLAND
GRISONS
SAVOY
PIEMONT
TYROL
GULF OF GENOA
GULF OF LYON
MEDITERRANEAN SEA
CORSICA
SARDINIA
ETRURIA
STATE OF VENICE

PRUSSIAN DISMEMBERMENT
POLAND
SILESIA
AUSTRIAN
GALICIA
DISMEMBERMENT
MORAVIA
HUNGARY
TRANSYLVANIA
SLAVONIA
CROATIA
BOSNIA
SERVIA
WALACHIA
BULGARIA
DALMATIA
TURKEY
ROMELIA
ALBANIA
VENICE OR ADRIATIC SEA
NAPLES

Figure 58

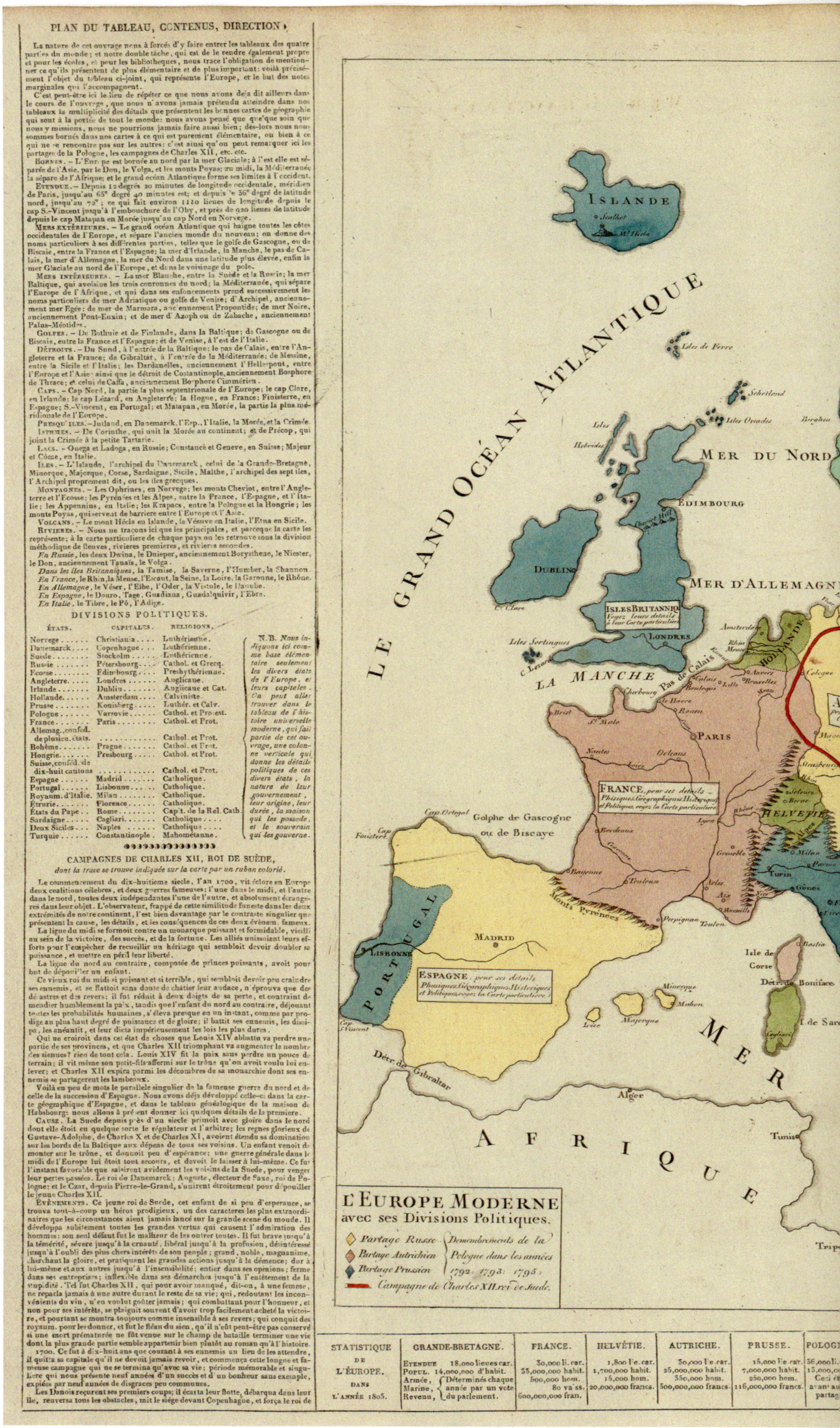
PLAN DU TABLEAU, CONTENUS, DIRECTION
DIVISIONS POLITIQUES.
CAMPAGNES DE CHARLES XII, ROI DE SUÈDE,
L'EUROPE MODERNE
avec ses Divisions Politiques.
Partage Russe
Partage Autrichien
Partage Prussien
Demembrements de la Pologne dans les années 1772. 1793. 1795.
Campagne de Charles XII roi de Suède.
STATISTIQUE DE L'EUROPE. DANS L'ANNÉE 1805.
GRANDE-BRETAGNE.
FRANCE.
HELVÉTIE.
AUTRICHE.
PRUSSE.
LE GRAND OCÉAN ATLANTIQUE
ISLANDE
MER DU NORD
MER D'ALLEMAGNE
EDIMBOURG
DUBLIN
ISLES BRITANNIQ
LONDRES
LA MANCHE
PARIS
FRANCE
HELVETIE
Golphe de Gascogne ou de Biscaye
MADRID
ESPAGNE
LISBONNE
PORTUGAL
Monts Pyrenees
Isle de Corse
Alger
Tunis
AFRIQUE
MER

N. 9.
CONTINUATION DE LA COLONNE A GAUCHE.
MER GLACIALE
Monts Poyas
MER BLANCHE
SUEDE
GOLPHE DE BOTHNIE
Golphe de Finlande
PETERSBOURG
STOCKOLM
MER BALTIQUE
MOSKOW
RUSSIE
PRUSSE
POLOGNE
VARSOVIE
BOHÊME
Prague
Cracovie
VIENNE
Monts Crapacks
HONGRIE
MER D'AZOF
CUBAN
CRIMÉE
MER NOIRE
TURQUIE D'EUROPE
CONSTANTINOPLE
Detr. de Constantinople
TURQUIE D'ASIE
ARCHIPEL
MORÉE
Candie
Chypres
SYRIE
MEDITERRANÉE
MER ADRIATIQUE
REPUBL. DES 7 ISLES
PARTAGES DE LA POLOGNE indiqués sur la carte.
A FLORENCE CHEZ MOLINI, LANDI ET C.°

Figure 59

The
RUSSIAN EMPIRE,
in
EUROPE and ASIA.
NORTHERN
OCEAN
NOVA ZEMBLA
C. Zelandia
North Cape
BOTHNIA GULF
BALTIC SEA
BLACK SEA
CASPIAN SEA
Aral
GEORGIA
PERSIA
ELUTHS
Orenburg
Astracan
Kalgonev I.
Karakoe Sea
Woguliţzians
Wercoturian Mountains
British Statute Miles 69 to a Degree
30 120 240 360 480 600
137 23
147 23
157 23 Longitude East 167 23 from

39
de East from London
EN OCEAN
OR
SEA
Land seen in 1778
BEERING'S STRAITS
SEA OF ANADIR
N. Cape
Jukagiri
Lena River
OF Yakutzk
Tunguſi
SEA OF KAMTSCHATKA
Bay of Penshink
Okotskoi Ostrog
KURILE I.
Saghalien Ula
Merghen
Tsitsicar
PEKIN
Yellow R.
SANDY DESART
JESO I.
GULF
COREA
NIPHON I.
SIKOKO I.
KIUSU I.
Irkutsk
Baikal L.
177 25
187 25
197 25
207 25

Figure 60

POLAND.

SITUATION, BOUNDARIES, AND EXTENT.

POLAND, situated between 15° 30′ and 35° East longitude, and between 47° 30′ and 56° 30′ North latitude, is bounded, on the North, by the Baltic and by Prussia; on the East, by Russia; on the South, by Hungary and Turkey; and on the West, by Germany. Its length is 700 miles, and breadth 600.

DIVISION.

Before the dismemberment of this country by Russia, Austria, and Prussia, it was divided into thirteen provinces, which were subdivided into palatinates: but it is now divided into fourteen provinces; of which nine belong to Russia, three to Prussia, and two to Austria.

N.B. In the following division, the cities distinguished by capital letters are chief cities of the new provinces.

RUSSIAN DOMINIONS.—(See Map, No. 54.)

ANCIENT PROVINCES.	PALATINATES.	CHIEF CITIES.	NEW PROVINCES.
COURLAND		MITTAU	COURLAND.
SAMOGITIA		Rosienie	
GRAND DUCHY OF LITHUANIA.	Wilna	WILNA	WILNA.
	Troki	Troki	
	Livonia	Duneburgh	POLOCK.
	Polock	POLOCK	
	Witepsk	Witepsk	MOHILOW.
	Mscislaw	MOHILOW	
	Minsk	MINSK	MINSK.
	Novogrodek	Novogrodek	
POLESIA	Brzesc	Brzesc	SLONIM.
		SLONIM	
VOLHYNIA	Volhynia	Luck	VOLHYNIA.
		PINSK	
UKRAINE	Kiow	KIOW	KIOW.
PODOLIA	Kaminiec	KAMINIEC	PODOLIA.
	Braclaw	Braclaw	

PRUSSIAN DOMINIONS.

PODLAKIA	Podlakia	Bielsk	NEW EAST PRUSSIA.
MAZOVIA	Mazovia	Warsaw	
	Polock	POLOCK	
POLISH PRUSSIA	Pomerelia	DANTZIC	WEST PRUSSIA.
	Marienburgh	Marienburgh	
	Culm	Culm	
GREAT POLAND	Poznan	Poznan	SOUTH PRUSSIA.
	Inoroclaw	Inoroclaw	
	Brzesc	Brzesc	
	Kalisz	Kalisch	
	Siradia	Siradia	
	Lenczica	Lenczica	
	Rawa	Rawa	
		WARSAW*	

* By the treaty of Tilsit, in 1807, Warsaw and a certain extent of the surrounding territory was erected into a Duchy by Buonaparte, and given to Saxony.

AUSTRIAN DOMINIONS.

LITTLE POLAND	Cracow	CRACOW	NEW GALLICIA.
	Sandomir	Sandomir	
	Lublin	Lublin	
RED RUSSIA	Russia	LEMBERG	GALLICIA.

In the following description, the new division of Poland has been preferred; to which however we shall prefix a short account of the ancient government.

ANCIENT CONSTITUTION OF POLAND.

The constitution of Poland differed little from an aristocracy, whence it has been indifferently called *a kingdom* and *a commonwealth*. The King, who was at the head of the republic, was elected by the nobility and clergy in the plains of Warsaw. The election always took place on horseback; and when the minority proved refractory, the majority had no other control over them, than to cut them to pieces with their sabres; but if the minority were sufficiently strong, a civil war ensued. Immediately after his election, the King signed the *pacta conventa* of the kingdom, by which he engaged that the crown should be elective; that his successor should be appointed during his life; that the diets should assemble every two years, that every nobleman or gentleman in the realm should have a vote in the diet of election; and that in case the King should infringe the laws and privileges of the nation, his subjects should be absolved from their allegiance. In fact, the King was no more than president of the senate, which was composed of the Primate of Gnesna, the Archbishop of Lemberg, fifteen Bishops, and 130 Laymen, consisting of the great officers of state, the Palatines, and Castellans The Palatines were governors of provinces, who held their offices for life. The office of the Castellans in time of peace was almost nominal; but when military or feudal services were required, they were the lieutenants of the Palatines, and commanded the troops of their several districts The diets of Poland where either ordinary or extraordinary: the former met once in two, or sometimes three years; the latter when summoned by the King, upon critical emergencies, but could not sit longer than a fortnight; and one dissenting voice rendered all their deliberations void.

The population of the whole of Poland before its first dismemberment in 1772, was estimated at 13,404,000.

AUSTRIAN DOMINIONS. (GALLICIA.)

CLIMATE, SOIL, AND PRODUCE.

The climate of Gallicia is the mildest of the late kingdom of Poland; and the air, though rather cold, is very pure. The soil is very productive; and, except in a few sandy tracts, is extremely fertile, both in corn and pasturage.

Gallicia produces, plentifully, grain of every kind, as also hemp, flax, tobacco; honey and wax are obtained in great abundance. The woods abound in oak, beech, pine, and fir trees.

NATURAL HISTORY.

MOUNTAINS.—Gallicia is, in general, a level country, except towards the South, on which side it is separated from Hungary by the Carpathian mountains.

RIVERS.—The principal rivers are—the Vistula, which rising in the Carpathian Mountains, traverses the country, and falls into the Baltic Sea near Dantzic, after a course of about 450 miles;—the Bug, which divides Gallicia from Russian and Prussian Poland, and falls into the Vistula below Warsaw;—the Dniester, and the Pruth, both which rise in South Gallicia, and after watering Moldavia, fall into the Black Sea.

METALS AND MINERALS.—Gallicia contains mines of iron, lead, and copper; and quarries of marble. Near Olkusch, are mines of silver and lead. The salt-mine of Wieliczka, near Cracow, are the richest and most productive in Europe. Out of some mines at Itza, about 70 miles North-east of Cracow, are dug several kinds of earth, excellently adapted to the potter's use, from which all the country is supplied with earthenware.

RELIGION.

The established religion of Gallicia is the Roman Catholic; but Lutherans, Calvinists, and Greeks, are tolerated; and the Jews, who are very numerous, enjoy considerable privileges. There is one archbishopric at Lemberg, and three bishoprics at Cracow, Chelm, and Luckow.

UNIVERSITIES, LITERATURE, &c.

The university of Cracow, founded in 1364, consists of eleven colleges, and had the superintendance of fourteen grammar-schools, dispersed through the city. The number of students, in 1778, amounted to six hundred. Though Copernicus, the great restorer of the *true* astronomical system, and some other learned men, were natives of Poland, yet many circumstances in this country are far from being favourable to literature. The contempt in which learning is held by the nobility, who place their chief importance in the privileges of their rank; the servitude of the lower people; and the universal superstition which pervades all ranks, have wonderfully retarded the progress of letters in this country. Of late, however, a taste for science has spread itself among the nobles, and they begin to regard it as an accomplishment.

COMMERCE AND MANUFACTURES.

The manufactures of Gallicia are not very considerable, being confined to articles of immediate necessity; they are, however, greatly favoured by the Austrian government, and are increasing and improving. The commerce is principally carried on by the Jews. Salt is the most important article of the exports, which consist besides of corn, tobacco, cattle, wool, skins, tallow, bristles, honey, and wax.

GEOGRAPHICAL AND STATIS

BATTLES AND SIEGES IN POLAND.

A.D.

1146 *Posnan:* besieged by Wladyslaus II. who is defeated and dethroned .... *South Prussia.*

1345 *Cracow:* besieged by John of Luxemburgh, King of Bohemia, who is forced to raise the siege .... *New Gallicia.*

1563 *Polock:* besieged and taken by the Russians *Polock.*

1576 *Polock:* retaken by Stephen Batthori ..... *Polock.*

1611 *Smolensk:* besieged and taken by Sigismund III. *Russia.*

1621 *Choczim:* the Turks defeated by the Poles . *Moldavia.*

1632 *Smolensk:* besieged by the Russians, who are forced to raise the siege by Wladyslaus VII. *Russia.*

1637 *Korsun:* the Poles defeated by the Cossacks *Kiow.*

1651 *Beresteczko:* the Cossacks and Tartars deft. by John Casimir............ *Volhynia.*

1656 *Warsaw:* taken by the Swedes, after a battle of three days............ *South Prussia.*

1671 *Bar, Niemirow,* and *Braclaw:* taken from the Cossacks by John Sobiesky .... *Podolia.*

1672 *Kaminiec:* taken by Mahomet IV....... *Podolia.*

1672 *Lemberg:* taken by Mahomet IV. .... *Gallicia.*

1673 *Choczim:* taken by John Sobiesky, after a signal victory gained over the Turks .... *Moldavia.*

1676 *Zurano:* Sobiesky besieged in his camp by the Turks............ *Gallicia.*

1684 *Buckovina Forest:* the Turks defeated by the Poles under Jablonowski............ *Buckovina.*

1700 *Riga:* besieged by Frederick Augustus I. who is forced to raise the siege by Charles XII. of Sweden .... *Livonia.*

1706 *Frawstad:* Schalemberg, General of Saxons, defeated by the Swedes u Renschild ....

1739 *Choczim:* The Turks defeated by the sians, under Munich............

1758 *Konigsberg:* taken by the Russians....

1788 *Choczim:* taken from the Turks by Russians ....

1794 *Warsaw:* taken by the Polish insur after they had defeated the Russians.

1794 *Cracow:* taken by the Prussians, after ing defeated Kosciusko ....

1794 *Warsaw:* besieged by the Prussians, raise the siege after two months .....

1794 *Warsaw:* taken and plundered by Russians, under Suwarow..........

1806 *Thorn:* taken by the French under Ma Ney............

1807 *Prussich Eylau:* the Russians defeate the French............

1807 *Dantzic:* capitulates to the French, General Lefevre............

1807 *Friedland:* the Russians completely feated by Buonaparte............

1807 *Konigsberg:* taken by the French...

1812 *Smolensk:* taken by the French, aft battle gained over the Russians ....

…AL MAP OF POLAND AND HUNGARY.

Figure 61

POLAND. No. 52.

PRUSSIAN DOMINIONS.

The Prussian dominions, besides the three provinces of New East, West, and South Prussia, dismembered from the late kingdom of Poland, consist of the ancient kingdom of Prussia, enclosed in Poland, which now forms the province of East Prussia; chief city, Konigsberg.

CLIMATE, SOIL, AND PRODUCE.

The climate of Prussia is more damp and raw, and the cold of longer duration, than in Germany, but the air is salubrious, and the natives frequently attain to a considerable age. The soil is tolerable, and produces corn, flax, hemp, hops, tobacco, various kinds of fruits, and timber in abundance.

NATURAL HISTORY.

The kingdom of Prussia is, in general, a level country; there are no mountains.

Lakes.—Prussia abounds with lakes; the principal of which are, the Sperling See, the Mauer See, and the Gneserich Lake. The first of these is twenty miles long and as many broad. In East Prussia, there are reckoned three hundred small lakes, and one hundred and sixty in West Prussia.

Rivers.—The chief rivers are, the Vistula, the Memel or Niemen, the Pregel, which falls into the Black Sea below Konigsberg, the Warta, and the Netze. Frederick's and Bromberg Canals unite the Netze with the Vistula.

At the mouths of the rivers Vistula and Memel are irregular havens, called by the Germans *haffs*, formed by long narrow tracts of land. That at the mouth of the Vistula is called *Friesch-haff*, and that at the mouth of the Memel, *Carisch-haff*.

GOVERNMENT.

His Prussian Majesty is absolute through all his dominions. The government of this kingdom is by a regency of four Counsellors of State: *viz.* the Great Master, the Great Burggrave, the Great Chancellor, and the Great Marshal. There are also some other councils, and 37 bailiwicks. The States consist of Counsellors of State, of deputies from the nobility, and of deputies from the commons. Besides these institutions, Frederick II. erected a board for commerce and navigation.

RELIGION.

The religion of Prussia is very tolerant. The established religions are the Lutheran and Calvinist, chiefly the former; but almost all other sects are tolerated.

UNIVERSITIES, LITERATURE, &c.

The university of Konigsberg was founded in 1544: to this must be added the Polish university of Posen or Poznan, which has now become subject to Prussia. There are many schools and seminaries of education in the kingdom of Prussia, but they have produced few men of eminence.

COMMERCE AND MANUFACTURES.

The Prussian manufactures, which are not inconsiderable, consist of glass, iron-work, paper, gunpowder, copper and brass mills; manufactures of cloth, camlet, linen, silk-stockings, and other articles. The inhabitants export a variety of naval stores, amber, linseed and hempseed, oatmeal, fish, mead, tallow, and caviare; with which it is said that 500 ships are loaded every year, chiefly from Konigsberg.

HUNGARY.

SITUATION, BOUNDARIES, AND EXTENT.

Hungary Proper, situated between 16° and 25° East longitude, and between 46° and 49° 30′ North latitude, is bounded, on the North, by Gallicia; on the East, by the Buckovina and Transylvania; on the South, by Sclavonia and Croatia; and on the West, by the circle of Austria and Moravia. Its length is 320 miles, and breadth 210; containing 59,500 square miles, with 105 inhabitants to each.

DIVISION.

Hungary is divided into Upper and Lower Hungary, situate on opposite sides of the Theiss. These are again each divided into two circles, *viz.* in Lower Hungary, the Country within the Danube (chief city, Presburgh,) and the Country beyond the Danube (chief city, Buda;) in Upper Hungary, the Country within the Theiss (chief city, Caschaw,) and the Country beyond the Theiss (chief city, Groswardein.) The province of Temeswar, called Bannat, has been considered as distinct from Hungary, because it was formerly governed by an independent King, and it has several times been in possession of the Turks, but the Austrians gaining possession of it, it was incorporated with the kingdom of Hungary.

CLIMATE, SOIL, AND PRODUCE.

The climate of the Southern parts of Hungary is unhealthy, owing to the numerous lakes, stagnated water, and marshes; but in the Northern part, which is mountainous and barren, the air is pure and wholesome.

No country in the world can boast a richer soil than the plain, which extends three hundred miles, between Presburgh and Belgrade.

The chief productions of Hungary are, corn, grass, tobacco, saffron, asparagus, melons, hops, pulse, millet, buck-wheat, delicious grapes, and fruits of various kinds. The vines of Hungary deserve particular attention, for the excellent wine afforded by their grapes, particularly the celebrated Tokay. Hungary is also remarkable for a fine breed of horses, generally mouse-coloured, and highly esteemed by military officers, so that great numbers of them are exported.

NATURAL HISTORY.

Mountains.—The Carpathian Mountains, which divide Hungary from Poland on the North, are the principal, though many detached mountains are found in the country. Their tops are generally covered with wood, and on their sides grow the richest grapes in the world.

Lakes.—Hungary contains several lakes, particularly the Platten Sea, near Stuhlweissenburgh, in Lower Hungary, about forty-six miles long and eight broad, abounding in fish; and the Newiedler Sea, near the Western frontiers, at a little distance from the right bank of the Danube.

Rivers.—The principal rivers are—the Danube, which entering Hungary near Presburgh, after watering that city as well as Buda, Peterwardein, &c. enters Turkey in its course to the Black Sea;—the Raab, which rises in Austria, and falls into the Danube below Raab;—the Vag, which rises out of the Carpathian Mountains, and falls into the Danube, not far from the mouth of the Raab;—the Theiss, which rises among the same mountains, waters Tokay, and falls into the Danube above Belgrade;—the Drave or Drau, which rises in Austria, separates Lower Hungary from Croatia and Sclavonia, and falls below Essek into the Danube;—the Save or Sau, which rises in Austria, separates Sclavonia from Turkey, and falls into the Danube near Belgrade;—the Marosh, which rises in the Carpathian Mountains, to the East of Transylvania, separates the Bannat from Upper Hungary, and falls into the Theiss.

Mineral Waters.—The Hungarian baths and mineral waters are esteemed the most sovereign of any in Europe; but their magnificent buildings, raised by the Turks when in possession of the country, particularly those of Buda, are suffered to go to decay.

Metals and Minerals.—Hungary contains mines of gold, silver, copper, very rich antimony, coal, salt, and alum. The gold mines are at Kremnitz, and the silver at Schemnitz, both in the North of Lower Hungary. Several kinds of precious stones are found in Hungary, particularly the opal, which hitherto has been discovered in no other country in the world.

GOVERNMENT.

The Hungarians have a diet, or parliament, consisting of two houses; the first composed of Magnates, or the great officers of the crown, Princes, Counts, Barons, and Archbishops; and the second of the Abbots, Prelates, and deputies from the chapters and each of the 52 counties, into which the kingdom is divided. These houses, however, form but one body, as their votes are taken together. There is likewise a Hungary-office, resembling the English chancery, which resides at Vienna; a Palatine's council, somewhat similar to the British privy council, but possessing a municipal jurisdiction, which resides at Presburgh. Every royal town has its senate, and the counties have magistrates, similar to British Justices of the Peace. There is also an exchequer, nine chambers, and other subordinate courts.

RELIGION.

The established religion of Hungary is the Roman Catholic; but the Protestants, who are very numerous, enjoy complete toleration. There are two Archbishoprics, and eleven Bishoprics. The Archbishop of Gran is Lord-Lieutenant of the country, Primate and Chancellor of Hungary; and has the exclusive right of officiating at the coronation of the King.

UNIVERSITIES.

The principal university of Hungary is that of Buda, which has an annual income equal to £20,000 sterling, of which, £4000 is allotted to pay the salaries of the professors. Here is a large library, and an observatory. Raab and Caschaw are likewise styled universities, but they are rather academies, founded by the Jesuits. The university of Tarnau has within these few years been removed to Pest, a city on the left bank of the Danube, opposite Buda, of which it may be considered a suburb.

Printed by J. Barfield, Wardour-Street, Printer to His Royal Highness the Prince Regent.

…field No. 91, Wardour Street, 1813. J. Walker Sculpt.

BATTLES AND SIEGES IN HUNGARY.

A.D.		
1044	*Raab*; Aba, King of Hungary defeated and slain by the Emperor Henry III.	Lower Hungary.
1389	*Caschaw*; the Hungarians defeated by Amurath I. who is slain after the battle by a wounded Christian	Upper Hungary.
1448	*Caschaw*; Hunniades defeated by Amurath II.	Upper Hungary.
1456	*Belgrade*; Mahomet II. defeated, and forced to raise the siege, by Hunniades	Servia.
1526	*Mohatz*; Louis II. defeated and slain by Solyman II.	Lower Hungary.
1527	*Tokay*; the army of John Zapolski defeated by that of his competitor, Ferdinand I.	Upper Hungary.
1541	*Buda*; the Hungarians completely defeated by Solyman II.	Lower Hungary.
1593	*Sissig*; the Turks cut to pieces or drowned, by the Archduke Ernest and Montecuculli. *Stahlweissemburg*; the Turks defeated by the Hungarians	Croatia.
1596	*Agria*; capitulates to Mahomet III.	Lower Hungary.
1598	*Raab*; retaken from the Turks by Baron Schwartzenberg	Lower Hungary.
1660	*Wardein*; taken by the Turks, after a siege of 47 days	Upper Hungary.
1663	*Neuhausel*; taken by the Turks, after a siege of 36 days, with a loss of 15,000 men	Lower Hungary.
1664	*St. Gothard*; the Vizir Mahomet Cuprogli defeated by Montecuculli	Lower Hungary.
1683	*Barkan*, near *Gran*; John Sobiesky defeated by the Vizir Kara Mustapha	Lower Hungary.
1684	*Watzen*; 15,000 Turks defeated by the Duke of Lorraine	Lower Hungary.
1687	*Mohats*; the Turks defeated by the Dukes Lorraine and Bavaria	Lower Hungary.
1688	*Munkatsh*; the Princess Tekeli capitulates, after a blockade of several years	Upper Hungary.
1688	*Stahlweissenburg*; taken from the Turks by Count Caraffe	Lower Hungary.
1688	*Belgrade*; taken, sword-in-hand, by the Elector of Bavaria	Servia.
1690	*Belgrade*; retaken by the Turks	Servia.
1691	*Salenkemin*; bloody battle fought by the Prince of Baden against the Turks; the success dubious	Sclavonia.
1697	*Zenta* (five miles south of Theresiople on the Theiss): signal victory gained by Prince Eugene, in which the Vizir, 17 Pachas, and 20,000 Turks, were slain, 10,000 drowned, and 3000 taken prisoners	Lower Hungary.
1708	*Trentchin*; Racoczi, a pretender, defeated by General Heister	Lower Hungary.
1716	*Peterwardein*; the Turks defeated by Prince Eugene	Sclavonia.
1716	*Temeswar*; taken from the Turks by Prince Eugene	Upper Hungary.
1717	*Belgrade*; the Turks defeated, and the city taken, by the Imperialists	Servia.

A NEW MAP OF POLAND, AND THE GRAND DUCHY OF LITHUANIA,
Shewing their DISMEMBERMENTS and Divisions BETWEEN AUSTRIA, RUSSIA AND PRUSSIA, ACCORDING TO THE CONGRESS OF VIENNA FROM THE LATEST AUTHORITIES:
By JOHN CARY, Engraver.
1819.
BALTIC OR EAST SEA
GULF OF DANTZICK
SCALES.
German and Polish Miles 15 to a Degree
Prussian Miles 20 to a Degree
British Statute Miles 69½ to a Degree
Common French Leagues 25 to a Degree

The Dates in the body of the Map signify the Years in which the Divisions took place.
The Colours, the Powers to which they now belong.
Rußia .... Blue
Prußia .... Red
Austria .... Yellow
Longitude East from Greenwich
LIVONIA
NOWOGOROD
RUSSIA
LITHUANIA
MOLDAVIA
WILNO
NOWOGRODEK
BRACLAW
KIOW
PODOLIA
WOLHINIA
Marshes Rokitno
Smolensk
Witepsk
Polock
Mohilow
1772
1793
1795

Figure 62

RUSSIA IN EUROPE
FINMARK
10°
15°
20°
25°
30°
35°
40°
45°
55°
60°
English Miles
STOCKHOLM
Upsala
Hudiksvall
BALTIC SEA
Gulf of Finland
Gottland
Oeland
ESTLANDIA
LIEFLAND
COURLAND
VILNA
GRODNO
WARSAW
POLAND
PRUSSIA
Thorn
Memel
Danzig
TVER
MOSCOW
SMOLENSK
MOGHILEV
MINSK
KALOUGA
TOOLA
ORLOV
KOURSK
KIEV
PODOLIA
VOLINSK
CHERSON
EKATERINOSLAV
POLTAVA
CRIMEA
Sea of Azov
MOLDAVIA
Jassy
WALACHIA
Bukarest
R. Danube
Nicopoli
Shumla
BLACK SEA
Sevastopol
Anapa
Kherson
Odessa
Lemberg
GALICIA
HUNGARY
Hermanstadt

Nova Zembla
SEA OF KARA
SEA OF OBE
ARCHANGEL
TOBOLSK
CASPIAN SEA
SEA OF ARAL
ASTRAKHAN
Country of Gog and
Magog or Majuje & Majuje
Kirgees Middle Horde
Gog and Magog
Kirgees Little Hord
Kirgees Great Horde
Orenburg
Tcheliabinsk
Ekaterinburg
Stavropol
Tzaritzin
Turkestan
Omsk
Tara
Solikamsk
45° from Greenwich
50°
55°
60°
65°
70°
75°
80°

Figure 63

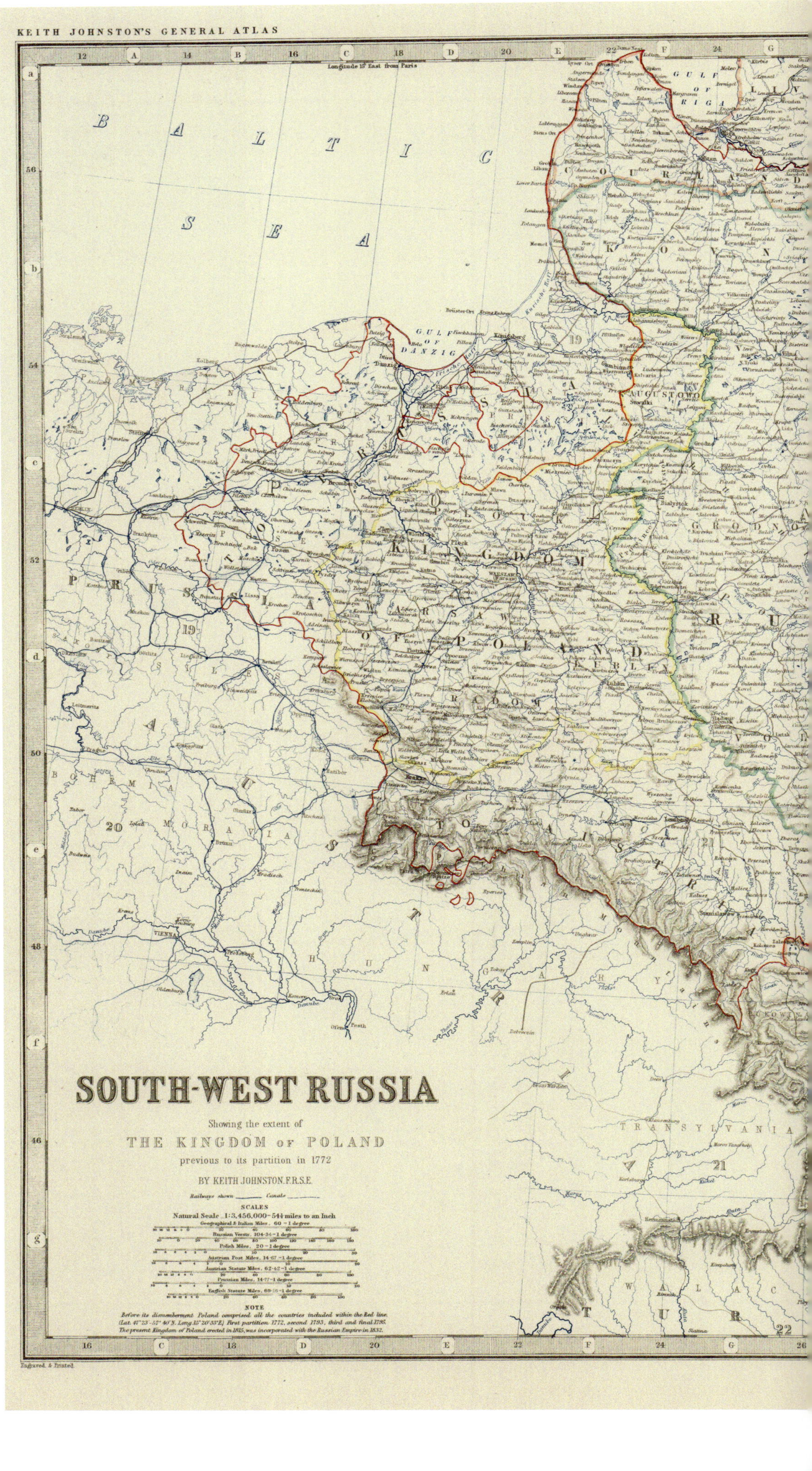
KEITH JOHNSTON'S GENERAL ATLAS
BALTIC SEA
GULF OF RIGA
GULF OF DANZIG
SOUTH-WEST RUSSIA
Showing the extent of
THE KINGDOM OF POLAND
previous to its partition in 1772
BY KEITH JOHNSTON F.R.S.E.
Railways shown Canals
SCALES
Natural Scale 1:3,456,000 = 54½ miles to an Inch
Geographical & Italian Miles. 60 = 1 degree
Russian Versts. 104·34 = 1 degree
Polish Miles. 20 = 1 degree
Austrian Post Miles. 14·67 = 1 degree
Austrian Statute Miles. 62·42 = 1 degree
Prussian Miles. 14·77 = 1 degree
English Statute Miles. 69·16 = 1 degree
NOTE
Before its dismemberment Poland comprised all the countries included within the Red line.
(Lat. 47°23'–57°40' N. Long. 15°20'33" E.) First partition 1772, second 1793, third and final 1795.
The present Kingdom of Poland erected in 1815, was incorporated with the Russian Empire in 1832.

27
MOSCOW
SEA OF AZOV
PEREKOP GULF
BLACK
SEA
Longitude East from Greenwich
W. & A.K. Johnston, Edinburgh
ood & Sons, Edinburgh & London.

Figure 64

Historisch - statistisch - geographische Darstellung des
KOENIGREICHS POLEN,
1770.
1773, 1793, 1795, 1807, 1815
MANNHEIM, Steindruckerey von Rudolph Schlicht. 1831.

NOWOGROD
LIEFLAND
GITIEN
1773
LITHAUEN
SMOLENSK
RIGA
Dünaburg
Polock
Witepsk
Kauen
WILNO
Minsk
Mohilow
Grodno
Białystock
1807
1795
1793
Pinsk
Brzesc
KLEIN RUSSLAND
RUSSISCHE UKRAINE
KIOW oder POLNISCHE UKRAINE
Kiow
Czernigow
Biała Cierkiew
WOLHYNIEN
PODOLIEN
BRACŁAW
Human
Lemberg
Chocim
MOLDAU

Figure 65

No. 43
EUROPAEISCHES
RUSSLAND
entworfen und gezeichnet
vom HAUPTM. RADEFELD.
1844.
MEYER'S - HANDATLAS

Figure 66

Sen. Ex Doc. No. 59_1st Sess. 36th Cong. Delafield's Report.
SEBASTOPOL and its ENVIRONS
showing the
Russian Defences
and the
APPROACHES AND OTHER WORKS OF ATTACK
of the
Allied Armies
during the Siege of 1854-55
from the Report of
Genl. Niel
Chief of the French Engrs.
Prepared to accompany the Report of Major Delafield on the art of War in Europe
in 1854-55, & 56.
NOTE. The references in figures on the horizontal curves are in metres above the level of the sea.
Volokhoff Tower (Telegraph)
Redoubt
NORTH FORT
Fort Constantine
Fort Michel
Battery No. 4 (Fort Catherine)
GREAT HARBOUR OF S
Fort Alexander
Fort Nicolas
Fort Paul
Battery No. 10 (Quarantine Fort)
QUARANTINE BAY
ARTILLERY BAY
SOUTH or INNER HARBOUR
KARABELNAYA BAY
SEBASTOPOL
KARABELNAYA SUBURB
DOCKS
FORT MALAKOFF
HOSPITAL
BASTION No. 2
BASTION No. 4
BASTION No. 5
Schwartz Red.
CITY
LABORATORY OR WORONZOFF RAVINE
FRENCH LEFT ATTACK
ENGLISH
Chapman Batteries
Note: Sebastopol and the ground of attack was surveyed and levelled by the officers of the engineer corps, the part north of the harbour is according to a Russian map found in the place after its capture. The trenches were drawn during the siege, and after the taking of the place a triangulation was made for rectifying the details.

Plate 8.
Redoubt
Redoubt
Redoubts
Batteries
Light House Batteries
ASTOPOL
Selinghinsk Rt
Volhynie Red.
MOUNT
SAPOUNE
RIGHT
FRENCH
English Battery
English Great Redoubt
Light House Redoubt
Lancaster Battery
Victoria Redoubt
Inkermann Bridge
Scale of 0m001 for 10 metres
Lith of Bowen & Co. Philad

Figure 67

THE
CRIMEA
BY
JAMES WYLD, GEOGRAPHER TO THE QUEEN & H.R.H. PRINCE ALBERT,
CHARING CROSS, EAST,
LONDON
FIFTH EDITION
JUNE 1855.
SCALES
English Miles
Russian Versts
Longitude East of Greenwich
GULF OF KARKINITE
EUXINE
KALAMITA BAY
SEVASTOPOL
BAKTCHI SARAI
SIMPHEROPOL
KARA SU BAZAR
PEREKOP
EUPATORIA
London. Published by James Wyld, Geographer to the Queen & H.R.H.

COPYRIGHT
LONGITUDE EAST OF GREENWICH
AUSTRIA
TRANSYLVANIA
RUSSIA
SEA OF AZOV
BLACK SEA
SEA OF AZOV
GULF OF ARABAT
Gulf of Kazandib
PENINSULA OF KERTCH
GULF OF TAMAN
STRAIT OF KERTCH
GULF OF KAFFA
BLACK SEA
EXPLANATION
CITY
TOWN
Village
Hamlet & House
Ancient Hamlets inhabited
Church
Monastery
Mosque
Fort
Post Road
Great Road
Cross Road
Lighthouse
Windmill
Water Mill
River
Water course
Ancient Names ... CIMMERIAN
The Range of the Guns are indicated thus
The Soundings are in Fathoms
GREAT HARBOUR
THE TOWN & HARBOUR OF SEVASTOPOL with the BATTERIES & APPROACHES
Geographic Miles
Russian Versts

Figure 68

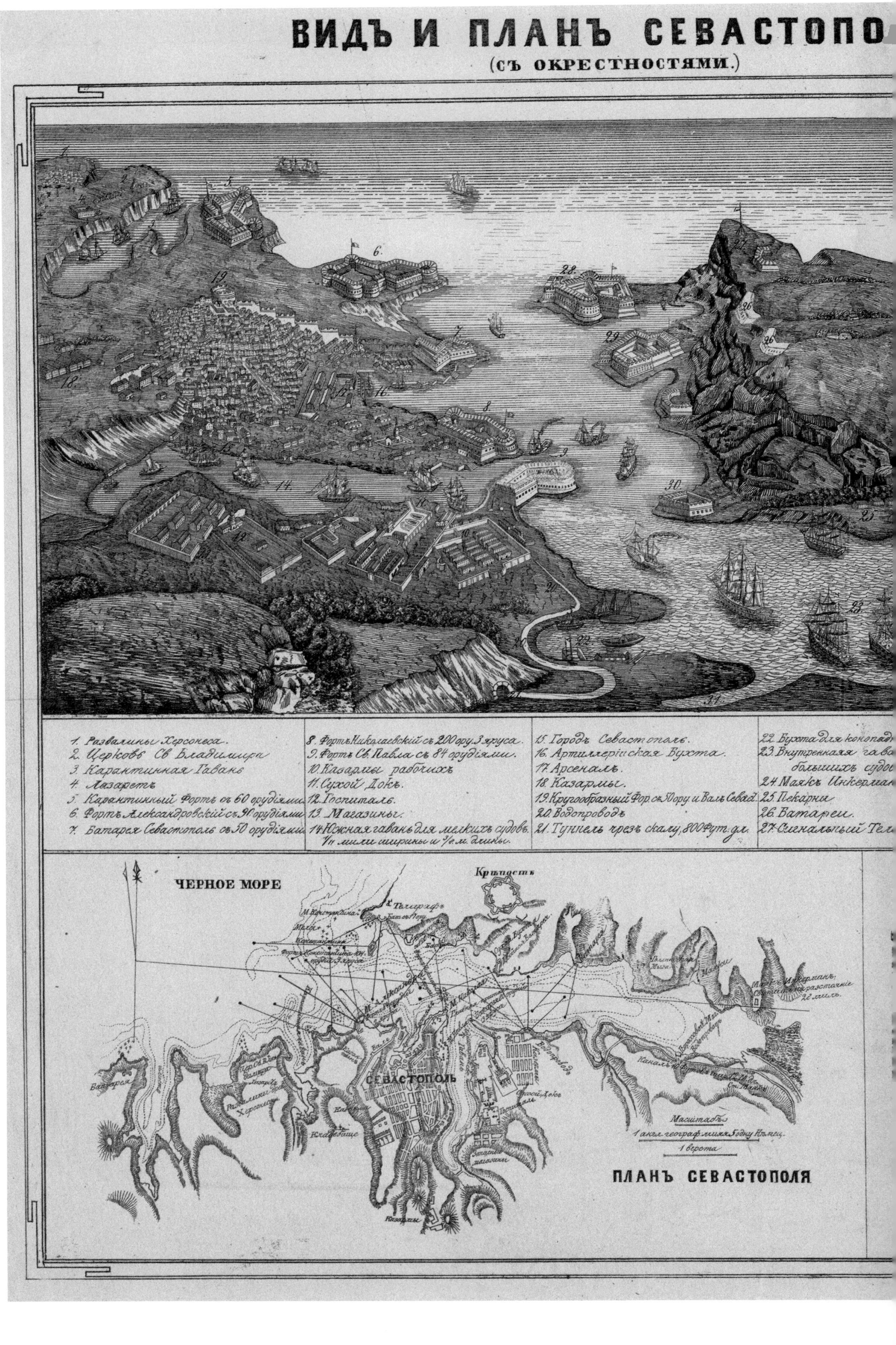
ВИДЪ И ПЛАНЪ СЕВАСТОПО
(съ окрестностями.)
1. Развалины Херсонеса.
2. Церковь Св Владимира
3. Карантинная Гавань
4. Лазаретъ
5. Карантинный Фортъ съ 60 орудіями.
6. Фортъ Александровскій съ 90 орудіями
7. Батарея Севастополь съ 50 орудіями.
8. Фортъ Николаевскій съ 200 ору. 3 яруса.
9. Фортъ Св. Павла съ 84 орудіями.
10. Казармы рабочихъ
11. Сухой Докъ.
12. Госпиталь.
13. Магазины.
14. Южная гавань для мелкихъ судовъ. 1/11 мили ширины и 3/5 м. длины.
15. Городъ Севастополь.
16. Артиллерiйская Бухта.
17. Арсеналъ.
18. Казармы.
19. Кругообразный Фор. съ 50 ору. и Валъ Севаст.
20. Водопроводъ
21. Туннель чрезъ скалу, 800 Фут. дл.
23. Внутренняя гав
24. Маякъ Инкерма
25. Пекарня
26. Батареи.
ЧЕРНОЕ МОРЕ
Крѣпость
Телеграфъ
СЕВАСТОПОЛЬ
Кладбище
Масштабъ
1 верста
ПЛАНЪ СЕВАСТОПОЛЯ

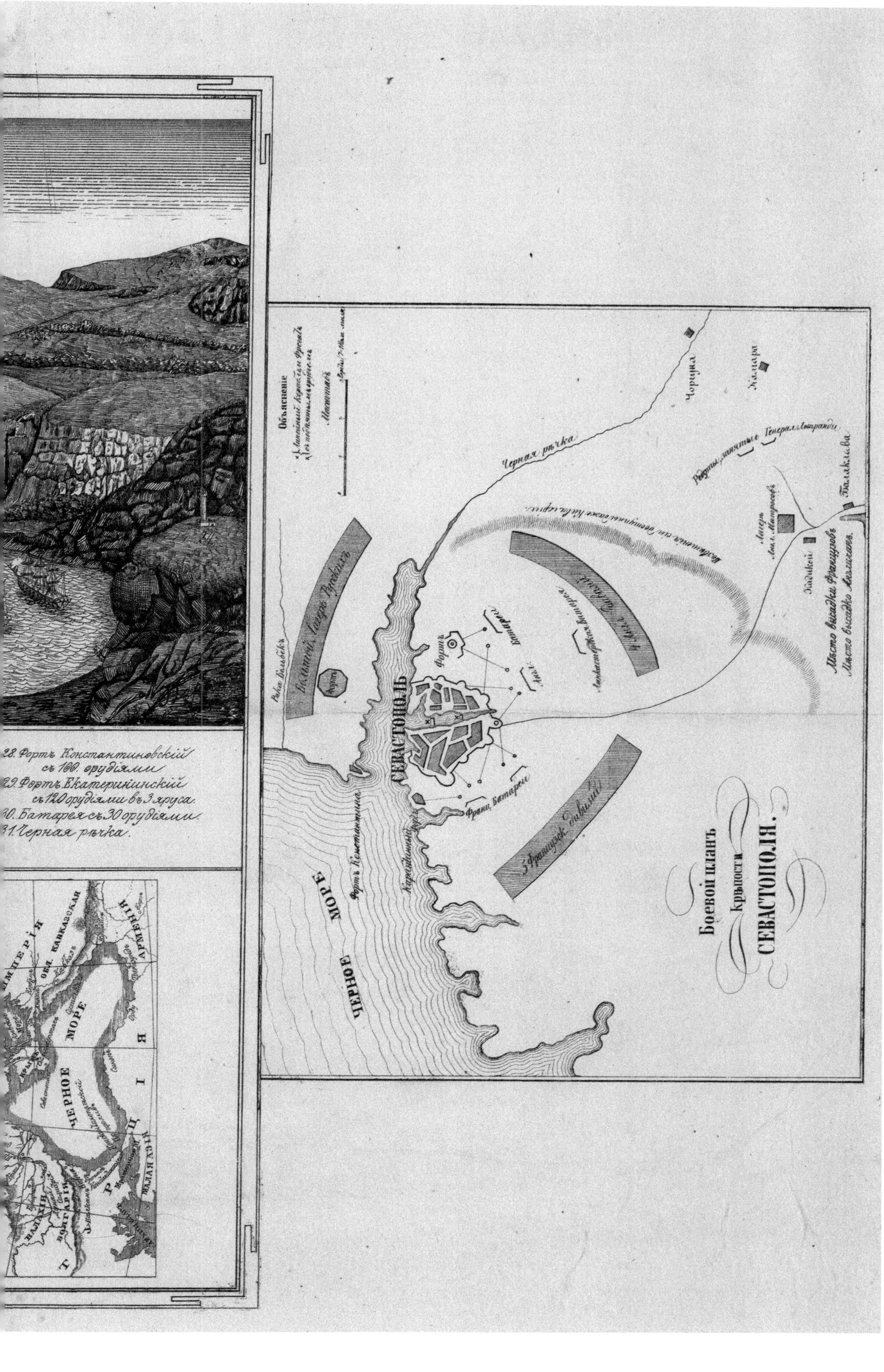
Боевой планъ
Крѣпости
СЕВАСТОПОЛЯ.
СЕВАСТОПОЛЬ
ЧЕРНОЕ МОРЕ

Figure 69

A COMPLETE
MAP OF THE
CRIMEA
Showing the
MILITARY AND CARRIAGE ROADS
with Distances
from
Various Points of the Western Coast.
From English & Russian Surveys.
PHILADELPHIA
PUBLISHED BY CHARLES DESILVER 253 MARKET ST
KEEN & LEE 148 LAKE ST CHICAGO.
SCALES:
English State Miles
Russian Versts
Military Roads
Post Stations
Light Houses
Fortified Places
GULF OF PEREKOP
KALAMITA BAY
CRIMEA
BLACK
Karkinet Bay
Perekop
Chaplinka
Kalanchak
Korga
Krasnaia
Khorli
Dgarilgatch Bay
Dgarilgatch Spit
Ak Metchit
Tarakhanskii Kut
Kosloy or Kupatoria
Lake Sasik
Simferopol
Sebastopol
Balaklava
Yalta
Alushta
Kara Subazar
Ichongar Pen.
Cenitchi
Azovskaya
Orta Ablan
Diurman Naiman
Saribulat
Kobchak
Landing of the Anglo French Army 14 September 1854.
Battle of the Alma 20th September 1854
Entered according to Act of Congress in the year 1856 by Charles Desilver in the Clerks office of the District Court of the Eastern District of Pennsylvania

Map Showing the Relative Position of the Crimea with the Seat of War.
AUSTRIA
HUNGARY
TRANSYLVANIA
RUSSIA
PODOLIA
BESSARABIA
MOLDAVIA
WALLACHIA
BOSNIA
SERVIA
BULGARIA
ROUMELIA
TURKEY
ASIA MINOR
GREECE
ITALY
CORSICA
SARDINIA
SICILY
Vienna
Rome
Naples
BLACK SEA
CRIMEA
CAUCASIA
GEORGIA
CASPIAN SEA
SEA OF AZOV
OF AZOV
ARABAT BAY
KAZANDIB BAY
KERTCH PENINSULA
BAY OF KAFFA
KERTCH
STRAITS OF KERTCH
GULF OF TAMAN
TAMAN Iᴰ
TEMRIOUK BAY
Phanagoria
Bogaz
ANAPAN
SEBASTOPOL HARBOUR &c.
HARBOUR OF SEBASTOPOL
SEBASTOPOL
North Bay
Inkerman Bay
Careening Bay
Artillery Park
One Geographical Mile
One English Mile

Figure 70

CCLXXXXI
KIEW
Aus d. Kunstanst. d. Bibliogr. Inst. in Hildbh.
Eigenthum d. Verleger

Figure 71

DCXV
LEMBERG
Aus d. Kunstanst. d. Bibl. Instit. in Hildbh.
Eigenthum d. Verleger.

Figure 72

DAS
EUROPÄISCHE
RUSSLAND
Revidirt von C. Gräf 1857.
DAS EIS MEER

DAS KASPISCHE MEER
DAS SCHWARZE MEER
WEIMAR: GEOGRAPHISCHES INSTITUT

Figure 73

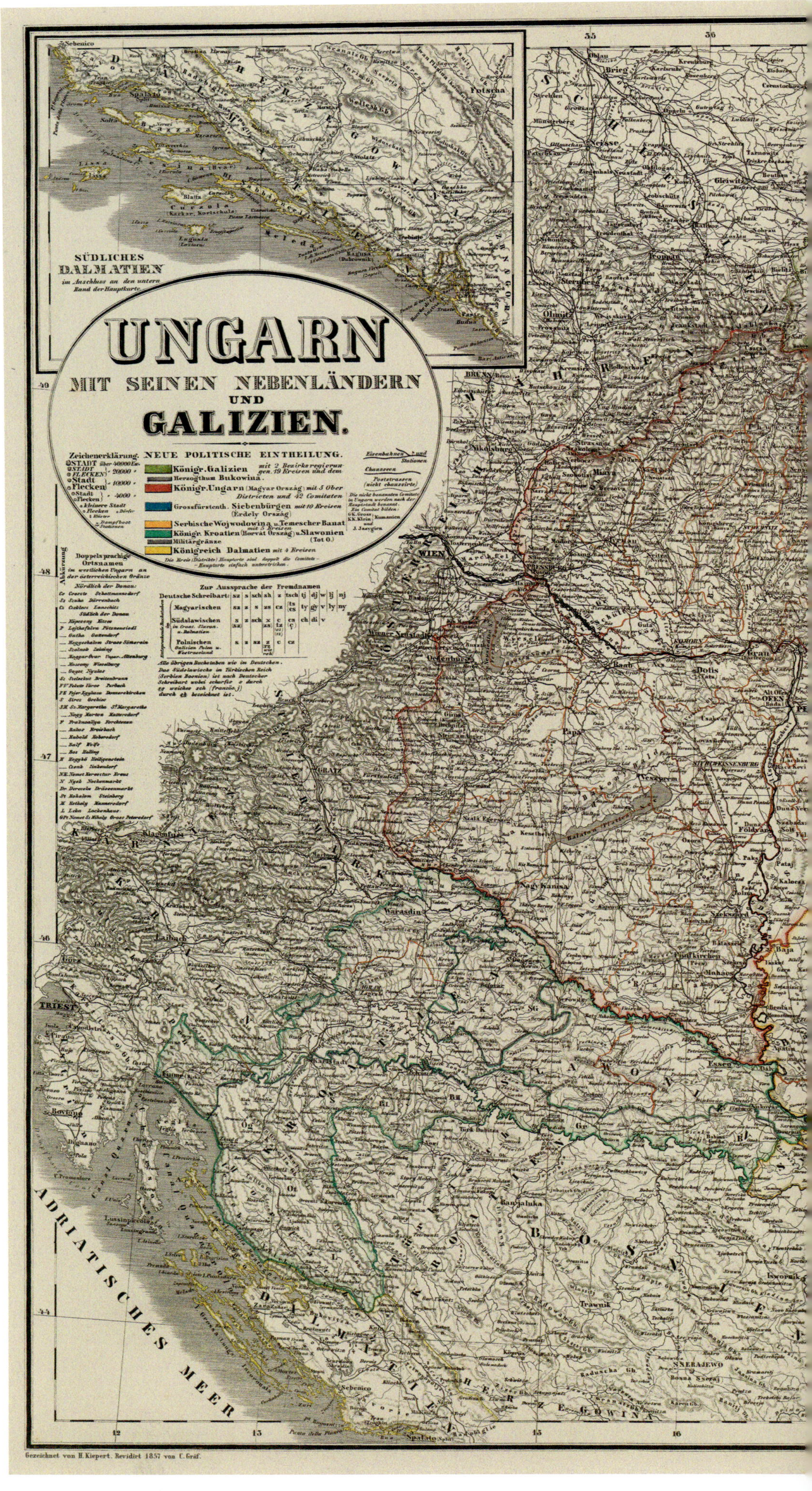
UNGARN
MIT SEINEN NEBENLÄNDERN
UND
GALIZIEN.
SÜDLICHES
DALMATIEN
im Anschluss an den untern Rand der Hauptkarte.
NEUE POLITISCHE EINTHEILUNG.
Zeichenerklärung.
Königr. Galizien
Herzogthum Bukowina
Königr. Ungarn
Grossfürstenth. Siebenbürgen
Serbische Wojwodowina u. Temescher Banat
Königr. Kroatien u. Slawonien
Militärgränze
Königreich Dalmatien
Doppelsprachige Ortsnamen
Zur Aussprache der Fremdnamen
ADRIATISCHES MEER
WIEN
TRIEST
Gezeichnet von H. Kiepert. Revidirt 1857 von C. Gräf.

Abgekürzte Bergnamen im Tatra Gebirge:
Abgekürzte Ortsnamen in der Zips:
LEMBERG
Tarnopol
DEBRECZEN
BUKURESCHT
WIDDIN
Gränz Regimenter
Maassstäbe in 1:1,600000
Erklärung geographischer Benennungen
Magyarische Heiligen Namen
EOGRAPHISCHES INSTITUT.
Gestochen v. F. Eulenstein

Figure 74

Українці
Великоруси
Білоруси
Поляки й Кашуби
Лужицькі Серби
Чехи й Словаки
Словінці
Сербо-Хорвати
Болгари
ЧЕХИ
СЛОВАКИ
СЛОВІНЦІ
ХОРВАТИ
Берлін
Дрезден
Прага
Венеція
Рим
Дубровник
Катаро

ЕТНОҐРАФІЧНА КАРТА
СЛАВЯНЩИНИ
По Нідерле.
Виборг
Гельсінґфорс
Петроград
Вятка
Вольоґда
Кострома
Ревель
Новгород
Рибінськ
Ярославль
Нижній Новгород
Казань
Юріїв
Псков
Твер
Владимір
Рига
Москва
Муром
Двинськ
Вязьма
Рязань
Симбірськ
Самара
Полоцьк
Вітебськ
Смоленськ
Калуґа
Тула
Пенза
Николаєвськ
Ковно
Вільна
Борисів
Мінськ
Могилів
Тамбов
Уральськ
Ґродно
Гомель
Бобруйськ
Брянськ
Орел
Саратов
Воронїж
Курськ
Острогожськ
Пинськ
Берестє
Камишин
Білгород
Олексіївка
Харків
Суми
Ромни
Ніжин
Ковель
Холм
Луцьк
Рівне
Житомир
Київ
Переяслав
Царицин
Броди
Кременець
Бердичів
Черкаси
Полтава
Кременчуг
Славянськ
Луганськ
Тернопіль
Проскурів
Вінниця
Жмеринка
Бахмут
Новочеркаськ
Камянець Под.
Елисаветград
Катеринослав
Олександрівськ
Таганрог
Ростов н. Д.
Астрахань
Коломия
Чернівці
Могилів
Ольвіопіль
Николаїв
Маріуполь
Азов
Балта
Миколаїв
Бердянськ
Ейськ
Кишинів
Ясі
Тирасполь
Херсон
Азівське М.
Ставрополь
Бендери
Одеса
Георгіевськ
Акерман
Катеринодар
Майкоп
Кронштадт
Галац
Євпаторія
Феодосія
Анапа
П'ятигорськ
Владикавказ
Браїла
Симферопіль
Новоросійськ
Севастопіль
Ялта
Букарешт
Констанца
Тифліс
ЧОРНЕ МОРЕ
Відін
Терново
Ниш
Софія
БОЛГАРИ
Скопле
Адріянопіль
Салунь
УКРАЇНЦІ
БІЛОРУСИ
ВЕЛИКОРУСИ
Волга
Дон
Дніпро
Дністер
Десна
Прип'ять

Figure 75

ЭТНОГРАФИЧЕСКАЯ
КАРТА
РОССІЙСКОЙ
ИМПЕРІИ.
Балтійское Море
Черное Море
Азовское Море
Каспійское Море
Аральское Море
ПЕРСІЯ
(Изъ энциклопедическаго Словаря изданія Брангаузъ-Ефрона)

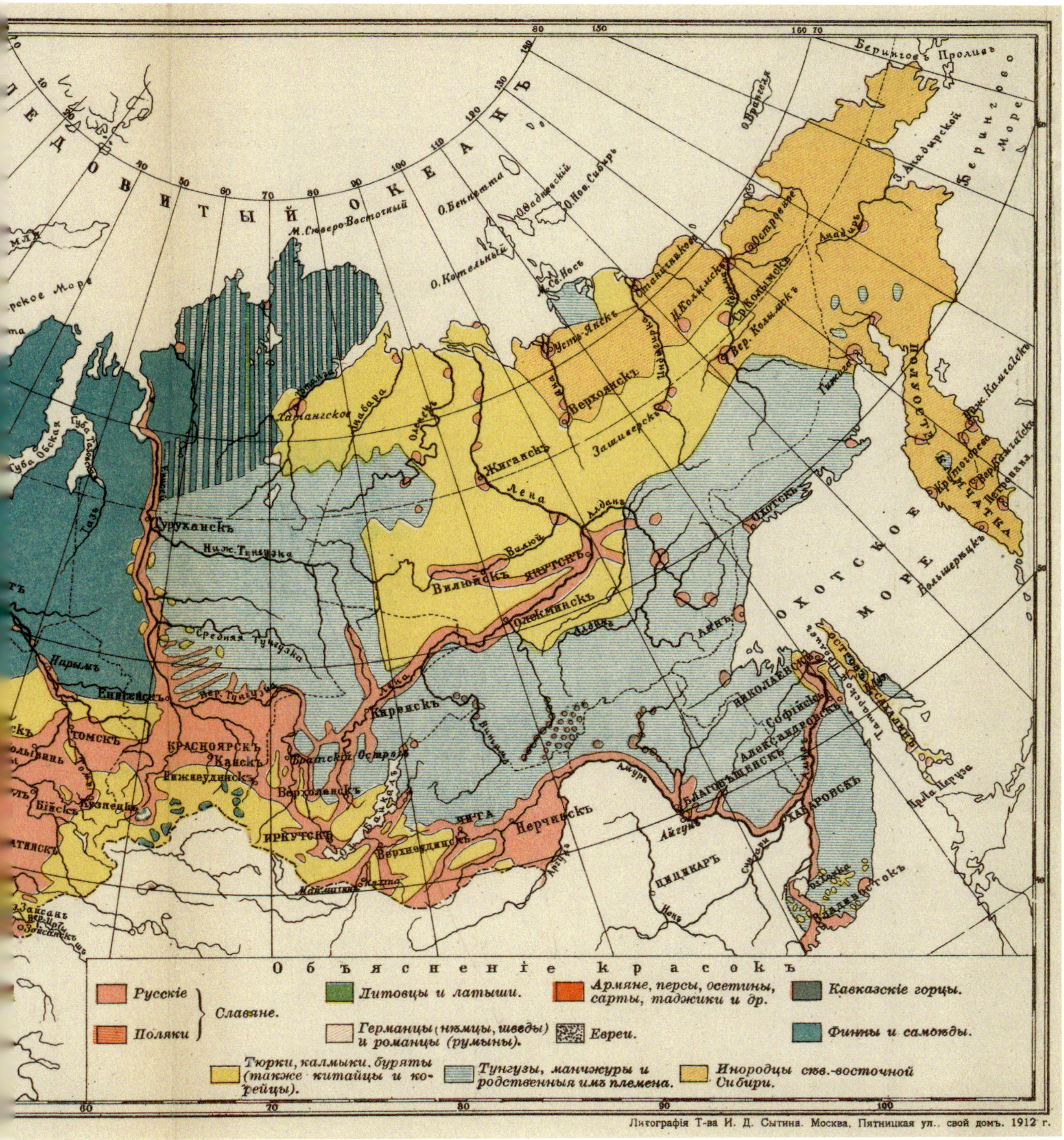
Объясненіе красокъ
Русскіе
Поляки
Славяне.
Литовцы и латыши.
Германцы (нѣмцы, шведы) и романцы (румыны).
Армяне, персы, осетины, сарты, таджики и др.
Евреи.
Кавказскіе горцы.
Финны и самоѣды.
Тюрки, калмыки, буряты (также китайцы и корейцы).
Тунгузы, манчжуры и родственныя имъ племена.
Инородцы сѣв.-восточной Сибири.
Литографія Т-ва И. Д. Сытина. Москва, Пятницкая ул., свой домъ. 1912 г.
Охотское Море
Туруханскъ
Якутскъ
Верхоянскъ
Красноярскъ
Иркутскъ
Томскъ
Нерчинскъ
Благовѣщенскъ
Хабаровскъ
Николаевскъ

Figure 76

A

DESCRIPTION

OF

UKRAINE,

by

Guillaume le Vasseur Sieur de Beauplan

New York – 1959

GUILLAUME LE VASSEUR SIEUR DE BEAUPLAN'S "DESCRIPTION OF UKRAINE" AND HIS MILITARY MAPS OF UKRAINE. — *By Bohdan Krawciw*

THE name *Rus'* (in Latin *Russia, Ruthenia*) which during the 10th and 12th centuries referred to the Kievan State — at first only to the Kiev land and later to the whole territory from the Syan and Buh rivers in the West to the Volga in the East — toward the end of this state's existence and particularly after the rise of the Halych-Volhynian State (1199) and the destruction of the Kievan Empire by the Mongols (1240), was attached to the western lands of Ukraine—Halych land and Volhyn.

Established on these territories during the 13th century and the first half of the 14th century, this name remained even later, after the invasion of Halych land by the Polish king Casimir. In the understanding of West European geographers and cosmographers of that day and of later times, as well as of Polish chroniclers and cartographers, *Rus'*, although it lay within the Polish boundaries, was a separate political, ethnic and cultural territory. In their writings and maps they distinguished it from Poland proper, from Polish ethnographic territories. For instance, on Marcus Beneventanus first map of Central-Eastern Europe entitled *Tabvla Moderna Polonie. Vngarie. Boemie. Germanie. Rvssie. Lithvanie.* which was included in Ptolemy's *Geography,* published in 1507 (Rome, Bernhardus Vinetus de Vitalibus), or on the map of the eminent scholar and geographer of the early 16th century, Martin Waldseemueller, entitled *Tabvla Moderna Sarmatie Evr. sive Hvngarie, Polonie, Rvssie, Prvssie et Valachie* (Johannes Schott, Strassburg) found in Ptolemy's *Geography* published in 1513, *Rus'* was defined in the titles of the maps as a separate territory, equal not only to Poland, Lithuania and Wallachia but to Germany, Hungary and Bohemia as well.

Both maps, as is verified by contemporary researchers (Birkenmajer, Kordt, Piekarski, Buczek, Chowaniec) were published on the basis of maps and information compiled by the Cracow canon and historian Bernard Wapowski († 1535), who at the beginning of the 16th century lived in Rome and enjoyed friendly and academic association with Marcus Beneventanus in Rome and Martin Waldseemueller in Saint-Die, author of several maps in the Strassburg edition of Ptolemy's *Geography.*

Rus' was treated—according to tradition—as a separate political and ethnic territory by the well-known cosmographer Sebastian Munster (1489-1552), publisher of several issues of Ptolemy's *Geography* (beginning in 1540 in Basel) and author of the widely-known *Cosmography* published first in 1542 and reprinted more than a score of times during the 16th and 17th centuries in German, Latin, Italian and other languages. In all these publications the description of *Rus'* (*Russia*), also referred to as *Ruthenia* and *Podolia,* is differentiated not only from the description of Poland but also from the description of Muscovy (*Moscowiters Lands*). Even Polish cartographers of the second half of the 16th century defined Rus' (*Russia*) on their maps as a separate territory within the borders of Poland, with distinct boundaries, as for instance Waclaw Grodecki in a map published in 1558 and included in 1570 in Abraham Ortelius' atlas entitled *Poloniae finitimarumque locorum descriptio. Auctore Veceslao Godreccio Polone.*

In accordance with such traditional treatment and with such an understanding of the name and concept of Rus' (Russia), all cartographers and publishers of atlases at the end of the 16th century and of the 17th century—Gerard Mercator, Jodocus and Henry Hondius, Johannes Blaeu, Jan Janson, Moses Pitt and others, if they published maps of Poland in their atlases, showed only its ethnographic territory—without Rus' (Russia) and Lithuania. Rus' was treated by them as a separate country together with Lithuania, which for a time (beginning in 1370 until 1569) encompassed Ukrainian lands—Volhyn, Podilya and Kiev province. In all these atlases there were separate maps for Lithuania and Rus'. Rus' was also treated as a separate land—administratively and ethnically.

In the descriptions of Poland of the Polish writer and historian Szymon Starowolski (1588-1656) printed in various publications and reprinted in all four editions of Joannes Blaeu's *Atlas Maior* the description of Rus' is differentiated from the description of Poland and Lithuania. In the Spanish translation of this description from the Spanish edition of Blaeu's *Atlas Mayor* (the description is in the collection of this writer), S. Starowolski, dividing Rus' into *Russia Roxa o Roxolania* and *Russia la Blanca,* includes in the first place all Ukrainian territory from Lancut, Ryashiv, Syanok, Krosno and Zamostya in the West to Kiev in the East. That is, he includes in Rus' the territories of Lviv, Kholm and Belz as well as Volhyn, Podilya and all of Kiev land. All these lands became known at the end of the 16th century under the name *Ukraina,* first applied on the large map of Lithuania of Tomasz Makowski (1575-1620?), published by Hessel Gerritsz in 1613 *cura et impensis* of the Lithuanian-Rus' prince Mikolaj Christophor Radziwill, in the inscription *"Volinia Vlterior quae tum Vkraina tum Nis ab aliis vocitatur."*

Wider knowledge of Ukraine as a territory subject to the personal authority of Polish kings but ethnically and culturally separate was brought into world geography and history by the geographer, cartographer and military engineer-architect Guillaume le Vasseur de Beauplan with his *Description d'Ukraine* and his general and special maps

Figure 77

Das Wohngebiet der UKRAINER in Europa.
Nördlicher Polarkreis
Färöer
Shetland-I.
Bergen
Stavanger
BRITISCHE INSELN
Hebriden
SCHOTTLD.
GROSZBRITANNIEN
Edinburg
NORD-SEE
Dublin
IRLAND
Irische S.
Liverpool
ENGLAND
LONDON
Amsterdam
Haag
Bremen
Kanal
Brüssel
BELGIEN
Köln
Brest
Havre
Frankfurt
PARIS
Nantes
Orléans
FRANKREICH
Lyon
Bern
Mont Blanc
Genua
Golf v. Biscaya
Santander
Bordeaux
Pyrenäen
Marseille
Porto
Valladolid
Duero
Madrid
Tajo
PORTUGAL
SPANIEN
Lissabon
Barcelona
Korsika
Valencia
Balearen
Sardinien
Sevilla
Cadiz
Gibraltar (Brit.)
Tanger
Ceuta (Span.)
MITTEL
Oran
Algier
Fes
Tunis
MAROKKO
ALGERIEN
30
20
10
0
60
50
40
Östl. Lä
Verlag der „Ukrainischen Rundschau“, Wien.

Archangelsk
Kotlas
Wjatka
Perm
Jekaterinburg
Tscheljabinsk
Tobolsk
Ufa
Orenburg
Samara
Kasan
Nowgorod
Jaroslawl
Moskau
St PETERSBURG
Stockholm
Helsingfors
Reval
Riga
Witebsk
Smolensk
Tula
Pensa
Orel
Saratow
Zarizyn
Astrachan
Tschernigow
Kijew
Schitomir
Charkow
Poltawa
Jekaterinoslaw
Rostow
Lemberg
Krakau
Warschau
Breslau
Danzig
Königsbg.
Brest Litowsk
Kamenez Podolsk
Czernowitz
Odessa
Cherson
Asow
Stawropol
Batum
Tiflis
Baku
Krasnowodsk
Bukarest
Belgrad
Sofia
Konstantinopel
Adrianopel
Smyrna
Angora
Brussa
Kreta
Rhodos
Cypern
Beirut
Bagdad
Teheran
SCHWARZES MEER
KASPISCHER SEE
Ust-Urt Plateau
FINNLAND
SCHWEDEN
RUMÄNIEN
BULGARIEN
SERBIEN
PERSIEN
KLEIN-ASIEN
TÜRKISCHES REICH
Kaukasus
Armenien
Ural-Gebirge
Wolga
Don
Dnjepr
Dnjestr
Donau
Petschora
Dwina
Ob
Tobol
Budapest
Szegedin
Agram
Bosnien
Sarajevo
Montenegro
Albanien
Saloniki
Kavala
GRIECHENLAND
Jonisches Meer
1 : 20,000.000.
0 200 400 600 800 1000 km
Kartogr. Anstalt G. Freytag & Berndt, Ges. m. b. H., Wien.

Figure 78

MAGYARORSZÁG NÉPRAJZI TÉRKÉPE
A NÉPSŰRŰSÉG ALAPJÁN
GRÓF TELEKI PÁL
Az 1910. évi népszámlálás alapján.
ETHNOGRAPHICAL MAP OF HUNGARY
BASED ON DENSITY OF POPULATION
BY
COUNT PAUL TELEKI
PROFESSOR OF GEOGRAPHY
Every square millimeter coloured indicates 100 inhabitants.
According to the census of 1910.
To Poland
To Czecho-Sloval
24.326 square miles 22·3 % of the whole territory of Hungary.
3·57 million inhabitants 19·5 % of the whole population of Hungary.
1·07 million magyar inhabitants 10·8 % of the whole magyar population
1·70 million slovac inhabitants 47·7 % of the total population of the ter
To Austria
Remained to Hungary
35.756 square miles 32·7 % of its territory.
7·60 million inhabitants 41·6 % of its population.
6·72 million magyars 67·5 % of the whole magyar population.
0·26 million slovaks, rumanians, serbs and croats.
Budapest
To Jugoslavia
To Jugoslavia
8.120 square miles 7·4 % of the whole territory of Hungary.
1·52 million inhabitants 8·3 % of the whole population of Hungary.
0·46 million magyar inhabitants 4·6 % of the whole magyar population of Hungary.
0·38 million serbian inhabitants (in the eastern territory) 24·9 % of the total population of the territory given to Jugoslavia.
0·15 million croatian and vend inhabitants (in the western territory) 10 % of population of the territory given to Jugoslavia.
MAGYARORSZÁG hegy- és vízrajzi térképe
MOUNTAINS AND PLAINS OF THE OLD KINGDOM OF HUNGARY
TO CZECHO-SLOVAKIA
TO RUMANIA
TO JUGOSLAVIA

Figure 79

(overlay shown)

MAGYARORSZÁG NÉPRAJZI TÉRKÉPE
A NÉPSŰRŰSÉG ALAPJÁN
SZERKESZTETTE
GRÓF TELEKI PÁL
Minden színezett négyszögmilliméter 100 lakost jelent.
Az 1910. évi népszámlálás alapján.
ETHNOGRAPHICAL MAP OF HUNGARY
BASED ON
DENSITY OF POPULATION
BY
COUNT PAUL TELEKI
PROFESSOR OF GEOGRAPHY
Every square millimeter coloured indicates 100 inhabitants.
According to the census of 1910.
MAGYARORSZÁG
hegy- és vízrajzi térképe
MOUNTAINS AND PLAINS OF THE OLD KINGDOM OF HUNGARY
Budapest

Figure 79

(overlay removed)

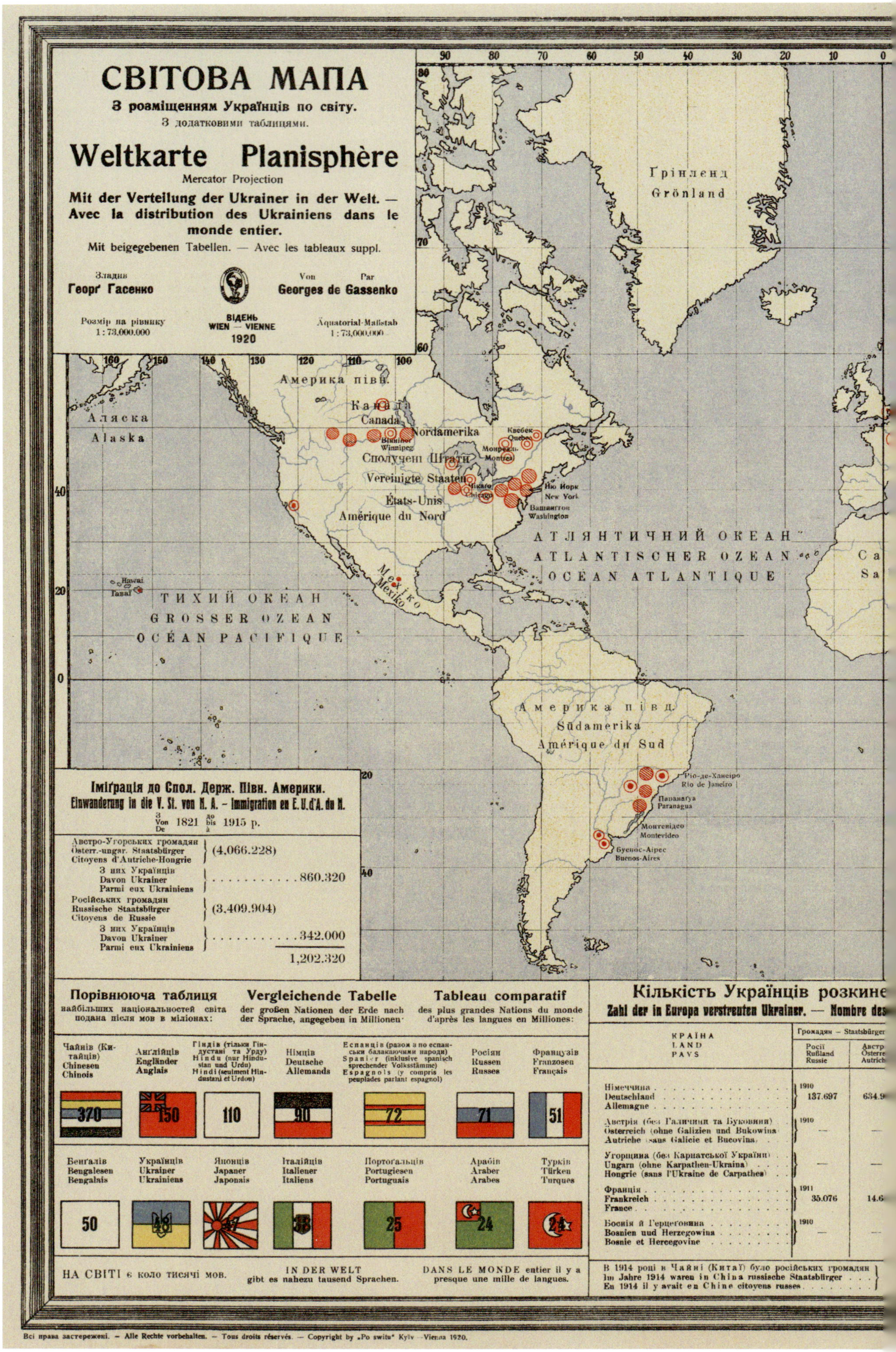

СВІТОВА МАПА
З розміщенням Українців по світу.
З додатковими таблицями.
Weltkarte Planisphère
Mercator Projection
Mit der Verteilung der Ukrainer in der Welt. — Avec la distribution des Ukrainiens dans le monde entier.
Mit beigegebenen Tabellen. — Avec les tableaux suppl.
Зладив Георґ Гасенко
Von Par Georges de Gassenko
Розмір па рівнику 1 : 73.000.000
ВІДЕНЬ WIEN — VIENNE 1920
Äquatorial-Maßstab 1 : 73.000.000
Ґрінленд
Grönland
Америка півн.
Канада
Canada
Аляска
Alaska
Вінніпеґ
Winnipeg
Nordamerika
Квебек
Quebec
Монреаль
Montreal
Сполучені Штати
Vereinigte Staaten
Чікаґо
Chicago
Ню Йорк
New York
Вашинґтон
Washington
États-Unis
Amérique du Nord
АТЛЯНТИЧНИЙ ОКЕАН
ATLANTISCHER OZEAN
OCÉAN ATLANTIQUE
Мексико
Mexiko
Гаваї
ТИХИЙ ОКЕАН
GROSSER OZEAN
OCÉAN PACIFIQUE
Америка півд.
Südamerika
Amérique du Sud
Ріо-де-Жанейро
Rio de Janeiro
Паранаґуа
Paranagua
Монтевідео
Montevideo
Буенос-Айрес
Buenos-Aires
Імігpація до Спол. Держ. Півн. Америки.
Einwanderung in die V. St. von N. A. — Immigration en E. U. d'A. du N.
З Von De 1821 до bis à 1915 р.
Австро-Угорських громадян / Österr.-ungar. Staatsbürger / Citoyens d'Autriche-Hongrie (4,066.228)
З них Українців / Davon Ukrainer / Parmi eux Ukrainiens 860.320
Російських громадян / Russische Staatsbürger / Citoyens de Russie (3.409.904)
З них Українців / Davon Ukrainer / Parmi eux Ukrainiens 342.000
1,202.320
Порівнююча таблиця найбільших національностей світа подана після мов в мілїонах:
Vergleichende Tabelle der großen Nationen der Erde nach der Sprache, angegeben in Millionen
Tableau comparatif des plus grandes Nations du monde d'après les langues en Millionen:
Чайнів (Китайців) Chinesen Chinois 370
Анґлійців Engländer Anglais 150
Гіндів (тільки Гіндустані та Урду) Hindu (nur Hindustan und Urdu) Hindi (seulment Hindustani et Urdou) 110
Німців Deutsche Allemands 90
Еспанців (разом з по еспанськи балакаючими народи) Spanier (inklusive spanisch sprechender Volksstämme) Espagnols (y compris les peuplades parlant espagnol) 72
Росіян Russen Russes 71
Французів Franzosen Français 51
Беньґалів Bengalesen Bengalais 50
Українців Ukrainer Ukrainiens 46
Японців Japaner Japonais 47
Італійців Italiener Italiens 38
Портоґальців Portugiesen Portuguais 25
Арабів Araber Arabes 24
Турків Türken Turques 24
НА СВІТІ є коло тисячі мов.
IN DER WELT gibt es nahezu tausend Sprachen.
DANS LE MONDE entier il y a presque une mille de langues.
Кількість Українців розкине
Zahl der in Europa verstreuten Ukrainer. — Nombre des
КРАЇНА LAND PAYS
Громадян — Staatsbürger
Росії Rußland Russie
Австр. Österr. Autrich.
Німеччина Deutschland Allemagne 1910 137.697
Австрія (без Галичини та Буковини) Österreich (ohne Galizien und Bukowina) Autriche (sans Galicie et Bucovina) 1910 —
Угорщина (без Карпатської України) Ungarn (ohne Karpathen-Ukraina) Hongrie (sans l'Ukraine de Carpathes) —
Франція Frankreich France 1911 35.076
Боснія й Герцеґовина Bosnien und Herzegowina Bosnie et Hercegovine 1910 —
В 1914 році в Чайні (Китаї) було російських громадян
Im Jahre 1914 waren in China russische Staatsbürger
En 1914 il y avait en Chine citoyens russes
Всі права застережені. — Alle Rechte vorbehalten. — Tous droits réservés. — Copyright by „Po switu" Kyiv — Vienna 1920.

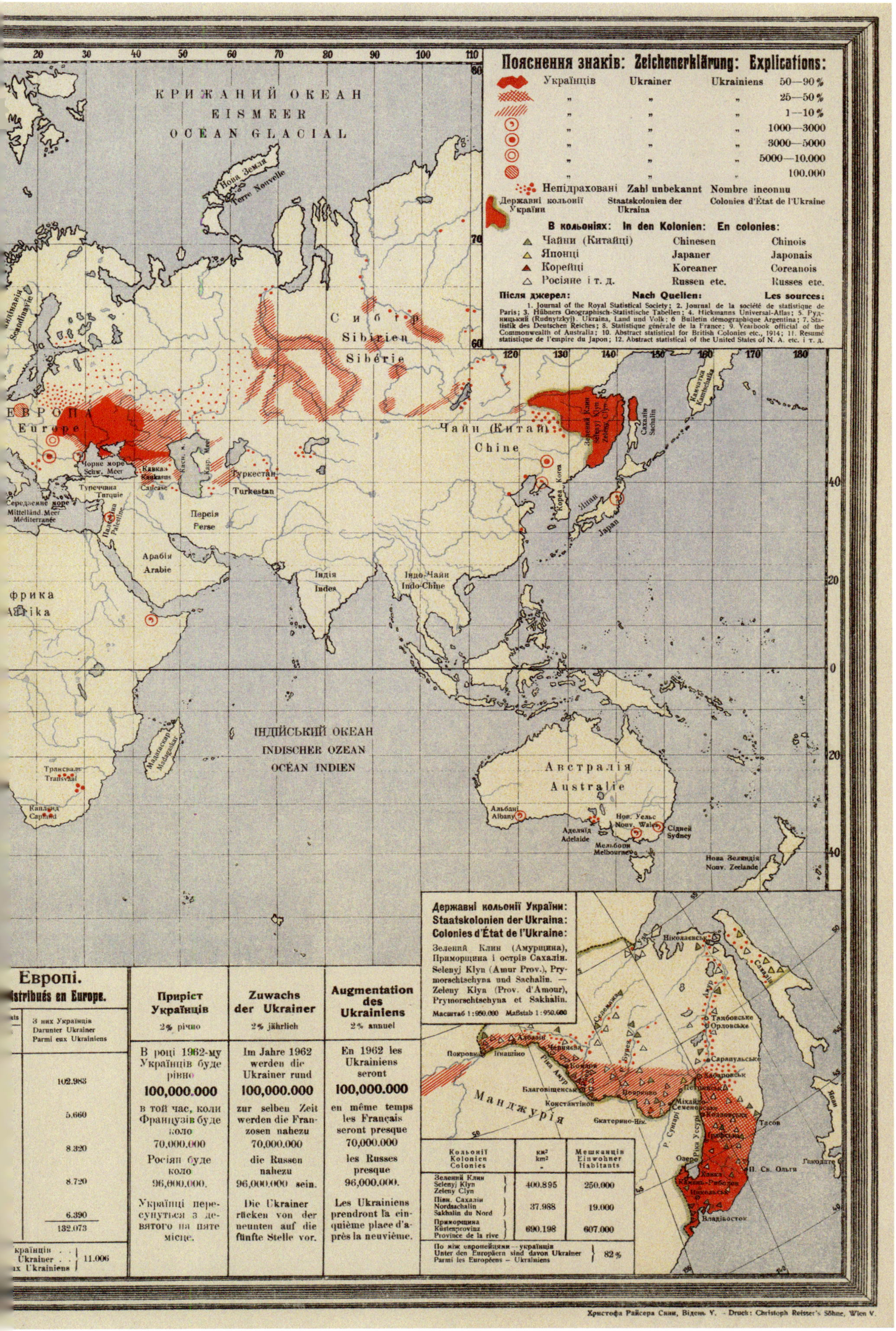
Пояснення знаків: Zeichenerklärung: Explications:
Українців Ukrainer Ukrainiens 50—90 %
25—50 %
1—10 %
1000—3000
3000—5000
5000—10.000
100.000
Непідраховані Zahl unbekannt Nombre inconnu
Державні кольонії України Staatskolonien der Ukraina Colonies d'État de l'Ukraine
В кольоніях: In den Kolonien: En colonies:
Чайни (Китайці) Chinesen Chinois
Японці Japaner Japonais
Корейці Koreaner Coreanois
Росіяне і т. д. Russen etc. Russes etc.
Після джерел: Nach Quellen: Les sources:
КРИЖАНИЙ ОКЕАН
EISMEER
OCEAN GLACIAL
Сибір
Sibirien
Sibérie
ЕВРОПА
Europe
Чайна (Китай)
Chine
Туркестан
Turkestan
Персія
Perse
Арабія
Arabie
Індія
Indes
Індо-Чайна
Indo-Chine
ІНДІЙСЬКИЙ ОКЕАН
INDISCHER OZEAN
OCÉAN INDIEN
Австралія
Australie
Альбані
Albany
Аделяїд
Adelaide
Мельбоурн
Melbourne
Сідней
Sydney
Нова Зеляндія
Nouv. Zeelande
Трансваль
Transvaal
Капланд
Capland
Африка
Afrika
Європі.
Приріст Українців 2% річно
Zuwachs der Ukrainer 2% jährlich
Augmentation des Ukrainiens 2% annuel
В році 1962-му Українців буде рівно 100,000.000 в той час, коли Французів буде коло 70,000.000 Росіян буде коло 96,000.000. Українці пересунуться з девятого на пяте місце.
Im Jahre 1962 werden die Ukrainer rund 100,000.000 zur selben Zeit werden die Franzosen nahezu 70,000.000 die Russen nahezu 96,000.000 sein. Die Ukrainer rücken von der neunten auf die fünfte Stelle vor.
En 1962 les Ukrainiens seront 100,000.000 en même temps les Français seront presque 70,000.000 les Russes presque 96,000.000. Les Ukrainiens prendront la cinquième place d'après la neuvième.
Державні кольонії України: Staatskolonien der Ukraina: Colonies d'État de l'Ukraine:
Зелений Клин (Амурщина), Приморщина і острів Сахалін. Selenyj Klyn (Amur Prov.), Prymorschtschyna und Sachalin. — Zeleny Klyn (Prov. d'Amour), Prymorschtschyna et Sakhalin.
Маньджурія
Владівосток
Кольонії Kolonien Colonies | км² km² | Мешканців Einwohner Habitants
Зелений Клин Selenyj Klyn Zeleny Clyn | 400.895 | 250.000
Півн. Сахалін Nordsachalin Sakhalin du Nord | 37.968 | 19.000
Приморщина Küstenprovinz Province de la rive | 690.198 | 607.000
По між европейцями — українців Unter den Europäern sind davon Ukrainer Parmi les Européens — Ukrainiens | 82 %
Христофа Райсера Сини, Відень V. — Druck: Christoph Reisser's Söhne, Wien V.

Figure 80

Україна
в числах зладив Г. ГАСЕНКО.
Українські національні барви
Die ukrainischen Nationalfarben
1593
1657
Гетьман Богдан Хмельницький
Hetjman Bohdan Chmelnyćkyj
Українська мова — є окремішньою славянською мовою,
Die ukrainische Sprache ist eine selbständige slawische Sprache,
ІСТОРІЯ.
УКРАЇНА
DIE UKRAIN
9. (22. Січня) 1918
Антропольоґічна табеля по акаде
Anthropologische Tabelle nach Aka
ТЕРИТОРІЯ УКРАЇНИ.
10 держав — взятих разом, займають такий великий простір, як одна Україна. Україна — є по Московщині найбільшим краєм Европи.
DAS GEBIET DER UKRAINA.
10 Staaten zusammen umfassen ein ebenso großes Gebiet wie die Ukraine allein. Die Ukraina ist nach Rußland das ausgedehnteste Land Europas.
УКРАЇНА
850.000 km²
UKRAINA
ФРАНЦІЯ
536.000 km²
FRANKREICH
Велико-Британія
315.000 km²
Großbritannien mit Irland
ІТАЛІЯ
287.000 km²
ITALIEN
Продукція збіжжя України й колишньої російської імперії — Getreideproduktion der Ukraina im Vergleiche mit derjenigen des ehemaligen Russischen Reiches (ohne die Ukraina)
УКРАЇНА - UKRAINA 44%
РОСІЯ (із ПОЛЬЩЕЮ)
RUSSLAND (MIT POLEN) 56%
850.000 km²
Простір України
Gebiet der Ukraina
22,360.000 km²
Простір колишньої Росії
Gebiet des ehemaligen Rußland
Україна — найбогатший у збіжжя край в Европі
Die Ukraina ist das getreidereichste Land von Europa
Офіціяльний державний б
Offizielles Staatsbudget des russ
10,833.000 Rubel
Прибуток - Gewinn
РОСІЯ з УКРАЇНОЮ
Продукція тютюну України й колишньої Росії.
Tabakproduktion der Ukraina im Vergleiche mit derjenigen im ehemaligen Russischen Reiche.
69%
УКРАЇНА - UKRAINA
31%
РОСІЯ (із Польщею)
Rußland (mit Polen)
Продукція соли України й колишньої Росії.
Salzproduktion der Ukraina im Vergleiche mit derjenigen im ehemaligen Russischen Reiche.
УКРАЇНА
UKRAINA
53%
47%
РОСІЯ (із Польщею)
Rußland (mit Polen)
Продукція антрациту України й колишньої Росії.
Anthrazitproduktion der Ukraina im Vergleiche mit derjenigen im ehemaligen Russischen Reiche.
УКРАЇНА - UKRAINA
99%
РОСІЯ (із Польщею)
Rußland (mit Polen)
1%
Продукція коксу України й колишньої Росії.
Koksproduktion in der Ukraina im Vergleiche mit derjenigen im ehemaligen Russischen Reiche.
УКРАЇНА - UKRAINA
99%
РОСІЯ (із Польщею)
Rußland (mit Polen)
1%
УКРАЇНА
Густота населення на 1 km².
Bevölkerungsdichte auf 1 km².
53 мешканців на 1 km² Einwohner auf 1 km²
УКРАЇНА - UKRAINA
74 мешканців на 1 km² Einwohner auf 1 km²
ФРАНЦІЯ - FRANKREICH
144 мешканців на 1 km² Einwohner auf 1 km²
В.-БРИТАНІЯ - GROSSBRITANNIEN
275 мешканців на 1 km² Einwohner auf 1 km²
БЕЛЬГІЯ BELGIEN
Вивіз збіжжя із
Getreideexport aus dem eh
Всі права застережені. – Alle Rechte vorbehalten. – Tous droits réservés. – Copyright by „Po switu" Kyiv – Vienna 1920.

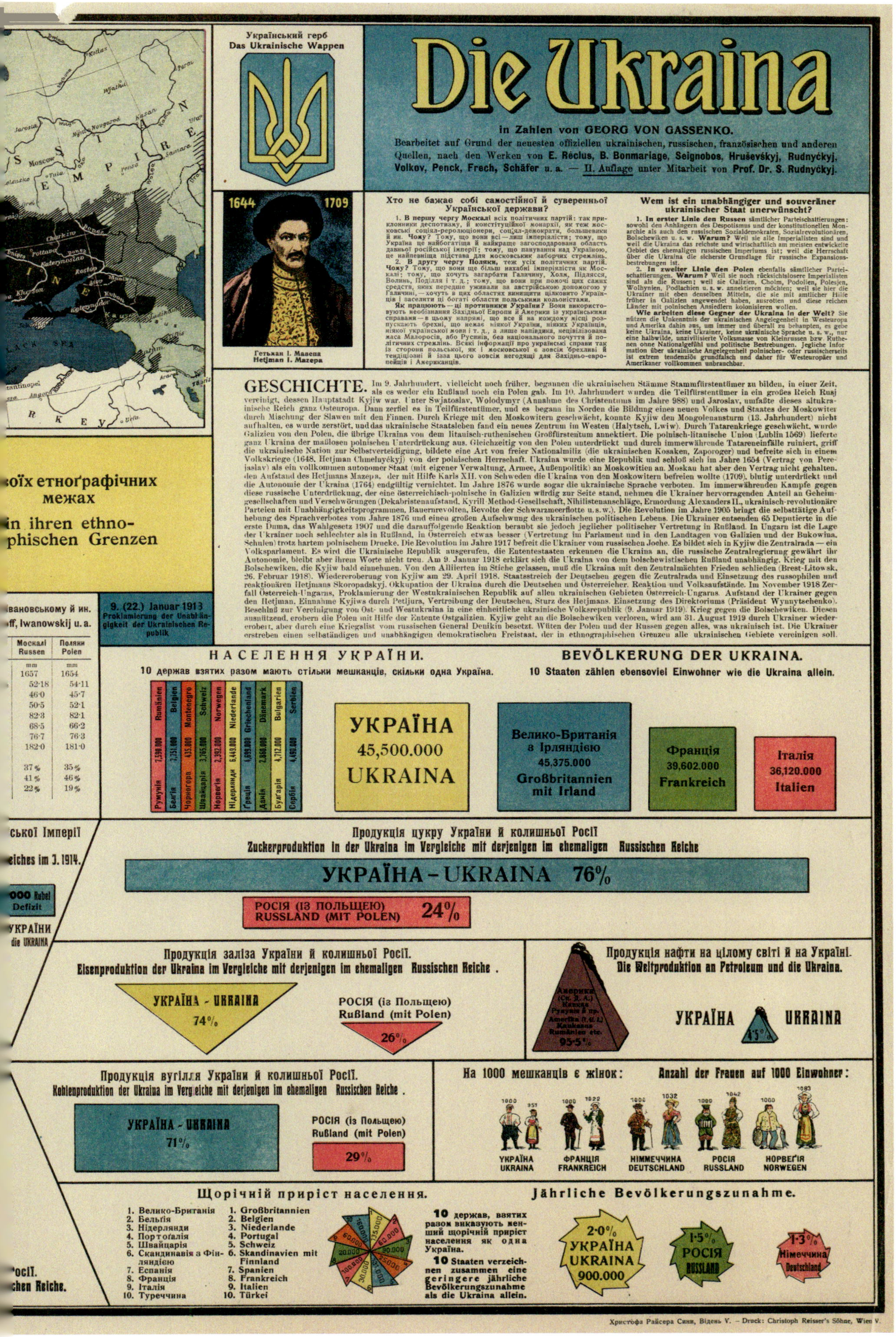
Український герб
Das Ukrainische Wappen
Die Ukraina
in Zahlen von GEORG VON GASSENKO.
1644 1709
Гетьман І. Мазепа
Hetjman I. Mazepa
Хто не бажає собі самостійної й суверенної Української держави?
Wem ist ein unabhängiger und souveräner ukrainischer Staat unerwünscht?
GESCHICHTE.
НАСЕЛЕННЯ УКРАЇНИ.
BEVÖLKERUNG DER UKRAINA.
10 Staaten zählen ebensoviel Einwohner wie die Ukraina allein.
УКРАЇНА
45,500.000
UKRAINA
Велико-Британія з Ірляндією
45.375.000
Großbritannien mit Irland
Франція
39,602.000
Frankreich
Італія
36,120.000
Italien
Продукція цукру України й колишньої Росії.
Zuckerproduktion in der Ukraina im Vergleiche mit derjenigen im ehemaligen Russischen Reiche
УКРАЇНА - UKRAINA 76%
РОСІЯ (ІЗ ПОЛЬЩЕЮ)
RUSSLAND (MIT POLEN) 24%
Продукція заліза України й колишньої Росії.
Eisenproduktion der Ukraina im Vergleiche mit derjenigen im ehemaligen Russischen Reiche.
УКРАЇНА - UKRAINA
74%
РОСІЯ (із Польщею)
Rußland (mit Polen)
26%
Продукція нафти на цілому світі й на Україні.
Die Weltproduktion an Petroleum und die Ukraina.
УКРАЇНА
UKRAINA
Продукція вугілля України й колишньої Росії.
Kohlenproduktion der Ukraina im Vergleiche mit derjenigen im ehemaligen Russischen Reiche.
УКРАЇНА - UKRAINA
71%
РОСІЯ (із Польщею)
Rußland (mit Polen)
29%
На 1000 мешканців є жінок:
Anzahl der Frauen auf 1000 Einwohner:
УКРАЇНА UKRAINA
ФРАНЦІЯ FRANKREICH
НІММЕЧЧИНА DEUTSCHLAND
РОСІЯ RUSSLAND
НОРВЕҐІЯ NORWEGEN
Щорічний приріст населення.
Jährliche Bevölkerungszunahme.
2·0%
УКРАЇНА
UKRAINA
900.000
1·5%
РОСІЯ
1·3%
Німеччина
Deutschland
Христофа Райсера Сини, Відень V. – Druck: Christoph Reisser's Söhne, Wien V.

Figure 81

371
UNION OF SOVIET SOCIALIST REPUBLICS IN EUROPE
THE GEORGE F. CRAM COMPANY
Atlas Headquarters—Est. 1867
INDIANAPOLIS
COPYRIGHT
SCALE OF MILES
ARCTIC OCEAN
KARA SEA
WHITE SEA
BALTIC SEA
GULF OF BOTHNIA
GULF OF FINLAND
GULF OF OB
RUSSIAN SOCIALIST F. S. R.
URAL MOUNTAINS
NOVAYA ZEMLYA
ARKHANGELSK
MURMANSK
KAREL
LOFOTEN ISLANDS
ESTONIA
LATVIA
Longitude East from Greenwich

CASPIAN SEA
(85.5 ft. below Sea Level)
BLACK SEA
SEA OF AZOV
TURKEY
ROMANIA
BULGARIA
POLAND
WHITE RUSSIAN S.S.R.
UKRAINIAN S.S.R.
TRANSCAUCASIAN S.F.S.R.
STALINGRAD
ASTRAKHAN
KALMYC
STAVROPOL
DAGESTAN
Mt. Ararat 16,925
Longitude East from Greenwich

Figure 82

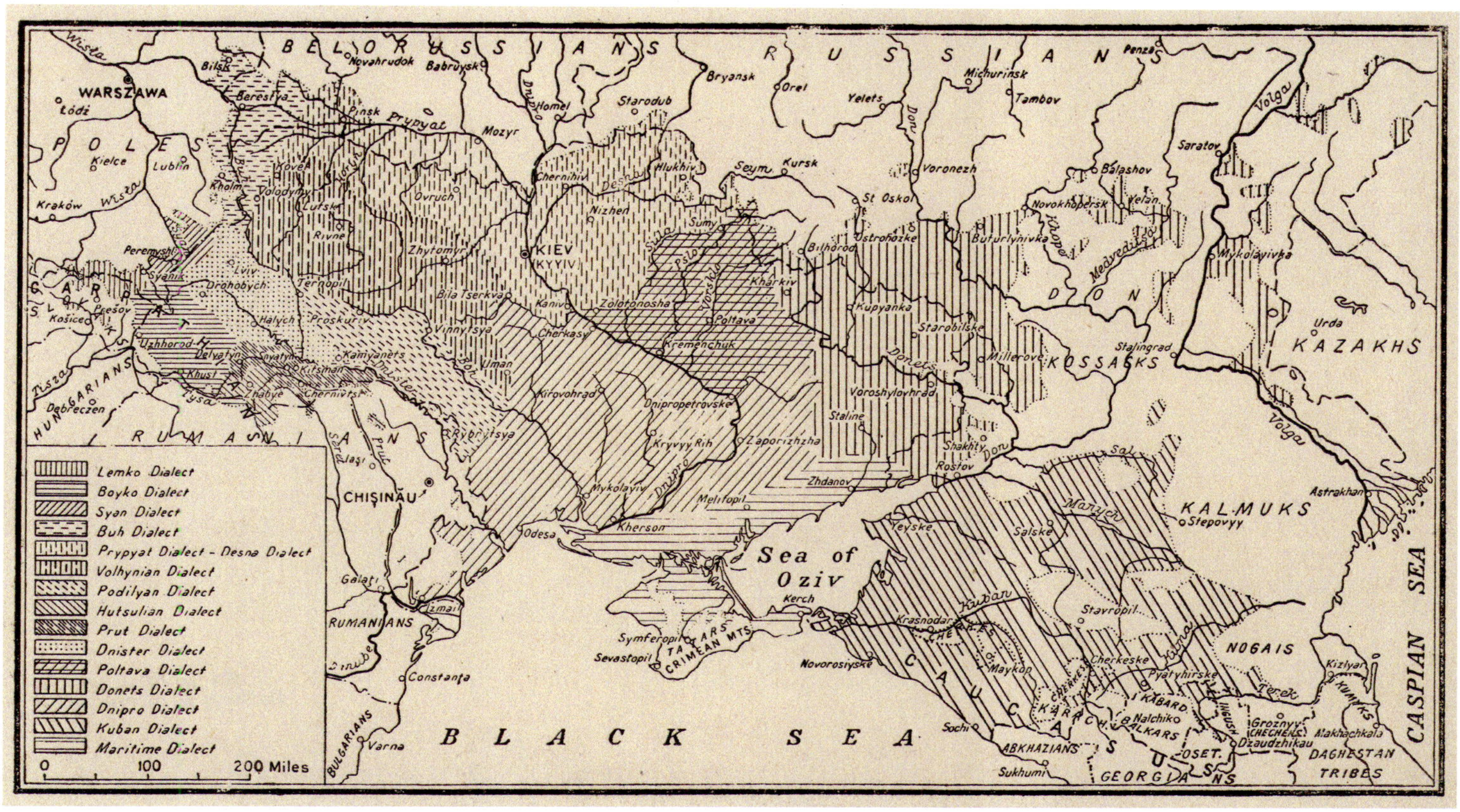
BELORUSSIANS
RUSSIANS
POLES
WARSZAWA
KIEV
KYYIV
KOSSACKS
KAZAKHS
KALMUKS
NOGAIS
RUMANIANS
HUNGARIANS
CHIŞINĂU
Sea of Oziv
BLACK SEA
CASPIAN SEA
CRIMEAN MTS.
TATARS
ABKHAZIANS
GEORGIANS
DAGHESTAN TRIBES
BULGARIANS
Lemko Dialect
Boyko Dialect
Syan Dialect
Buh Dialect
Prypyat Dialect - Desna Dialect
Volhynian Dialect
Podilyan Dialect
Hutsulian Dialect
Prut Dialect
Dnister Dialect
Poltava Dialect
Donets Dialect
Dnipro Dialect
Kuban Dialect
Maritime Dialect
0 100 200 Miles

Figure 83

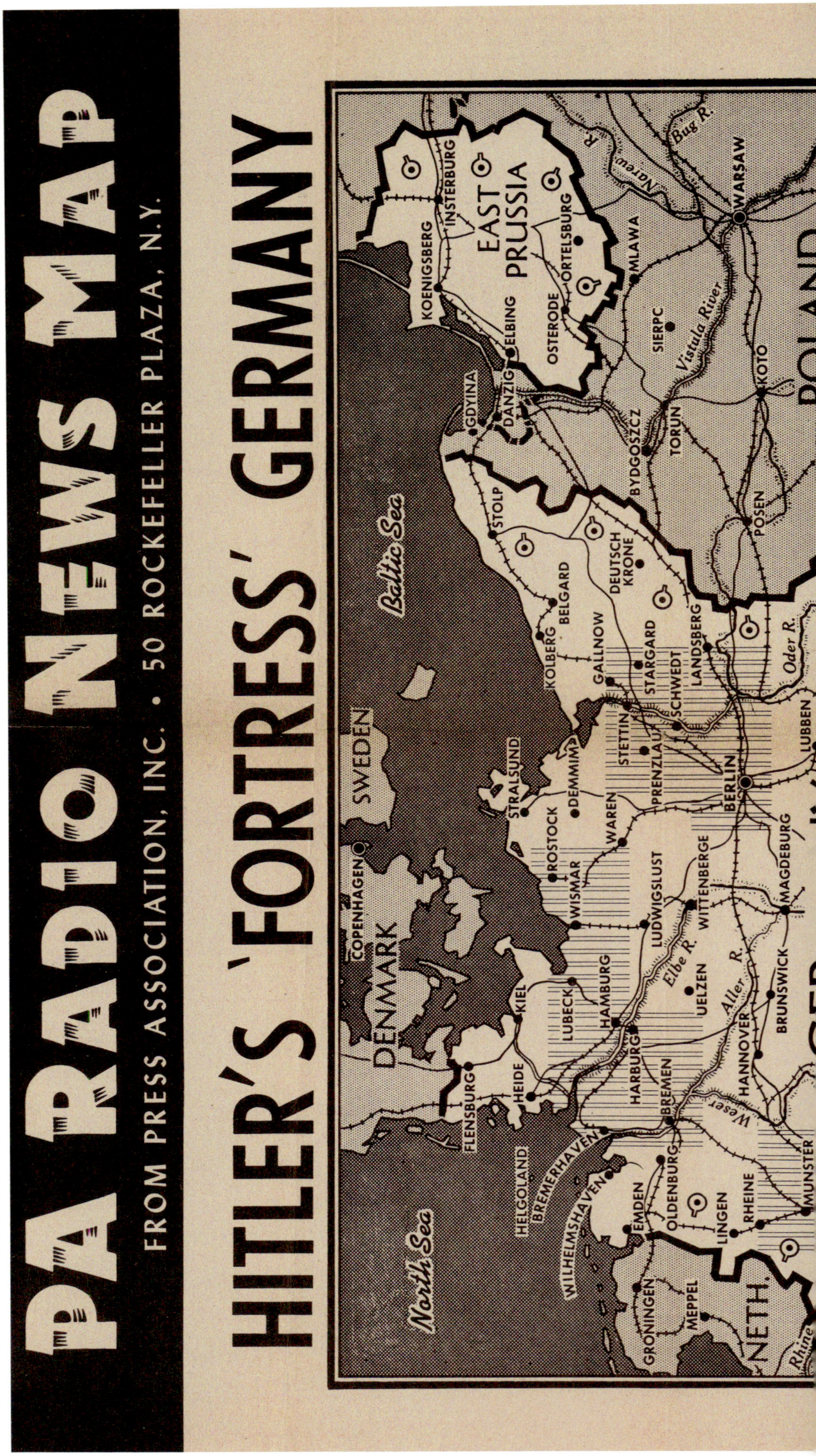
PA RADIO NEWS MAP
FROM PRESS ASSOCIATION, INC. • 50 ROCKEFELLER PLAZA, N.Y.
HITLER'S 'FORTRESS' GERMANY
North Sea
Baltic Sea
DENMARK
COPENHAGEN
SWEDEN
EAST PRUSSIA
POLAND
NETH.
KOENIGSBERG
INSTERBURG
ORTELSBURG
OSTERODE
ELBING
MLAWA
WARSAW
SIERPC
Vistula River
Narew R.
Bug R.
KOTO
TORUN
BYDGOSZCZ
POSEN
GDYINA
DANZIG
STOLP
DEUTSCH KRONE
BELGARD
KOLBERG
GALLNOW
STARGARD
LANDSBERG
SCHWEDT
Oder R.
STETTIN
PRENZLAU
LUBBEN
BERLIN
STRALSUND
DEMMIM
ROSTOCK
WAREN
WISMAR
LUDWIGSLUST
WITTENBERGE
MAGDEBURG
KIEL
LUBECK
HAMBURG
Elbe R.
UELZEN
Aller R.
BRUNSWICK
HARBURG
HANNOVER
FLENSBURG
HEIDE
BREMEN
Weser
HELGOLAND
BREMERHAVEN
WILHELMSHAVEN
EMDEN
OLDENBURG
LINGEN
RHEINE
MUNSTER
GRONINGEN
MEPPEL
Rhine

WITH Nazi Lebensraum shrinking daily, the Battle of Germany looms as perhaps the climactic episode of the Hitler saga. Battle lines already are at Germany's September 1939 boundaries—but with traffic headed the other way and the armies of the Reich as outclassed as were their opponents in the days of the surprise Blitzkrieg.

What are Germany's defensive resources?

In the first place, the Reich when shorn of Austria and Czechoslovakia, is smaller than Texas. It is almost totally without natural oil sources. Though highly industrialized, Germany's factories are concentrated in the Ruhr area, near the new western front, and in Silesia which is close to the advancing Russians. For nearly five years Reich cities in all sections have been subjected to air attacks.

DISREGARDING the apparent break between Hitler's Nazi party and the Junkers generals, another great cleavage within Germany argues against that nation's power to resist. This is represented by 14,000,000 foreign "slave" laborers impressed for war work. Poles, Russians, Frenchmen, Czechs—all with reason to hate their masters—they are the time bomb which showdown battles may touch off.

On Germany's western border the Siegfried Line offers little hope to the Herrenvolk behind it. That it should exceed in performance the French Maginot Line of 1940 does not seem likely in view of Allied strength. However, a major German stand is foreseen in this region of the great fortified lines—fortified in depth as contrasted with the narrow Maginot Line ramparts.

ON THE south, mountain barriers promise greater immunity from invasion. The Alps, broken by the heavily protected Brenner Pass, extend from Switzerland to the Danube, and the Carpathians offer a natural barrier along the old Czech border.

Germany's eastern frontier, however, is on the plain which has made Poland so frequently a battlefield. The "sacred soil" of East Prussia is defended by strong fortifications between the lakes of that region. But the frontier of Germany proper is long and without natural features. Here the Nazis have been reported strengthening old positions, though some experts believe they will depend chiefly on the line of the Oder River—if they hold out that long. The Oder is less than 50 miles from Berlin.

—*Robert N. Cool*

Figure 84

RUSSIA
IN EUROPE AND
ITS BORDER STATES
ARCTIC OCEAN
SIBERIA
NOVAYA ZEMLYA
Kara Sea
WHITE SEA
ESTONIA
GULF OF FINLAND
SWEDEN
NORWAY

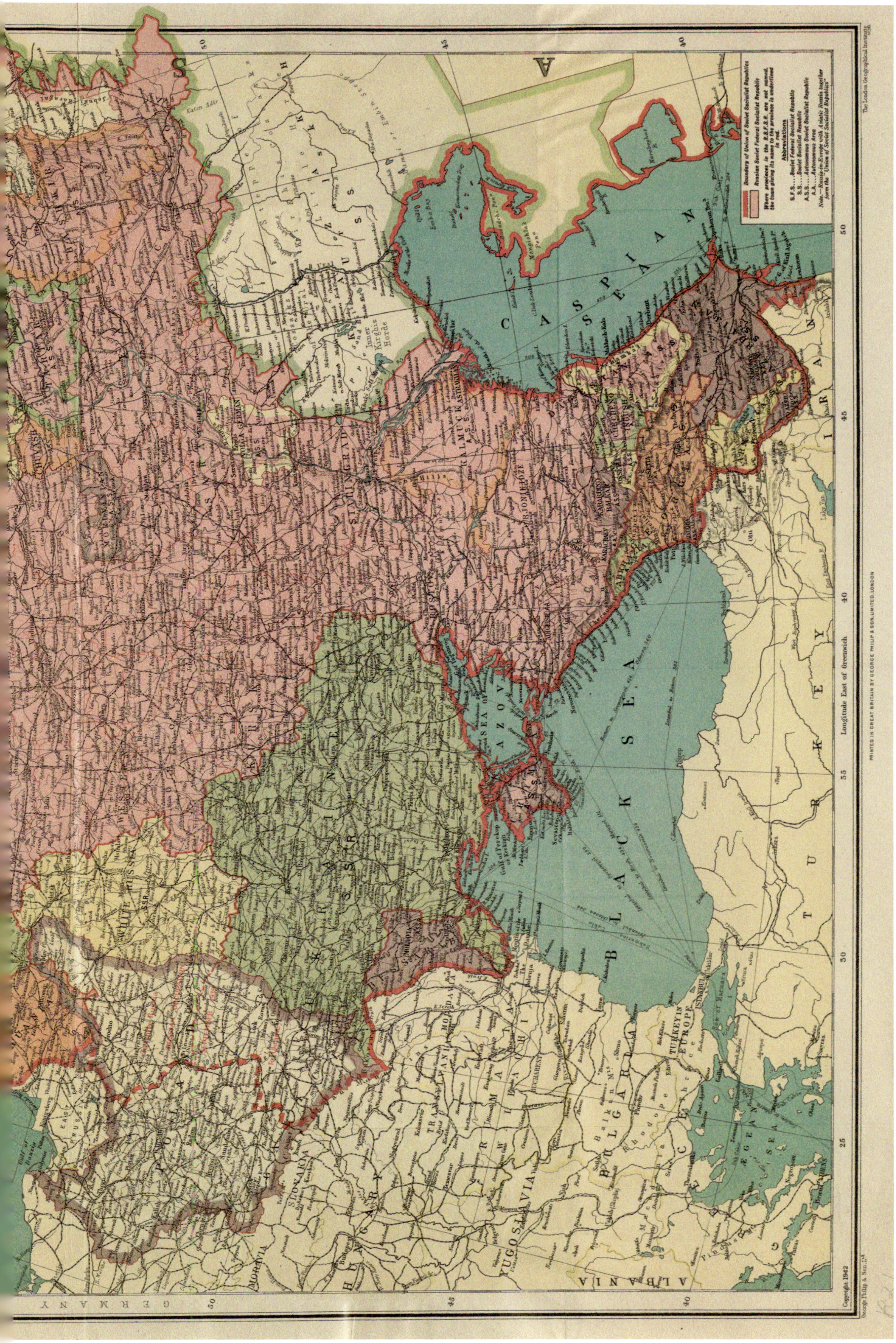
CASPIAN SEA
BLACK SEA
SEA OF AZOV
TURKEY
IRAN
BULGARIA
RUMANIA
YUGOSLAVIA
ALBANIA
HUNGARY
SLOVAKIA
MORAVIA
GERMANY
UKRAINE S.S.R.
WHITE RUSSIA S.S.R.
Longitude East of Greenwich
PRINTED IN GREAT BRITAIN BY GEORGE PHILIP & SON, LIMITED, LONDON

Figure 85

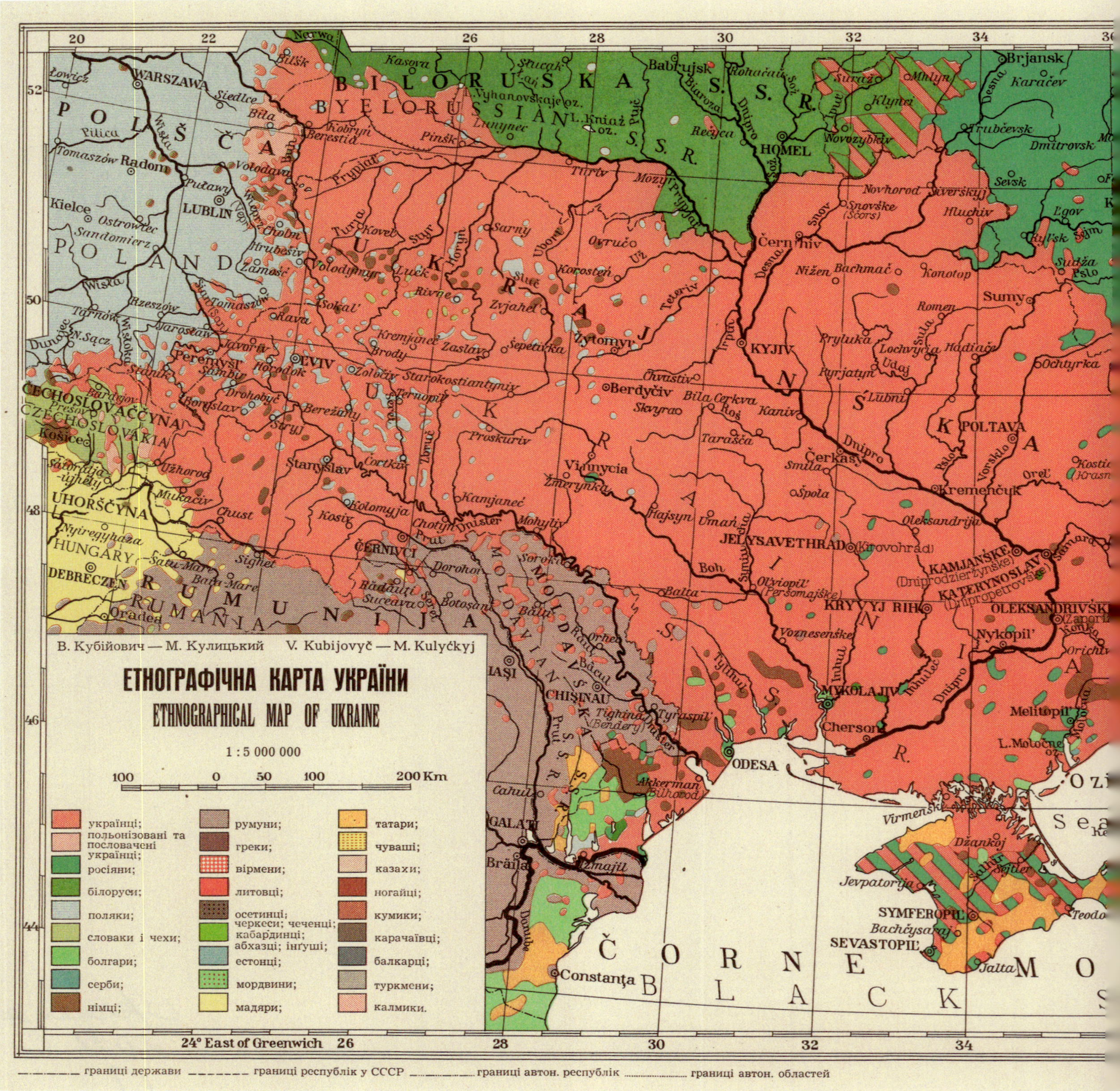
В. Кубійович — М. Кулицький V. Kubijovyč — M. Kulyćkyj
ЕТНОГРАФІЧНА КАРТА УКРАЇНИ
ETHNOGRAPHICAL MAP OF UKRAINE
1 : 5 000 000
100 0 50 100 200 Km
українці;
польонізовані та пословачені українці;
росіяни;
білоруси;
поляки;
словаки і чехи;
болгари;
серби;
німці;
румуни;
греки;
вірмени;
литовці;
осетинці;
черкеси; чеченці; кабардинці; абхазці; інгуші;
естонці;
мордвини;
мадяри;
татари;
чуваші;
казахи;
ногайці;
кумики;
карачаївці;
балкарці;
туркмени;
калмики.
24° East of Greenwich
границі держави
границі республік у СССР
границі автон. республік
границі автон. областей

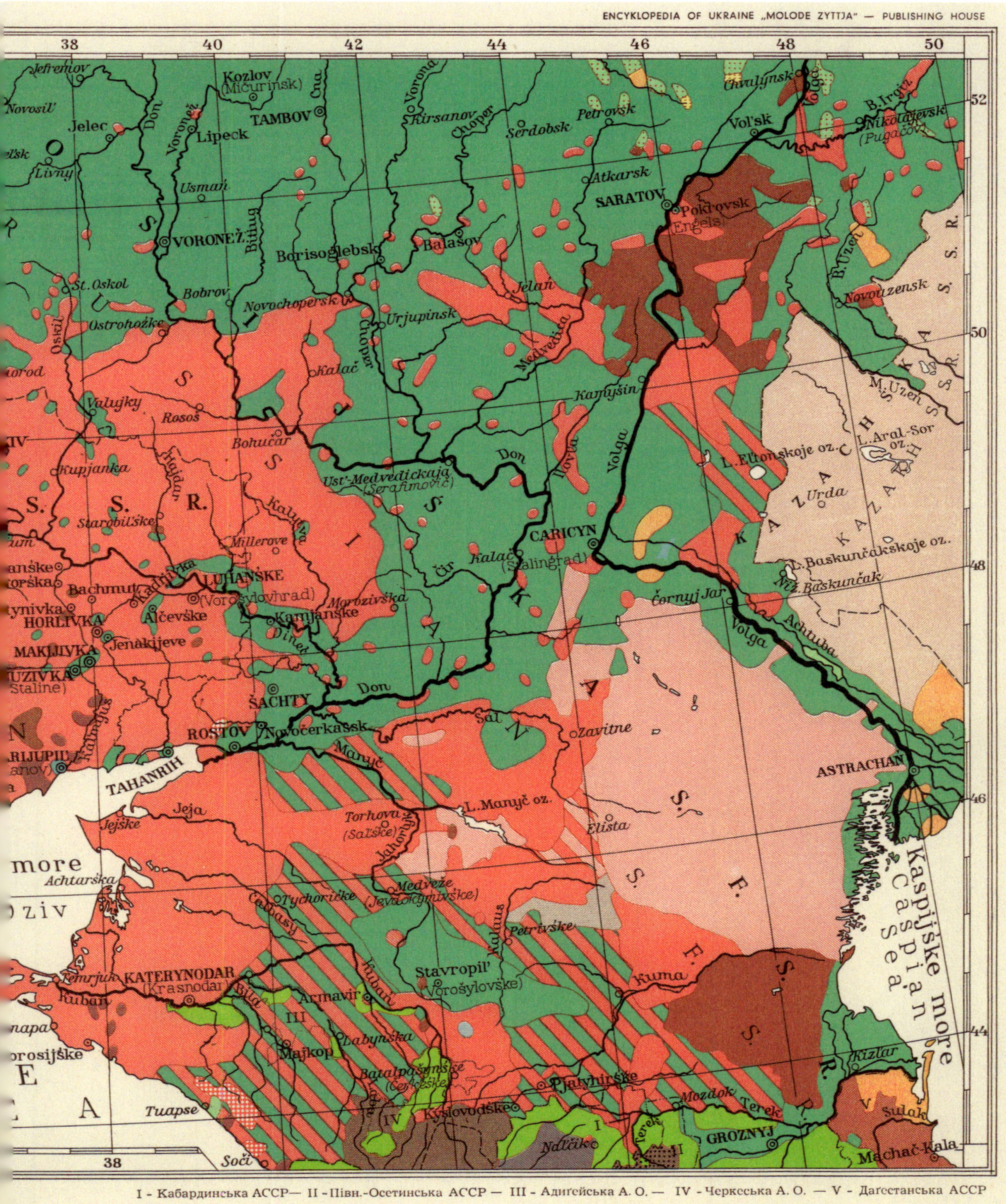
ENCYKLOPEDIA OF UKRAINE „MOLODE ZYTTJA" — PUBLISHING HOUSE
38
40
42
44
46
48
50
Jefremov
Kozlov (Mičurinsk)
Novosil'
Jelec
Lipeck
TAMBOV
Kirsanov
Serdobsk
Petrovsk
Chvalynsk
Vol'sk
Nikolajevsk (Pugačov)
Livny
Usmań
Atkarsk
SARATOV
Pokrovsk (Engels)
VORONEŽ
Borisoglebsk
Balašov
St. Oskol
Bobrov
Novochopersk
Jelań
Novouzensk
Ostrohožke
Urjupinsk
Kalač
Kamyšin
Valujky
Rosoš
Bohučar
Kupjanka
Ust'-Medvedickaja (Serafimovič)
L. Eltonskoje oz.
L. Aral-Sor oz.
Urda
Starobil'ške
Millerove
CARICYN (Stalingrad)
L. Baskunčakskoje oz.
Niž. Baskunčak
Bachmut
LUHANŠKE (Vorošylovhrad)
Černyj Jar
HORLIVKA
Alčevške
Kamjanške
MAKIJIVKA
Jenakijeve
JUZIVKA (Staline)
ŠACHTY
Don
Volga
Achtuba
ROSTOV
Novočerkassk
Zavitne
ASTRACHAN
TAHANRIH
L. Manyč oz.
Torhova (Sal'ške)
Elista
Jejške
Jeja
more
Achtarška
Medveže (Jevdokymivške)
Tychorićke
Petrivške
Kaspijške more
Caspian Sea
KATERYNODAR (Krasnodar)
Armavir
Stavropil' (Vorošylovške)
Kuma
Majkop
Labynska
Batalpašynške (Čerkeške)
Pjatyhirške
Kizlar
Tuapse
Kyslovodške
Mozdok
Terek
Sulak
GROZNYJ
Nal'čik
Soči
Machač-Kala
52
50
48
46
44
I - Кабардинська АССР— II - Півн.-Осетинська АССР — III - Адигейська А. О. — IV - Черкеська А. О. — V - Дагестанська АССР

Figure 86

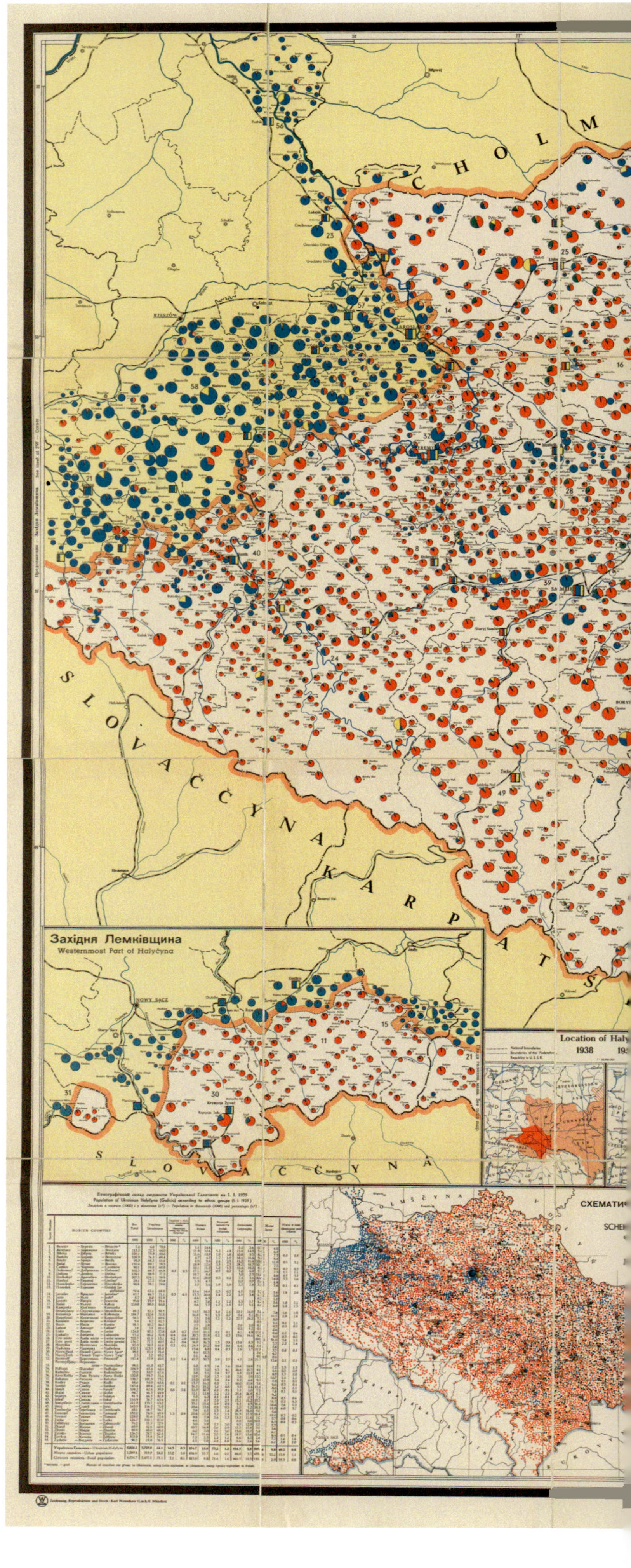
C H O L M
RZESZÓW
S L O V A Č Č Y N A
K A R P A T S
Західня Лемківщина
Westernmost Part of Halyčyna
NOWY SACZ
S L O V A Č Č Y N A
Location of Haly
1938
СХЕМАТИ
SCHE

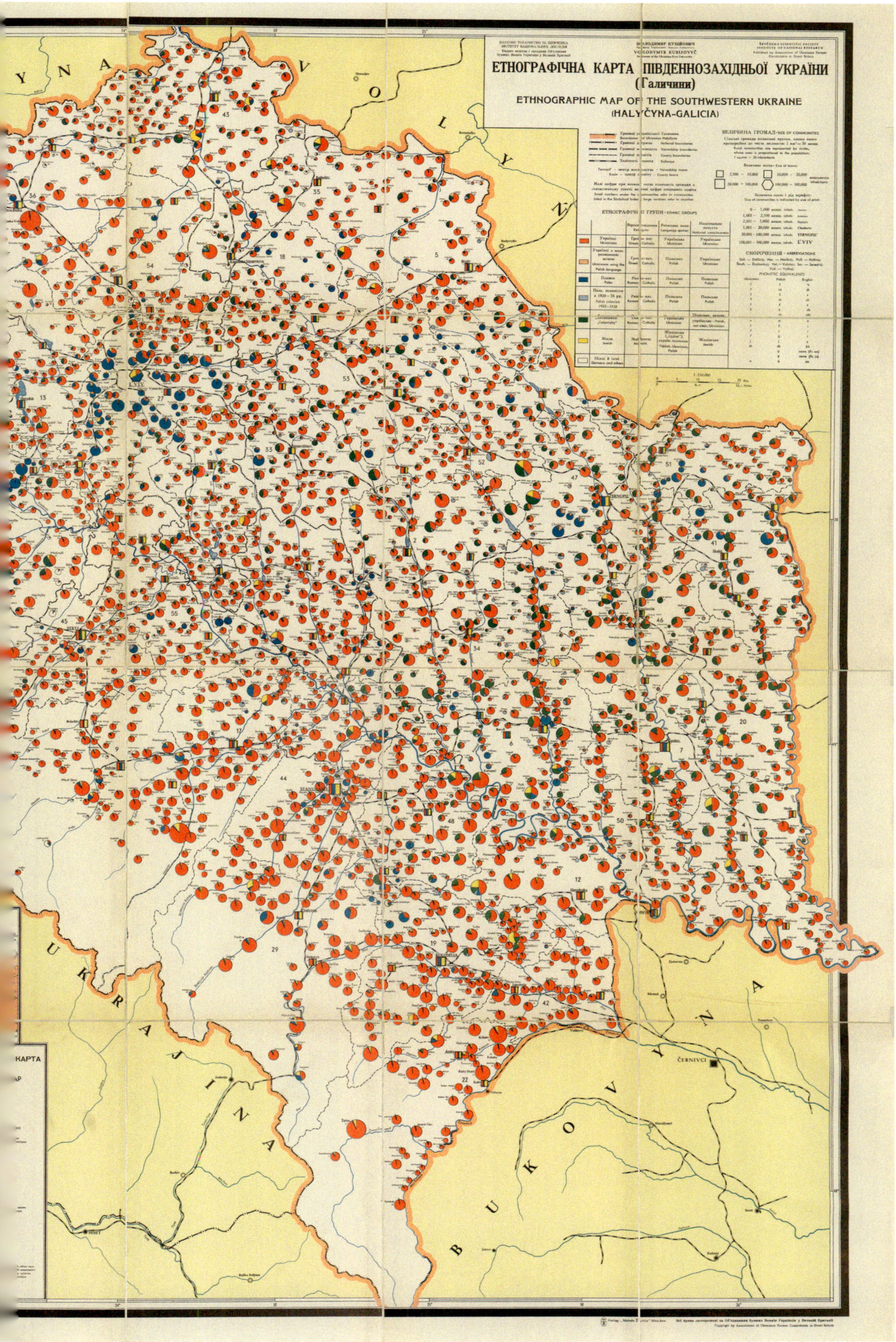
ЕТНОГРАФІЧНА КАРТА ПІВДЕННОЗАХІДНЬОЇ УКРАЇНИ
(Галичини)
ETHNOGRAPHIC MAP OF THE SOUTHWESTERN UKRAINE
(HALYČYNA-GALICIA)
V O L Y N
U K R A J I N A
B U K O V Y N A
ČERNIVCI

Figure 87

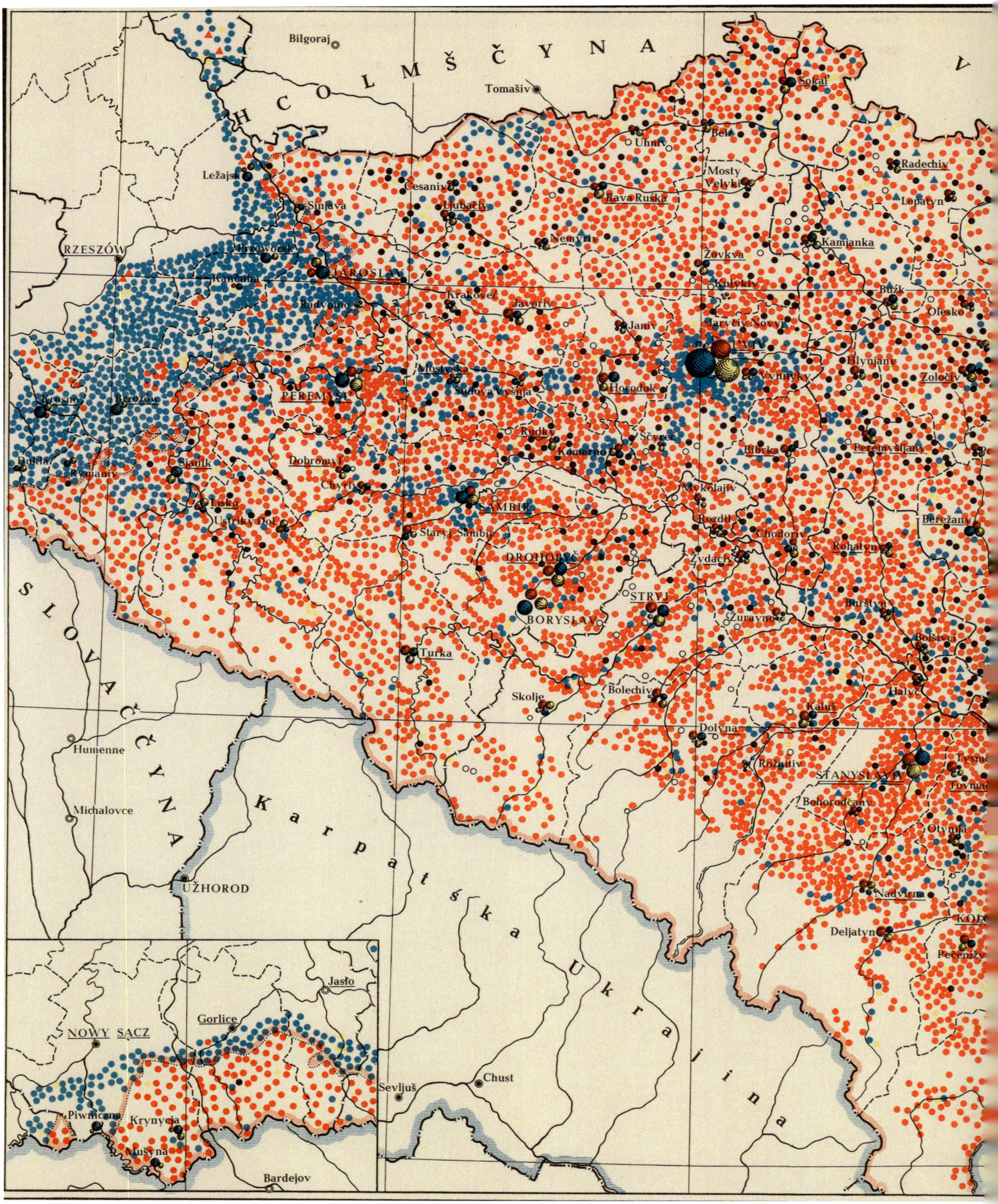
Bilgoraj
CHOLMŠČYNA
Tomašiv
Ležajsk
Sinjava
Cesaniv
Ljubačiv
Rava Ruška
Belz
Uhniv
Sokal
Mosty Velyki
Radechiv
Lopatyn
RZESZÓW
Przeworsk
JAROSLAV
Nemyriv
Kamjanka
Žovkva
Radymno
Krakovec
Javoriv
Janiv
Kulykiv
Busk
Olesko
LVIV
Vynnyky
Hlynjany
Zoločiv
Mostyska
Sudova Vyšnja
Horodok
PEREMYŠL
Krosno
Brzozów
Rudky
Komarno
Ščyrec
Bibrka
Peremyšljany
Dukla
Rymanów
Sjanik
Dobromyl
Chyriv
Lisko
Ustryky Dol.
SAMBIR
Mykolajiv
Rozdil
Chodoriv
Berežany
Staryj Sambir
Žydačiv
Rohatyn
DROHOBYČ
STRYJ
BORYSLAV
Žuravno
Burštyn
Turka
Bolechiv
Skolje
Kaluš
Halyč
Dolyna
Rožnjativ
STANYSLAVIV
Bohorodčany
Tysmenycja
Tovmač
Otynja
Nadvirna
Deljatyn
Pečenižyn
SLOVAČČYNA
Humenne
Michalovce
UŽHOROD
Karpatśka Ukrajina
Chust
Sevljuš
Jasło
Gorlice
NOWY SACZ
Piwniczna
Krynycja
Mušyna
Bardejov

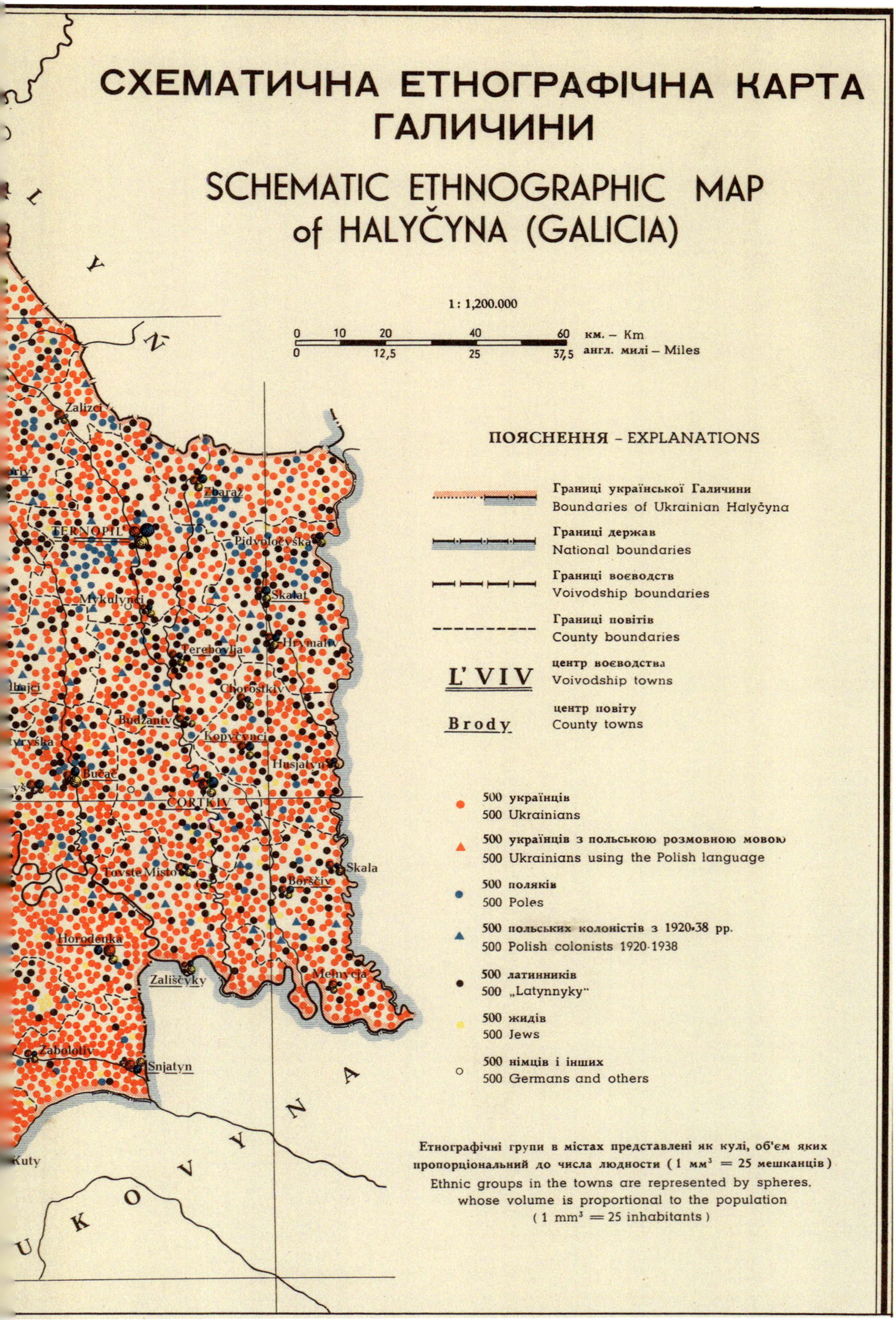

Figure 87

(detail)

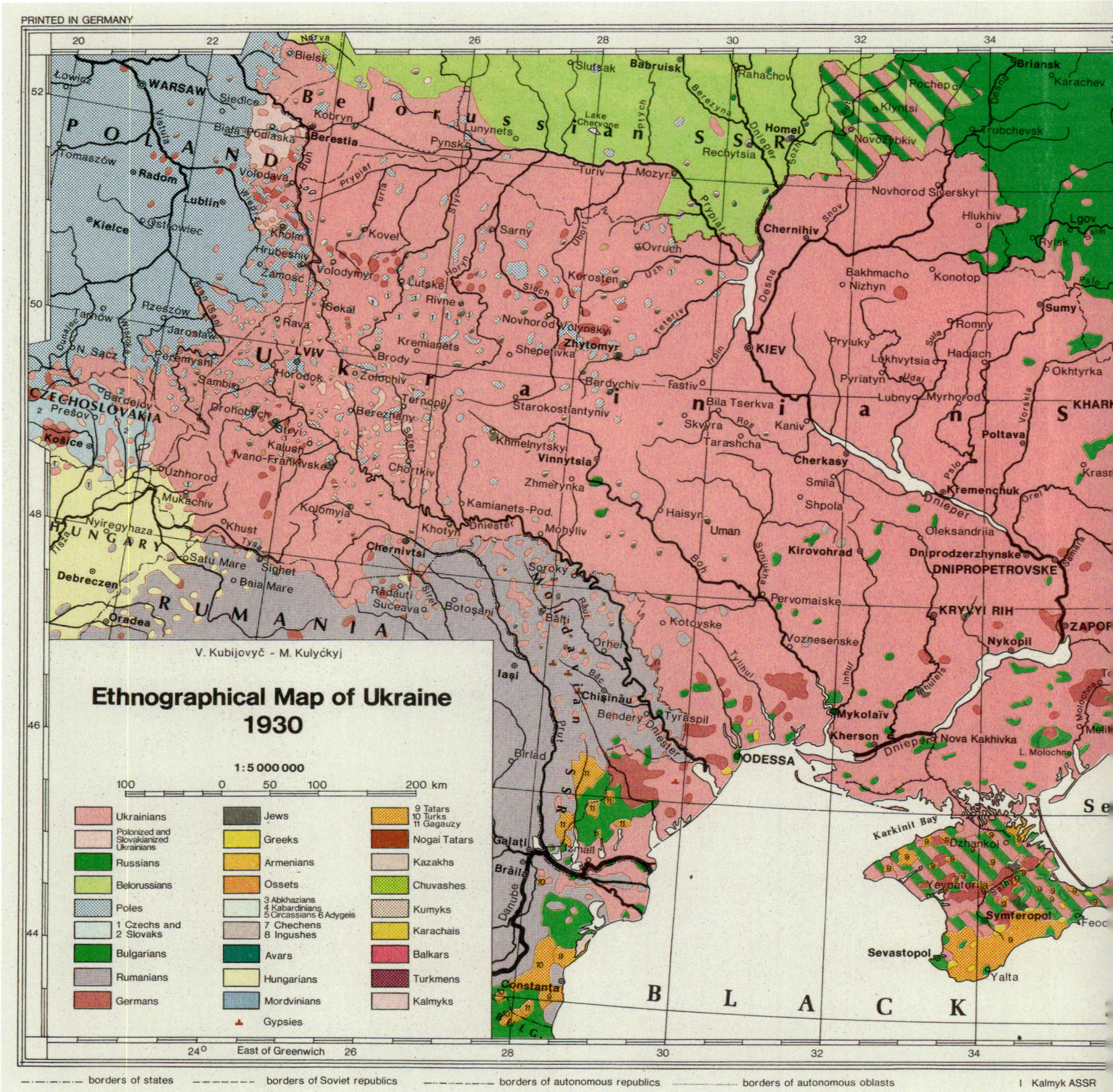
PRINTED IN GERMANY
V. Kubijovyč - M. Kulyćkyj
Ethnographical Map of Ukraine
1930
1:5 000 000
100 0 50 100 200 km
Ukrainians
Polonized and Slovakianized Ukrainians
Russians
Belorussians
Poles
1 Czechs and 2 Slovaks
Bulgarians
Rumanians
Germans
Jews
Greeks
Armenians
Ossets
3 Abkhazians 4 Kabardinians 5 Circassians 6 Adygeis
7 Chechens 8 Ingushes
Avars
Hungarians
Mordvinians
Gypsies
9 Tatars 10 Turks 11 Gagauzy
Nogai Tatars
Kazakhs
Chuvashes
Kumyks
Karachais
Balkars
Turkmens
Kalmyks
POLAND
Belorussian SSR
Ukrainians
CZECHOSLOVAKIA
HUNGARY
RUMANIA
BLACK
East of Greenwich
borders of states
borders of Soviet republics
borders of autonomous republics
borders of autonomous oblasts
I Kalmyk ASSR

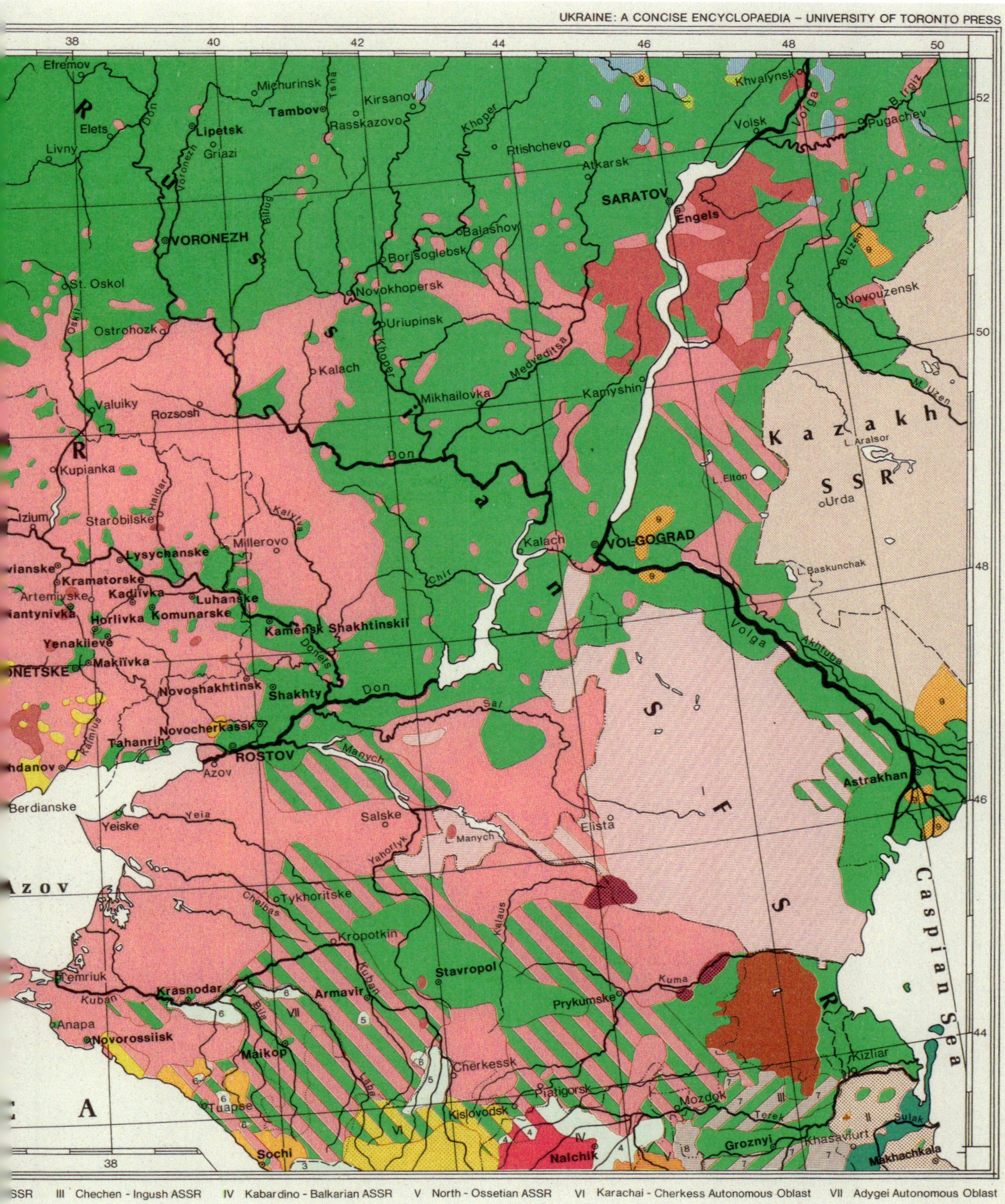
UKRAINE: A CONCISE ENCYCLOPAEDIA – UNIVERSITY OF TORONTO PRESS
R U S S I A N
S F S R
Kazakh SSR
Caspian Sea
A Z O V
Efremov
Michurinsk
Tambov
Kirsanov
Rasskazovo
Khvalynsk
Volsk
Pugachev
Elets
Lipetsk
Griazi
Livny
Rtishchevo
Atkarsk
SARATOV
Engels
VORONEZH
Balashov
Borisoglebsk
Novokhopersk
St. Oskol
Novouzensk
Uriupinsk
Ostrohozk
Kalach
Mikhailovka
Kamyshin
Valuiky
Rozsosh
Kupianka
Izium
Starobilske
Millerovo
Lysychanske
Kramatorske
Artemivske
Kadiivka
Luhanske
Horlivka
Komunarske
Kamensk Shakhtinskii
Yenakiieve
Makiivka
Novoshakhtinsk
Shakhty
Novocherkassk
Tahanrih
ROSTOV
Azov
Berdianske
Yeiske
Salske
Elista
L. Manych
Tykhoritske
Kropotkin
Stavropol
Temriuk
Krasnodar
Armavir
Anapa
Novorossiisk
Maikop
Cherkessk
Piatigorsk
Kislovodsk
Nalchik
Mozdok
Groznyi
Khasaviurt
Kizliar
Makhachkala
Tuapse
Sochi
Prykumske
VOLGOGRAD
Kalach
L. Elton
L. Aralsor
Urda
L. Baskunchak
Astrakhan
Don
Volga
Akhtuba
Khoper
Medveditsa
Manych
Kuban
Kuma
Terek
Sulak
III Chechen - Ingush ASSR IV Kabardino - Balkarian ASSR V North - Ossetian ASSR VI Karachai - Cherkess Autonomous Oblast VII Adygei Autonomous Oblast

Figure 88

ЗАГАЛЬНА
КАРТА
УКРАЇНИ
Зладив М. ДЯЧИШИН
ЗАХОДОМ І НАКЛАДОМ ЧАС. „СВОБОДА", ОРҐАНУ УКРАЇНСЬКОГО НАР. СОЮЗА В АМЕРИЦІ
ПОЯСНЕНЯ:
Етноґрафічна границя України (Др. Величко).
Границя після мир. договору (9 ·II. 1918.).
Границя державна (перед війною).
Границя ґубернїяльна і краєва.
Міста понад 100 тисяч населеня
" від 50 до 100 " "
" " 10 " 50 " "
" менше 10 " "
Висота суші:
до 200 метрів
від 200 " 500 "
Желїзниці. (Начерк в приближеню).
від 500 до 1500 метрів
понад 1500 метрів
Розмір: 1: 2,580,000
Східна довжина від Ґрінїч
МІНСЬКА ГУБ.
МОГИЛІВСЬКА ГУБ.
Городненська ґуб.
ЧЕРНИГІВЩИНА
Полїська низина
Волинська височина
Подільська плита
Правобічна височина
Наддніпрянська низина
Семигородська Верховина
Каркинитський залив
КРИМ
Яйла
ЧОРНЕ МО

ПЕНЗАНСЬКА ГУБ.
СІМБІРСЬКА ГУБ.
ТАМБОВСЬКА ГУБ.
САРАТІВСЬКА ГУБ.
САМАРСЬКА ГУБ.
ГУБЕРНІЯ
ВОРОНІЖСЬКА ГУБ.
ВОРОНІЖ
НОВОХОПЕРСЬК
САРАТІВ
ЦАРИЦИН
АСТРАХАНСЬКА ГУБ.
АСТРАХАН
ЗЕМЛЯ ДОНСЬКОГО ВІЙСЬКА
ДОНСЬКИЙ КРЯЖ
РОСТІВ
СТАВРОПОЛЬСЬКА ГУБ.
СТАВРОПОЛЬ
КУБАНСЬКЕ ВІЙСЬКО
ТЕРСЬКЕ ВІЙСЬКО
КАВКАЗ
КАСПІЙСЬКЕ МОРЕ
К-І-Р-Г-І-З-И
Т-А-Т-А-Р-И
НОГАЙЦІ
К-А-Л-М-У-К-И
ЕРГЕНІ ГОРИ
НОВОРОСІЙСЬК
ЧОРНОМОРСЬКИЙ
© RAND McNALLY & CO., N.Y.

Figure 89

THE MAP OF
UKRAINE
Sponsored by the LEAGUE of AMERICANS of UKRAINIAN DESCENT,
State Organization of UKRAINIAN CONGRESS COMMITTEE of AMERICA.
Chicago, Ill.
John Duzansky, President
Compiled and drawn by WILLIAM KARPA
Scale 1:1,350,000
20 10 0 20 40 60 80 100 Miles
Towns & Villages: Highways: 1st class roads 2nd class roads Railroads: Double track Single
International Boundaries County Boundaries Canals Swamps
Capitals:
KYIV
LVIV
ODESA
CHERNIVTSI
BUCURESTI
IAȘI
GALAȚI
Lublin
Homel
Chernyhiv
Zhytomyr
Berdychiv
Vynnytsya
Rivne
Luts'k
Kovel'
Ternopil'
Stanyslaviv
Uzhhorod
Mukachiv
Kamyanets' Podil's'kyy
Kolomyya
Drohobych
Boryslav
Peremyshl'
Uman'
Bila Tserkva
Kishyniv
Bendery
Bilhorod Dnistrovyy
Ol'viopil'
BL
(CHO
MOUTHS OF THE DUNAY
COPYRIGHT 1965 WILLIAM KARPA PUBLISHED BY THE LEAGUE OF AMERICANS OF UKRAINIAN DESCENT, INC. CHICAGO, ILL

TERRITORY SETTLEMENT OF UKRAINIAN PEOPLE
White Ruthenians
Russians
Rumanians
Don Cossacks
BLACK SEA
Sea of Oziv
(Ozivs'ke More)
KHARKIV
POLTAVA
SUMY
LUHANS'KE
HORLIVKA
MAKIYIVKA
YUZIVKA
MARIUPIL'
ROSTOV
TAHANRIH
KATERYNODAR
SYMFEROPIL'
Sevastopil'
Kherson
Melitopil'
Berdyanka
Zaporizhzhia
KRYVYY RIH
Kremenchuk
SICHESLAV
Kamyans'ke
Oleksandriya
Romen
Okhtyrka
Kerch
KARKINITS'KA ZATOKA (GULF)
EVPATORIYS'KA ZATOKA (GULF)
TEMRYUKS'KA ZATOKA
SEA
MORE)
PRINTED BY PHOTOPRESS, 1918 CHICAGO

Figure 90

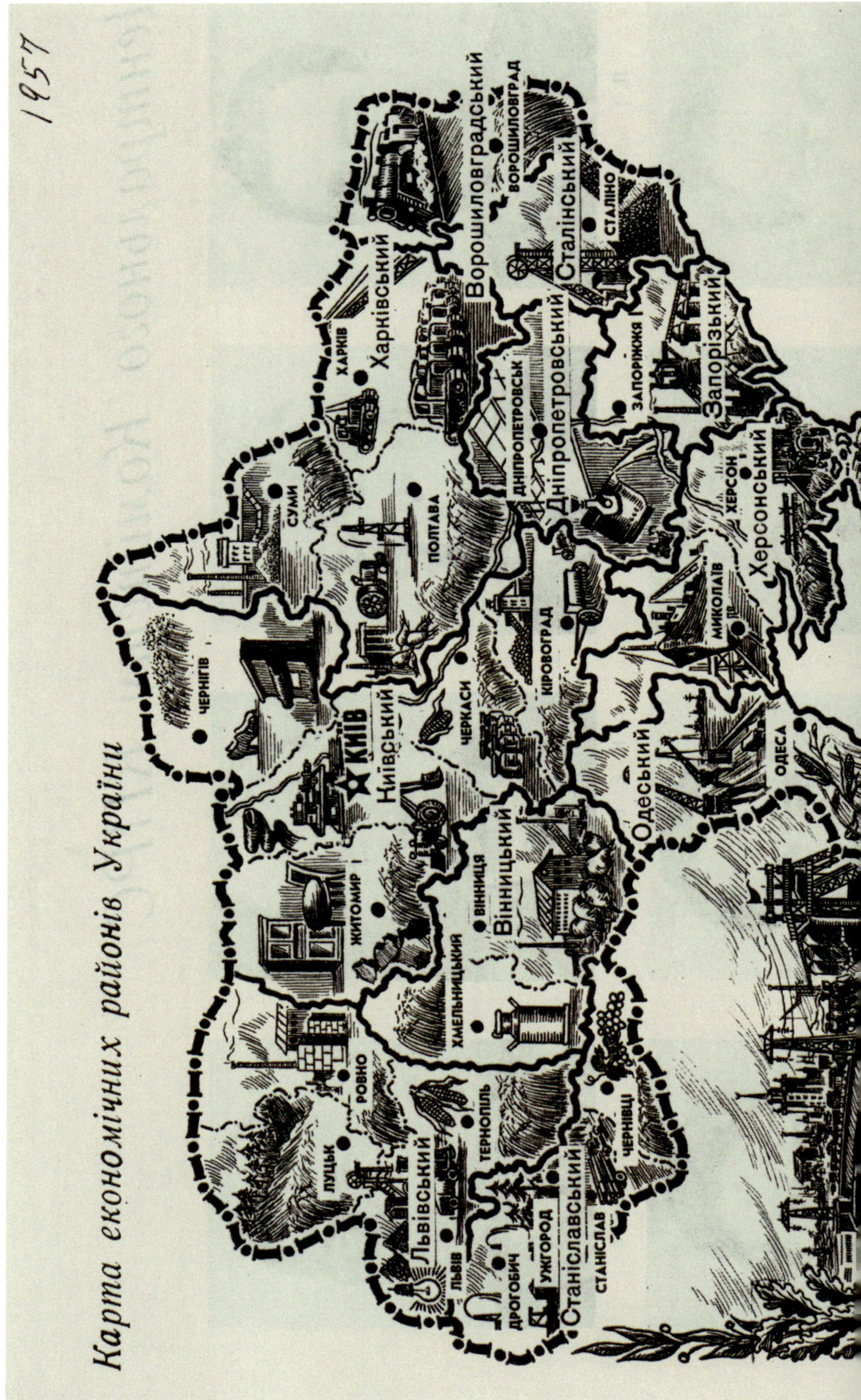
Карта економічних районів України
1957
Харківський
Харків
Ворошиловградський
Ворошиловград
Сталінський
Сталіно
Дніпропетровський
Дніпропетровськ
Запорізький
Запоріжжя
Херсонський
Херсон
Миколаїв
Одеський
Одеса
Кіровоград
Полтава
Суми
Чернігів
Київський
Київ
Черкаси
Вінницький
Вінниця
Житомир
Хмельницький
Львівський
Львів
Луцьк
Ровно
Тернопіль
Дрогобич
Ужгород
Станіславський
Станіслав
Чернівці

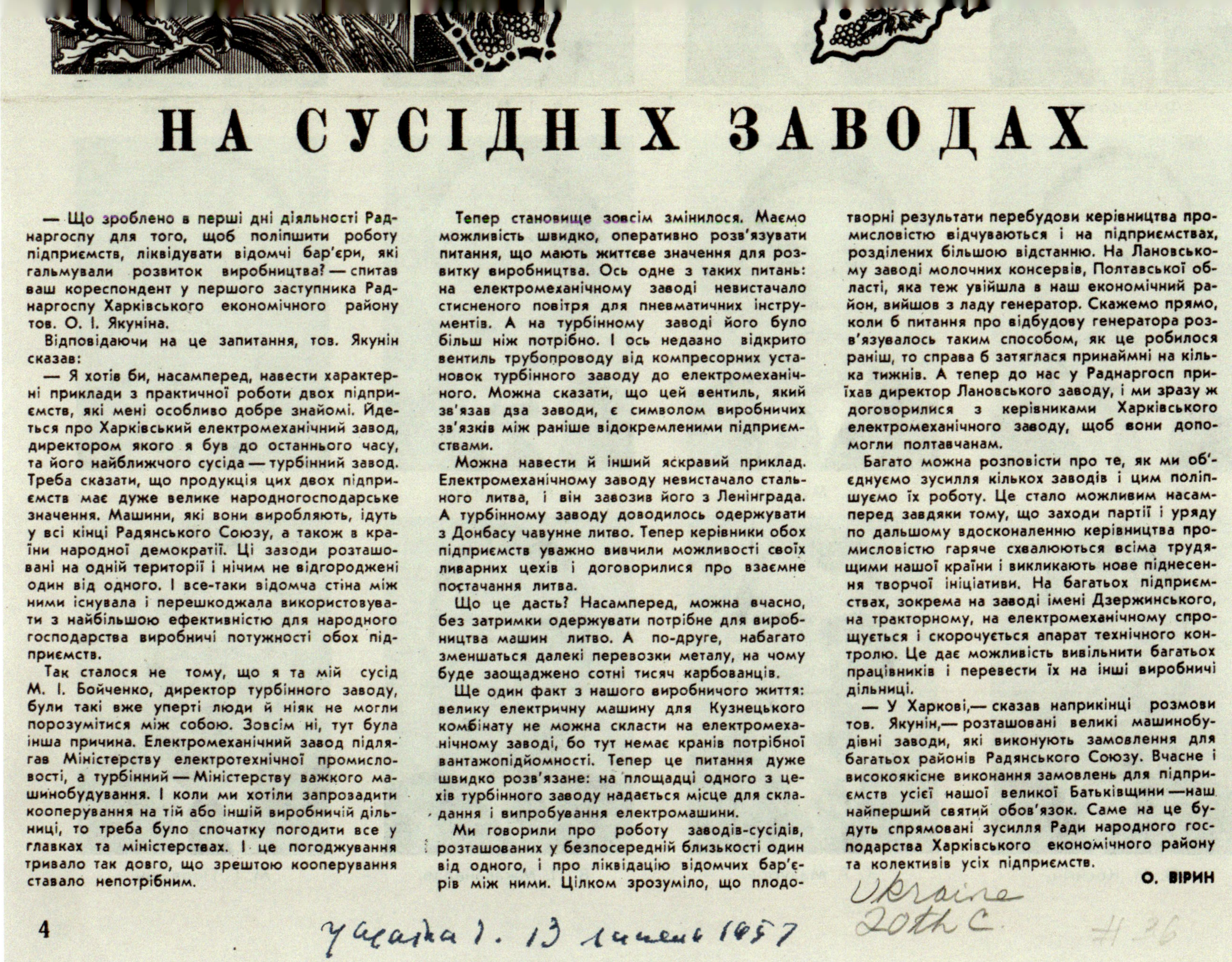

НА СУСІДНІХ ЗАВОДАХ

— Що зроблено в перші дні діяльності Раднаргоспу для того, щоб поліпшити роботу підприємств, ліквідувати відомчі бар'єри, які гальмували розвиток виробництва? — спитав ваш кореспондент у першого заступника Раднаргоспу Харківського економічного району тов. О. І. Якуніна.

Відповідаючи на це запитання, тов. Якунін сказав:

— Я хотів би, насамперед, навести характерні приклади з практичної роботи двох підприємств, які мені особливо добре знайомі. Йдеться про Харківський електромеханічний завод, директором якого я був до останнього часу, та його найближчого сусіда — турбінний завод. Треба сказати, що продукція цих двох підприємств має дуже велике народногосподарське значення. Машини, які вони виробляють, ідуть у всі кінці Радянського Союзу, а також в країни народної демократії. Ці заводи розташовані на одній території і нічим не відгороджені один від одного. І все-таки відомча стіна між ними існувала і перешкоджала використовувати з найбільшою ефективністю для народного господарства виробничі потужності обох підприємств.

Так сталося не тому, що я та мій сусід М. І. Бойченко, директор турбінного заводу, були такі вже уперті люди й ніяк не могли порозумітися між собою. Зовсім ні, тут була інша причина. Електромеханічний завод підлягав Міністерству електротехнічної промисловості, а турбінний — Міністерству важкого машинобудування. І коли ми хотіли запровадити кооперування на тій або іншій виробничій дільниці, то треба було спочатку погодити все у главках та міністерствах. І це погоджування тривало так довго, що зрештою кооперування ставало непотрібним.

Тепер становище зовсім змінилося. Маємо можливість швидко, оперативно розв'язувати питання, що мають життєве значення для розвитку виробництва. Ось одне з таких питань: на електромеханічному заводі невистачало стисненого повітря для пневматичних інструментів. А на турбінному заводі його було більш ніж потрібно. І ось недавно відкрито вентиль трубопроводу від компресорних установок турбінного заводу до електромеханічного. Можна сказати, що цей вентиль, який зв'язав два заводи, є символом виробничих зв'язків між раніше відокремленими підприємствами.

Можна навести й інший яскравий приклад. Електромеханічному заводу невистачало стального литва, і він завозив його з Ленінграда. А турбінному заводу доводилось одержувати з Донбасу чавунне литво. Тепер керівники обох підприємств уважно вивчили можливості своїх ливарних цехів і договорилися про взаємне постачання литва.

Що це дасть? Насамперед, можна вчасно, без затримки одержувати потрібне для виробництва машин литво. А по-друге, набагато зменшаться далекі перевозки металу, на чому буде заощаджено сотні тисяч карбованців.

Ще один факт з нашого виробничого життя: велику електричну машину для Кузнецького комбінату не можна скласти на електромеханічному заводі, бо тут немає кранів потрібної вантажопідйомності. Тепер це питання дуже швидко розв'язане: на площадці одного з цехів турбінного заводу надається місце для складання і випробування електромашини.

Ми говорили про роботу заводів-сусідів, розташованих у безпосередній близькості один від одного, і про ліквідацію відомчих бар'єрів між ними. Цілком зрозуміло, що плодотворні результати перебудови керівництва промисловістю відчуваються і на підприємствах, розділених більшою відстанню. На Лановському заводі молочних консервів, Полтавської області, яка теж увійшла в наш економічний район, вийшов з ладу генератор. Скажемо прямо, коли б питання про відбудову генератора розв'язувалось таким способом, як це робилося раніш, то справа б затяглася принаймні на кілька тижнів. А тепер до нас у Раднаргосп приїхав директор Лановського заводу, і ми зразу ж договорилися з керівниками Харківського електромеханічного заводу, щоб вони допомогли полтавчанам.

Багато можна розповісти про те, як ми об'єднуємо зусилля кількох заводів і цим поліпшуємо їх роботу. Це стало можливим насамперед завдяки тому, що заходи партії і уряду по дальшому вдосконаленню керівництва промисловістю гаряче схвалюються всіма трудящими нашої країни і викликають нове піднесення творчої ініціативи. На багатьох підприємствах, зокрема на заводі імені Дзержинського, на тракторному, на електромеханічному спрощується і скорочується апарат технічного контролю. Це дає можливість вивільнити багатьох працівників і перевести їх на інші виробничі дільниці.

— У Харкові, — сказав наприкінці розмови тов. Якунін, — розташовані великі машинобудівні заводи, які виконують замовлення для багатьох районів Радянського Союзу. Вчасне і високоякісне виконання замовлень для підприємств усієї нашої великої Батьківщини — наш найперший святий обов'язок. Саме на це будуть спрямовані зусилля Ради народного господарства Харківського економічного району та колективів усіх підприємств.

О. ВІРИН

4

Figure 91

Київський Державний музей українського мистецтва
Пам'ятник князю Володимиру
Українська Республіканська філармонія
Рада Міністрів Української РСР
Андріївська церква
Київський Державний історичний музей
Пам'ятник Богдану Хмельницькому
Державний історико-архітектурний заповідник-Софійський музей
Золоті ворота
Державний академічний театр опери та балету імені Т. Г. Шевченка
Київський зоопарк
Володимирський собор
Київський політехнічний інститут
Пам'ятник М. О. Щорсу
Академія наук Української РСР
КОСТЯНТИНІВСЬКА
ВУЛИЦЯ ФРУНЗЕ
КУРЕНІВКА
ЛУК'ЯНІВКА
ВУЛИЦЯ АРТЕМА
Пушкінський парк
БРЕСТЛИТОВСЬКЕ ШОСЕ

ПЛАН КИЄВА
НІ МІСЦЯ І ПАМ'ЯТНИКИ
Верховна Рада Української РСР
Пішоходний міст через Дніпро
Пам'ятник Слави
Пам'ятник М. Ф. Ватутіну
Будинок Міської Ради
Державний заповідник-музей Київо-Печерська лавра
Державний український драматичний театр імені Івана Франка
Державний російський драматичний театр імені Лесі Українки
Київський Державний музей Т. Г. Шевченка
Кінотеатр „Київ"
ий університет
Київський філіал центрального музею В. І. Леніна
Пам'ятник Т. Г. Шевченку
Республіканська виставка передового досвіду в народному господарстві
Київський Державний музей російського мистецтва
Стадіон імені М. С. Хрущова
ДНІПРО
НАБЕРЕЖНЕ ШОСЕ
Міський пляж
Аскольдова могила
ВУЛИЦЯ КІРОВА
ВУЛИЦЯ ЖОВТНЕВОЇ РЕВОЛЮЦІЇ
МОСКОВСЬКА ВУЛИЦЯ
Іподром
СТАРОНАВОДНИЦЬКА ВУЛИЦЯ
Ботанічний сад Академії наук Української РСР
АВТОСТРАДА
ХРЕЩАТИК
ПУШКІНСЬКА ВУЛИЦЯ
ЧЕРВОНОАРМІЙСЬКА ВУЛИЦЯ
ВУЛИЦЯ ТОЛСТОГО
САКСАГАНСЬКОГО
Батиєва гора
СОЛОМ'ЯНКА
ПРОСПЕКТ 40-річчя ЖОВТНЯ
Дорога на виставку
СТАЛІНКА
В

Figure 92

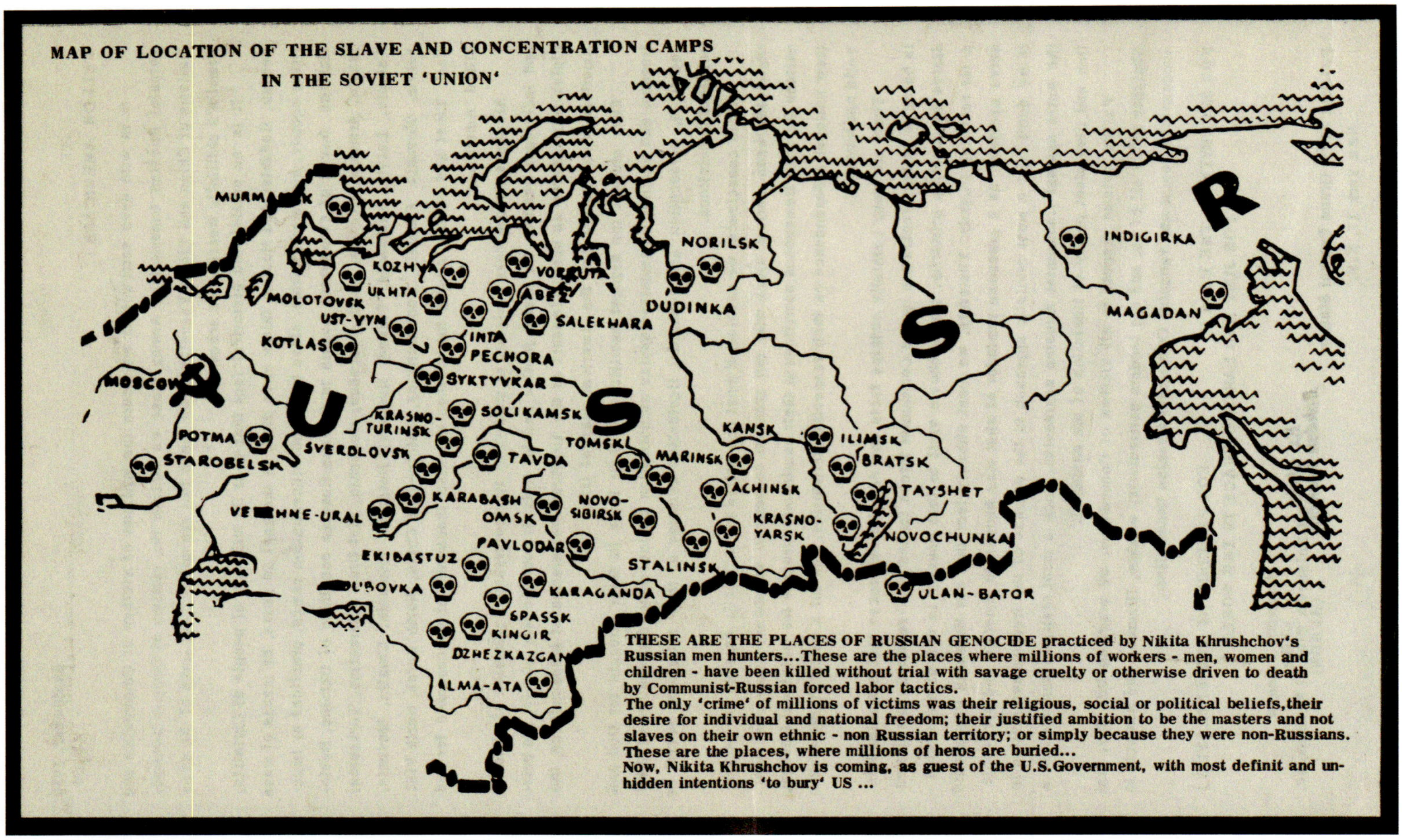
MAP OF LOCATION OF THE SLAVE AND CONCENTRATION CAMPS
IN THE SOVIET 'UNION'
U
S
S
R
KOZHVA
UKHTA
VORKUTA
MOLOTOVSK
ABEZ
UST-VYM
SALEKHARA
KOTLAS
INTA
PECHORA
SYKTYVKAR
KRASNO-TURINSK
SOLIKAMSK
POTMA
STAROBELSK
SVERDLOVSK
TAVDA
KARABASH
OMSK
NOVO-SIBIRSK
PAVLODAR
EKIBASTUZ
KARAGANDA
SPASSK
KINGIR
DZHEZKAZGAN
ALMA-ATA
NORILSK
DUDINKA
TOMSK
MARINSK
KANSK
ILIMSK
BRATSK
ACHINSK
TAYSHET
KRASNO-YARSK
NOVOCHUNKA
STALINSK
ULAN-BATOR
INDIGIRKA
MAGADAN
THESE ARE THE PLACES OF RUSSIAN GENOCIDE practiced by Nikita Khrushchov's Russian men hunters...These are the places where millions of workers - men, women and children - have been killed without trial with savage cruelty or otherwise driven to death by Communist-Russian forced labor tactics.
The only 'crime' of millions of victims was their religious, social or political beliefs,their desire for individual and national freedom; their justified ambition to be the masters and not slaves on their own ethnic - non Russian territory; or simply because they were non-Russians.
These are the places, where millions of heros are buried...
Now, Nikita Khrushchov is coming, as guest of the U.S.Government, with most definit and unhidden intentions 'to bury' US ...

Figure 93

THERE CAN BE NO PEACE IN EUROPE UNTIL UKRAINE IS REUNITED AND FORTY-FIVE MILLION UKRAINIANS ARE FREE AGAIN
Russian Bolsheviks occupied 833,570 sq. km. and enslaved 36,825,000 Ukrainians
Poland occupied 121,041 sq. km. and enslaved 6,743,000 Ukrainians
Czechoslovakia occupied 14,673 sq. km. with 472,000 Ukrainians
Roumania occupied 17,264 sq. km. and enslaved 1,270,000 Ukrainians
POLAND
RUSSIA
UKRAINE
KUBAN
BLACK SEA
SEA OF ASOV
CRIMEA
BESSARABIA
ROUMANIA
CZECHOSLOVAKIA
Warsaw
Cracow
Lublin
Peremysl
Lemberg
Kiev
Kharkov
Bila Tserkov
Kaminetz
Tschyhyryn
Ekaterinoslav
Odessa
Kherson
Rostov
Volga
Don
Donetz
Dniester
Dnieper
Bug
Kuban
Pripet

Figure 94

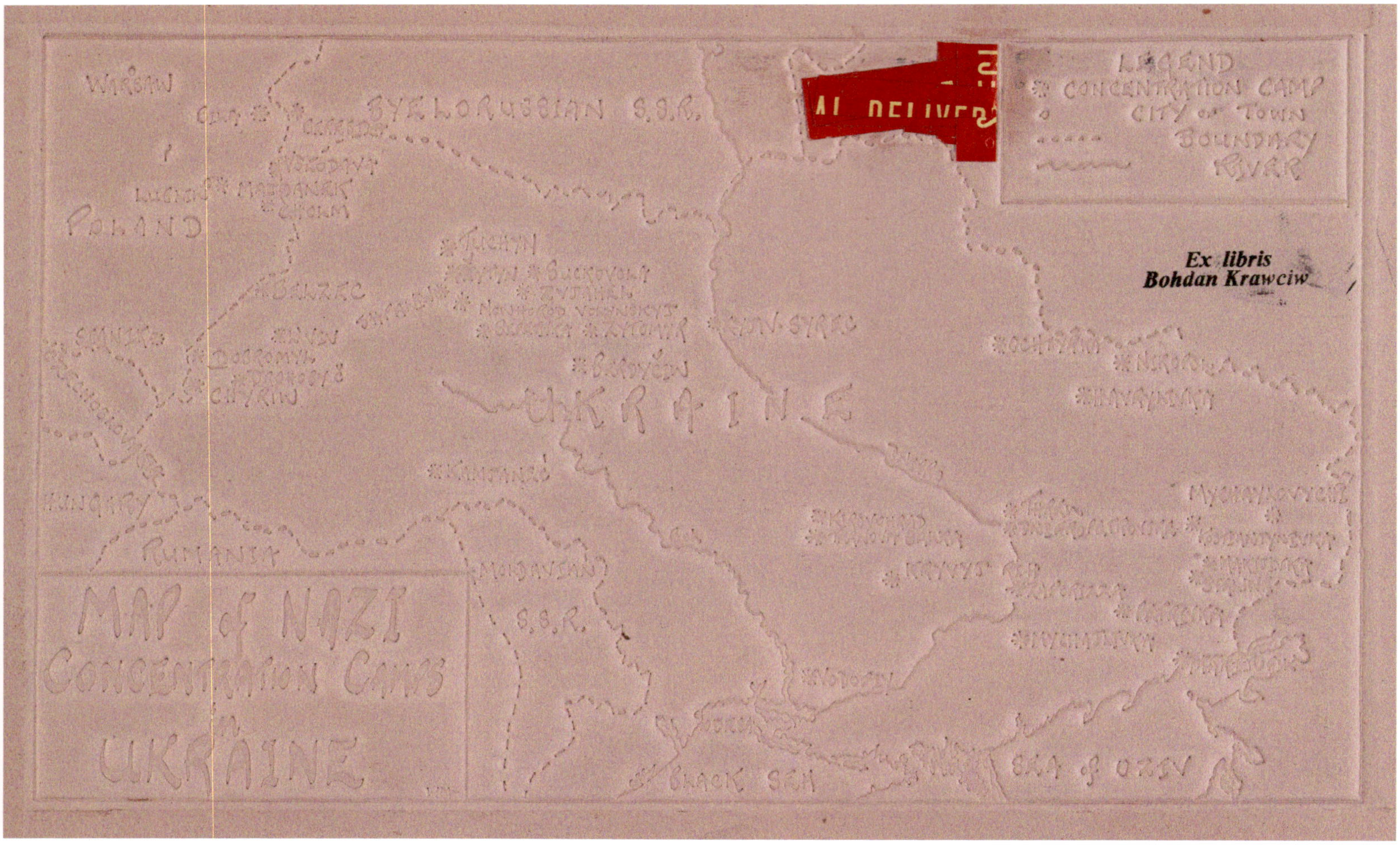

Figure 95

(unenhanced)

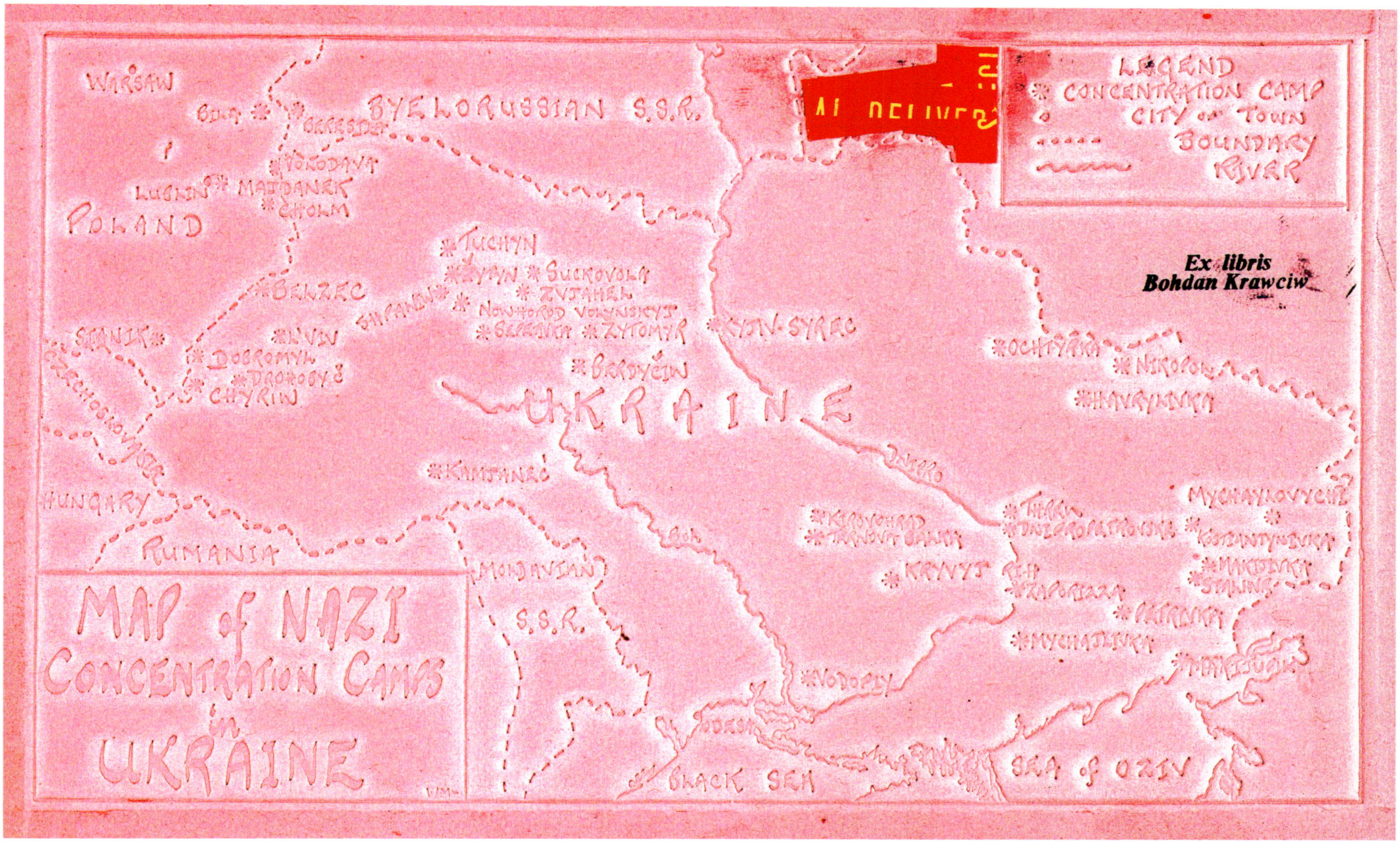

Figure 95

(contrast enhanced)

Appendix: All Maps in the Collection

PART ONE (BY CARTOGRAPHER)

Maps with an Author Indicated

Entry (No.): 1
Author: Aa, Pieter (Pierre, Petrus) van der (1659–1733)
Title: LA // HONGRIE, // SUIVANT LES // NOUVELLES OBSERVATIONS // DE // MESSrs DE L'ACADEMIE ROYALE // DES SCIENCES, ETC. // AUGMENTÉES DE NOUVEAU // A LEIDE, //CHEZ PIERRE VANDER AA // AVEC PRIVILEGE./
Year: ca. 1713–33
Dimensions: 37.5 x 30.0 (29.0 x 22.3) cm
Phys. Desc.: Engraved, Colored
Language(s): French

Entry (No.): 2
Author: Aa, Pieter (Pierre, Petrus) van der (1659–1733)
Title: CROSNO // PETITE VILLE DE LA RUSSIE ROUGE, EN POLOGNE. // A Leide, Chez Pierre vander Aa.
Year: 1729
Dimensions: 49.5 x 38.0 (36.0 x 29.0) cm
Phys. Desc.: Engraved, Colored
Language(s): French

Entry (No.): 3
Author: Aa, Pieter (Pierre, Petrus) van der (1659–1733)
Title: KAMINIEK, // VILLE CAPITALE DE LA HAUTE PÓDOLIE, //CONTRÉE DE LA POLOGNE. A Leide, Chez Pierre vander Aa.
Year: 1729
Dimensions: 49.5 x 38.0 (33.5 x 25.5) cm
Phys. Desc.: Engraved
Language(s): French

Entry (No.): 4
Author: Aa, Pieter (Pierre, Petrus) van der (1659–1733)
Title: LA // POLOGNE, // SUIVANT LES // NOUVELLES // OBSERVATIONS DE // MESSrs DE L'ACADEMIE // ROYALE // DES SCIENCES, ETC. // AUGMENTÉES DE NOUVEAU... // A LEIDE, // CHEZ P. VANDER AA. // AVEC PRIVILEGE/
Year: 1729
Dimensions: 49.0 x 37.0 (29.5 x 22.5) cm
Phys. Desc.: Engraved
Language(s): French

Entry (No.): 5
Author: Aa, Pieter (Pierre, Petrus) van der (1659–1733)
Title: LA GRANDE TARTARIE, SUIVANT LES NOUVELLES OBSERVATIONS DE MESSrs DE L'ACADEMIE ROYALS DES SCIENCES, ETC. Augmentées de Nouveau. A Leide, chez Pierre vander AA. Avec privilege.
Year: 1729
Dimensions: 49.0 x 37.0 (29.5 x 22.5) cm
Phys. Desc.: Engraved, Uncolored
Language(s): French

Entry (No.): 6
Author: Aa, Pieter (Pierre, Petrus) van der (1659–1733)
Title: La Moscovie Meridionale, Suivant les Nouvelles Observations de Messrs de L'Academie Royale des Sciences, etc. Augmentées de Nouveau. A Leide, chez Pierre vander AA. Avec Privilege.
Year: 1729
Dimensions: 37.9 x 26.0 cm
Phys. Desc.: Engraved, Uncolored
Language(s): French

Entry (No.): 7
Author: Aa, Pieter (Pierre, Petrus) van der (1659–1733)
Title: La Moscovie Septentrionale, Suivant les Nouvelles Observations de Messrs de L'Academie Royale des Sciences, etc. Augmentées de Nouveau. A Leide, chez Pierre vander AA. Avec Privilege.
Year: 1729
Dimensions: 37.9 x 25.7 cm
Phys. Desc.: Engraved, Uncolored
Language(s): French

Entry (No.): 8
Author: Aa, Pieter (Pierre, Petrus) van der (1659–1733)
Title: UKRAINE, // GRAND PAYS DE LA // RUSSIE ROUGE, // AVEC UNE PARTIE DE LA // POLOGNE, MOSCOVIE, BULGARIE, VALACHIE, // PODOLIE ET VOLHYNIE.

// TIRÉE DE DIFFERENTES CARTES, QUI ONT ÉTÉ // FAITES SUR LES LIEUX, ET DES RELATIONS LES PLUS //FIDELES, PRÉSENTEMENT MISE ET LUMIERE, PAR // PIERRE VANDER AA, // MARCHAND LIBRAIRE // A LEIDE, // AVEC PRIVILEGE.
Year: 1729
Dimensions: 49.6 x 36.4 cm
Phys. Desc.: Engraved
Language(s): French

Entry (No.): 9
Author: Aa, Pieter (Pierre, Petrus) van der (1659–1733)
Title: ZAMOS- // CIVM, NOUA // POLONIAE CIUI- // TAS, À MAGNIFICO // EIUS REGNI CAN- // CELLARIO À FUNDA- // MENTIS EXST- // RUCTA, // Se vend a Leide, Chez Pierre vander Aa. Avec Privilege.
Year: 1729
Dimensions: 49.5 x 38.0 (46.0 x 35.0) cm
Phys. Desc.: Engraved
Language(s): French

Entry (No.): 10
Author: Academia Scientiarum Petropolitana
Title: CHARTE / DER KRIEGS OPERATIONEN // AM DONN U. DNIEPER / IHRO RUSSISCH // KAYSERL. // MAJESTAET // GLORIEUSEN ARMEEN A. 1736. // GESTOCHEN BEY DER KAYSERL. ACAD. // DER WISSENSCH: IN St. PETERSBURG // ZU HABEN BEY // MATTHAEUS SEUTTER, //... / IN // AUGSBURG. /
Year: 1736
Dimensions: 61.5 x 53.7 (44.0 x 38.0) cm
Phys. Desc.: Engraved, Colored
Language(s): German

Entry (No.): 11
Author: Academia Scientiarum Petropolitana
Title: CHARTE // DER KRIEGS-OPERATIONEN // AM DONN UND DNIEPER / IHRO RUSSISCH KAYSERL // MAJESTAET // GLORIEUSEN ARMEEN A: 1736. // GESTOCHEN BEY DER // KAŸSERL: ACAD: // DER WISSENSCH: IN St. PETERSBURG.
Year: 1736
Dimensions: 60.7 x 49.0 (46.0 x 39.0) cm
Phys. Desc.: Engraved
Language(s): German

Entry (No.): 12
Author: Academia Scientiarum Petropolitana
Title: CHARTE // derer von der// RUSSISCH-KEYSER: ARMEE/ im Iahr 1736 // zwischen und an dem Dnieper und Donn // wieder die Turcken // und Tartarn // siegreich unternommenen // Kriegs-Operationen //
Year: 1736
Dimensions: 80.0 x 63.0 (75.0 x 58.0) cm
Phys. Desc.: Engraved
Language(s): German
Figure 46

Entry (No.): 13
Author: Academia Scientiarum Petropolitana
Title: MAPPA // GENERALIS // TOTIUS IMPERII // RUSSICI
Year: 1745
Dimensions: 103.0 x 63.0 (97.3 x 55.7) cm
Phys. Desc.: Engraved, Colored Outline
Language(s): Latin
Figure 47

Entry (No.): 14
Author: Academia Scientiarum Petropolitana
Title: TABULA GEOGRAPHICA // GUBERNIUM SMOLENSCENSE // CUM PARTIBUS // KIOVIENSIS // BELGORODENSIS ET // VORONICENSIS // GUBERNII COMPLECTENS //
Yer: 1745
Dimensions: 67.7 x 52.3 (55.0 x 48.2) cm
Phys. Desc.: Engraved, Colored Outline
Language(s): Latin

Entry (No.): 15
Author: Academia Scientiarum Petropolitana
Title: TATARIA // MINOR // CUM // ADIACENTIBUS // KIOVIENSI // ET BELGORO- // DENSI // GUBERNIIS. //
Year: 1745
Dimensions: 67.7 x 52.3 (55.3 x 48.3) cm
Phys. Desc.: Engraved, Colored Outline
Language(s): Latin

Entry (No.): 16
Author: Academia Scientiarum Petropolitana
Title: TERRITORIUM // PONTUM EVXINUM // ET MARE CASPIUM // INTERIACENS // CUBANIAE ET GEORGIAE // DELINEATIONEM GEOGRA- // PHICAM RELIQUAMQUE // PARTEM FLIVII VOLGAE // EIUSQUE OSTIUM // EXHIBENS.
Year: 1745
Dimensions: 67.7 x 52.3 (55.0 x 48.3) cm
Phys. Desc.: Engraved, Colored Outline
Language(s): Latin

Entry (No.): 17
Author: Adams, John
Title: Poland
Year: 1803
Dimensions: 22.2 x 18.3 (199 x 158) cm
Phys. Desc.: Engraved, Colored Outline
Language(s): English

Entry (No.): 18
Author: Adams, John
Title: Southern part of // Russia in // Europe.
Year: 1803
Dimensions: 22.2 x 18.3 (20.3 x 16.0) cm
Phys. Desc.: Engraved, Colored
Language(s): English

Entry (No.): 19
Author: Adams, John
Title: Turkey // in Europe // and Hungary.

Year: 1804
Dimensions: 22.2 x 18.3 (19.2 x 16.1) cm
Phys. Desc.: Engraved, Colored
Language(s): English

Entry (No.): 20
Author: Allard (Allardt), Carolus (Karel) (1648–ca. 1706)
Title: Regni // Poloniae, // Magni Ducatus Lithuaniae.
Year: 1667–68
Dimensions: 61.0 x 52.2 (59.2 x 50.0) cm
Phys. Desc.: Engraved, Colored
Language(s): Latin

Entry (No.): 21
Author: Allard (Allardt), Carolus (Karel) (1648–ca. 1706)
Author: Witsen, Nicolas (1641–1717)
Title: Tartaria, sive Magni Chami Imperium ex Credendis Amplissimi Viri D[ni] Nicolai Witsen, Cos: Amst: Aliorumq. probandorum, et hodie vigentium Geographorum Archetypis congestum, auctum, et in lucem editum a Carolo Allard, Amst: Bat: Cum Privilegio Potentissimorum D.D. Ordinum Hollandiae et Westfrisiae.
Year: 1667–68
Dimensions: 61.0 x 52.2 (59.2 x 50.0) cm
Phys. Desc.: Engraved, Colored
Language(s): Latin

Entry (No.): 22
Author: Allard (Allardt), Carolus (Karel) (1648–ca. 1706)
Title: Exactissima Asiae Delineatio, In Praecipuas regiones, Caeterasq. Partes divisa, et denuo in lucem edita per Carolum Allard, Amstero-Batavum; Cum Privilegio Potentissimorum D.D. Ordinum Hollandiae et Westfrisiae.
Year: ca. 1690–96
Dimensions: 63.8 x 53.5 (59.0 x 50.2) cm
Phys. Desc.: Engraved, Colored Outline
Language(s): Latin

Entry (No.): 23
Author: Allard (Allardt), Carolus (Karel) (1648–ca. 1706)
Title: Totius Regni Hungariae Maximaeque Partis Danubii Fluminis una cum adjacentibus et finitimis Regionibus Novissima Delineatio per Carolum Allard Amstelo Batarum.
Year: 1690
Dimensions: 82.0 x 53.0 (78.7 x 43.0) cm
Phys. Desc.: Engraved, Colored
Language(s): Latin

Entry (No.): 24
Author: Allard (Allardt), Carolus (Karel) (1648–ca. 1706)
Title: Regni Poloniae, Magni Ducatus Lithuaniae Coeterarumq, Regi Poloniae Subditarum Regionum Tabula; In Omnes suos Ducatus, Palatinatus &c. divisa et in lucem edita a Carolo Allard, Amstelo Batavo. Cum Privilegio Potentissimorum D.D. Ordinum Hollandiae et Westfr.
Year: 1696
Dimensions: 63.5 x 53.2 (59.0 x 44.3) cm
Phys. Desc.: Engraved, Colored
Language(s): Latin

Figure 23

Entry (No.): 25
Author: Andelfinger, Johann Jacob
Title: Plan de Constantinople de son Port Canal et son Environs.
Year: Unknown
Dimensions: 58.3 x 52.0 (54.8 x 48.0) cm
Phys. Desc.: Engraved, Colored Outline
Language(s): French

Entry (No.): 26
Author: Arrowsmith, Aaron (1750–1823)
Title: Europe
Year: 1798
Dimensions: 28.6 x 23.5 (25.0 x 20.3) cm
Phys. Desc.: Engraved, Colored
Language(s): English

Entry (No.): 27
Author: Arrowsmith, Samuel (1790–1873)
Title: Russia in Europe.
Year: 1828
Dimensions: 34.5 x 27.5 (29.6 x 23.0) cm
Phys. Desc.: Engraved, Colored
Language(s): English
Figure 63

Entry (No.): 28
Author: Aspin, Jehosaphat (worked 1805–40) (filed under Lavoisne by Krawciw)
Title: ASIA: // DRAWN // FROM THE BEST AUTHORITIES, FOR THE // ELUDICATION OF / LAVOISNE'S // GENEALOGICAL, HISTORICAL, CHRONOLOGICAL, // & GEOGRAPHICAL // ATLAS. // BY J. ASPIN // 1813.
Year: 1813
Dimensions: 53.2 x 42.5 (32.2 x 31.2) cm
Phys. Desc.: Engraved, Colored
Language(s): English

Entry (No.): 29
Author: Barbié du Bocage, Jean-Denis (1760–1825)
Title: CARTE DU PALUS MÉOTIDE ET DU PONT EUXIN, Pour le Voyage du Jeune Anacharsis, Reduite en grande partie d'après celle de M. D'ANVILLE. Par. M. BARBIÉ DU BOCAGE. MDCCLXXXI.
Year: 1781
Dimensions: 32.0 x 22.5 (23.6 x 17.5) cm
Phys. Desc.: Engraved, Uncolored
Language(s): French

Entry (No.): 30
Author: Beauplan, Guillaume le Vasseur de (1595–1685)
Title: Ukraine Pars-quae Pokutia Vulgo dicitur Per Guil le Vasseur de Beauplan S.R.M. Poloniae Archit.[m] militarem etc.
Year: 1650
Dimensions: 63.0 x 52.5 (52.5 x 44.0) cm
Phys. Desc.: Engraved, Colored Cartouches

Language(s): Latin

Entry (No.): 31
Author: Beauplan, Guillaume le Vasseur de (1595–1685)
Title: Nova totius Regni Poloniae, Magnia Ducatus Prussiae et Lithuaniae, cum suis Palatinatibus ac confinis." "Exacta delineato per G. le Vasseur de Beauplan S.R.M.[tis] Architectum milit. et Capitaneum. Amsterdami, Apud Danckerum Dankerts.
Year: 1651
Dimensions: 54.0 x 44.0 (50.0 x 38.5) cm
Phys. Desc.: Engraved, Uncolored Cartouche
Language(s): Latin

Entry (No.): 32
Author: Beauplan, Guillaume le Vasseur de (1595–1685)
Title: Ukrainae Pars, quae Kiovia Palatinatus Vulgo dicitur. Per Guilhelmum le Vasseur de Beauplan S.R.M. Poloniae Architectum militarem etc.
Year: 1659–72
Dimensions: 64.5 x 54.5 (56.7 x 40.5) cm
Phys. Desc.: Engraved, Colored Cartouche
Language(s): Latin
Figure 17

Entry (No.): 33
Author: Beauplan, Guillaume le Vasseur de (1595–1685)
Title: Description d'Ukranie, qui sont plusieurs Provinces du Royaume de Pologne. Contenves depuis les confins de la Moscovie, isques aux limites de la Transylvanie, Ensemble leurs moeurs, facons de viures et de faire la Guerre. Par le Sieur de Beauplan, A Roüen chez Jacques Cailloüé, dans la Cour du Palais. M.DC.LX.
Year: 1660
Dimensions: 31.6 x 20.5 cm
Phys. Desc.: text
Language(s): Latin
Figure 16

Entry (No.): 34
Author: Beauplan, Guillaume le Vasseur de (1595–1685)
Title: Chersonesvs Taurica et Tartaria Praepopensis, Vel Crimea.
Year: 1662
Dimensions: 64.2 x 54.9 (49.8 x 37.9) cm
Phys. Desc.: Engraved, Colored Cartouche
Language(s): Latin

Entry (No.): 35
Author: Beauplan, Guillaume le Vasseur de (1595–1685)
Title: Descriptio Borysthenis Fluvii vulgo Niepr, sive Dniepr dicti: simul + de Moribus cosacorum Zaporoviorum. Tractus Borysthenis Vulgo Dniepr et Niepr dicti à Bouzin usque ad Chortyca Ostrow.
Year: 1662–72
Dimensions: 64.0 x 55.0 (56.0 x 42.5) cm
Phys. Desc.: Engraved, Colored
Language(s): Latin

Entry (No.): 36
Author: Beauplan, Guillaume le Vasseur de (1595–1685)
Title: Descriptio Borysthenis Fluvii vulgo Niepr, sive Dniepr dicti: simul + de Moribus cosacorum Zaporoviorum. Tractus Borysthenis Vulgo Dniepr et Niepr dicti à chortika ostro ad Urbem Oczakow ubi in Pontum Euxium se exonerat.
Year: 1662–72
Dimensions: 64.0 x 55.0 (54.0 x 46.5) cm
Phys. Desc.: Engraved, Colored
Language(s): Latin

Entry (No.): 37
Author: Beauplan, Guillaume le Vasseur de (1595–1685)
Title: Descriptio Borysthenis Fluvii vulgo Niepr, sive Dniepr dicti: simul + de Moribus cosacorum Zaporoviorum. Tractus Borysthenis Vulgo Dniepr et Niepr dicti à chortika ostro ad Urbem Oczakow ubi in Pontum Euxium se exonerat.
Year: 1662–72
Dimensions: 64.0 x 55.0 (54.0 x 46.5) cm
Phys. Desc.: Engraved, Colored
Language(s): Latin

Entry (No.): 38
Author: Beauplan, Guillaume le Vasseur de (1595–1685)
Title: Descriptio Borysthenis Fluvii vulgo Niepr, sive Dniepr dicti: simul + de Moribus cosacorum Zaporoviorum. Tractus Borysthenis Vulgo Dniepr et Niepr dicti à chortika ostro ad Urbem Oczakow ubi in Pontum Euxium se exonerat.
Year: 1662–72
Dimensions: 64.0 x 55.0 (54.0 x 46.5) cm
Phys. Desc.: Engraved, Colored
Language(s): Latin

Entry (No.): 39
Author: Beauplan, Guillaume le Vasseur de (1595–1685)
Title: Basse Podolie ou Palatinat de Braclaw, tiré de la Grande Ukraine, du S.[r] le Vasseur de Beauplan. Par le S.[r] d'Abbeville Geogr. Ord.[re] du Roy. A Paris Chez Pierre Mariette, Rue St. Jacques a L'Esperance. Avec Privilege pour Vingt Ans. 1665.
Year: 1665
Dimensions: 56.0 x 39.0 (53.5 x 35.5) cm
Phys. Desc.: Engraved, Colored Cartouche
Language(s): French

Entry (No.): 40
Author: Beauplan, Guillaume le Vasseur de (1595–1685)
Title: Basse Volhynie, ou Palatinat de Kiow, tiré entirement de la grande Ukraine, du Sieur le Vasseur de Beauplan. Par le S[r]. Sanson d'Abbeville Geogr. ord[re] du Roy. A Paris. Chez P. Mariette, Rue S[t]. Jacques a L'Esperance avec Privilege pour Ving Ans.1665.
Year: 1665
Dimensions: 58.3 x 42.8 (57.3 x 41.8) cm
Phys. Desc.: Engraved, Colored
Language(s): French

Entry (No.): 41

Author: Beauplan, Guillaume le Vasseur de (1595–1685)
Title: Haute Podolie ou Palatinat de Kamieniec tirée entierem de la Gr. Ukranie, du S^{r}. le Vasseur de Beauplan" Par le S^{r}. Sanson d'Abb. Geogr. ord du Roy. A Paris. Chez Pierre Mariette Rue St. Jacques a l'Esperance. Avec Privilege de la Majesté pour vingt ans. 1665.
Year: 1665
Dimensions: 56.0 x 42.0 (52.9 x 40.2) cm
Phys. Desc.: Engraved, Colored
Language(s): French

Entry (No.): 42
Author: Beauplan, Guillaume le Vasseur de (1595–1685)
Title: Haute Volhynie, ou Palatinat de Lusuc; tire de la Grande Carte d'Ukraine, du S^{r}. le Vasseur de Beauplan. Par le S^{r}. Sanson d'Abbeville Geogr. ord^{re} du Roy. A Paris. Chez Pierre Mariette, Rue St. Jacques, a L'Esperance avec Privilege pour Vingt Ans. 1665.
Year: 1665
Dimensions: 59.0 x 40.5 (57.0 x 38.2) cm
Phys. Desc.: Engraved, Colored
Language(s): French

Entry (No.): 43
Author: Beauplan, Guillaume le Vasseur de (1595–1685)
Title: Russie Noire divisée en ses Palatinats + c. tirée pour la plus grande partie de la grande Carte de l'Ukraine, du S^{r}. le Vasseur de Beauplan. Par le S^{r}. Sanson d'Abbeville Geogr. ord^{re} du Roy. A Paris. Chez P. Mariette, rue Jacques a l'Esperance avec Privilege pour Vingt Ans. 1665.
Year: 1665
Dimensions: 54.5 x 45.5 (52.5 x 44.5) cm
Phys. Desc.: Engraved, Colored
Language(s): French

Entry (No.): 44
Author: Beauplan, Guillaume le Vasseur de (1595–1685)
Title: Tractus Borysthenis Vulgo Dniepr et Niepr dicti, à Kiovia ad Urbum Oczakow ubi in Pontium Euxinum se exonerat.
Year: 1680
Dimensions: 62.5 x 53.5 (53.5 x 45.5) cm
Phys. Desc.: Engraved, Colored
Language(s): Latin

Entry (No.): 45
Author: Beauplan, Guillaume le Vasseur de (1595–1685)
Title: Typus Generalis Ukrainae sive Palatinatuum Podoliae, Kiovensis et Braczlaviensis terras nova delineatione exhibens. Ex Officina Janssonio-Waesbergiana et Moses Pitt.
Year: 1680
Dimensions: 68.8 x 59.0 (54.0 x 42.4) cm
Phys. Desc.: Engraved, Colored
Language(s): Latin
Figure 18

Entry (No.): 46
Author: Beauplan, Guillaume le Vasseur de (1595–1685)
Title: Ukrainae pars, quae Podolia Palatinus Vulgo dicitur» Per Guil. le Vasseur de Beauplan S.R.M^{tis}. Poloniae Architectum militarem etc.
Year: 1680–83
Dimensions: 62.0 x 53.5 (53.5 x 44.5) cm
Phys. Desc.: Engraved, Colored
Language(s): Latin

Entry (No.): 47
Author: Beauplan, Guillaume le Vasseur de (1595–1685)
Title: Ukrainae Pars, que Braclavia Palatinatus Vulgo Dicitur. Per Guilhelmum le Vasseur de Beauplan S.R.M^{tis} Poloniae Architectum militarem etc.
Year: 1680–83
Dimensions: 62.0 x 51.5 (57.7 x 45.0) cm
Phys. Desc.: Engraved, Colored
Language(s): Latin

Entry (No.): 48
Author: Beauplan, Guillaume le Vasseur de (1595–1685)
Title: Taurica Chersonesus.
Year: 1701
Dimensions: 58.5 x 49.0 (49.8 x 39.0) cm
Phys. Desc.: Engraved, Colored
Language(s): Latin

Entry (No.): 49
Author: Beauplan, Guillaume le Vasseur de (1595–1685)
Title: A Description of Ukraine, Containing Several Provinces of the Kingdom of Poland, Lying between the Confines of Muscovy, and the Borders of Transylvania. Together with their Customs, Manner of Life, and how they manage their Wars. Written in French by the Sieur de Beauplan.
Year: 1704
Dimensions: 25.0 x 31.4 cm
Phys. Desc.: Without Cover
Language(s): English

Entry (No.): 50
Author: Beauplan, Guillaume le Vasseur de (1595–1685)
Title: The Empire of Moscovy
Year: 1704
Dimensions: 19.0 x 17.1 (17.3 x 14.0) cm
Phys. Desc.: Uncolored
Language(s): English

Entry (No.): 51
Author: Beauplan, Guillaume le Vasseur de (1595–1685)
Title: A Description of Ukraine by Guillaume le Vasseur Sieur de Beauplan, New York, 1959. Editorial board J.F. Vovchuk, J.T. Petryshyn, B. Krawciw, O. Sokolyshyn, V. Davydenko. Published by the Organization for the Defense of Four Freedoms of Ukraine, Inc.
Year: 1959
Dimensions: 27.3 x 36.0 cm
Phys. Desc.: Book with Maps
Language(s): English
Figure 77

Entry (No.): 52
Author: Bellin, Jacques Nicolas (1703–72)

Title: An Essay of a New and Compact Map Containing the known Parts of the Terrestrial Globe, by N. Bellin, Engineer of the Marine at the Hague by P. De Hondt. MDCCL.
Year: 1750
Dimensions: 68.6 x 55.0 (68.0 x 49.5) cm
Phys. Desc.: Engraved, Uncolored
Language(s): English, French, Dutch
Figure 33

Entry (No.): 53
Author: Bellin, Jacques Nicolas (1703–72)
Title: Carte de la Crimée.
Year: 1750
Dimensions: 23.5 x 32.0 (17.1 x 21.3) cm
Phys. Desc.: Engraved, Uncolored
Language(s): French

Entry (No.): 54
Author: Bellin, Jacques Nicolas (1703–72)
Title: Carte de la Mer Caspienne et ses Environs.
Year: 1750
Dimensions: 23.0 x 30.5 (17.2 x 21.5) cm
Phys. Desc.: Engraved, Uncolored
Language(s): French

Entry (No.): 55
Author: Bellin, Jacques Nicolas (1703–72)
Title: Carte de la Mer Noire.
Year: 1750
Dimensions: 23.0 x 30.5 (17.4 x 21.5) cm
Phys. Desc.: Engraved, Uncolored
Language(s): French

Entry (No.): 56
Author: Bellin, Jacques Nicolas (1703–72)
Title: Carte de la Moscovie Europeene.
Year: 1750
Dimensions: 43.5 x 32.0 (29.0 x 23.5) cm
Phys. Desc.: Engraved, Uncolored
Language(s): French

Entry (No.): 57
Author: Bellin, Jacques Nicolas (1703–72)
Title: Carte de la Moscovie Europeene.
Year: 1750
Dimensions: 43.5 x 32.0 (29.7 x 23.5) cm
Phys. Desc.: Engraved, Uncolored
Language(s): French

Entry (No.): 58
Author: Bellin, Jacques Nicolas (1703–72)
Title: Carte des // Embouchures // du Tanais. // Et des Environs depuis Taigan, // jusqu'a Nova-Krepa.
Year: 1750
Dimensions: 23.7 x 32.0 (16.7 x 23.4) cm
Phys. Desc.: Engraved, Uncolored
Language(s): French

Entry (No.): 59
Author: Bellin, Jacques Nicolas (1703–72)
Title: Carte Pour les Voyages de Rubruquis, Marco Polo, Jen-kin-son, + c. Inserès dans L'Histoire Generale des Voyages.
Year: 1750
Dimensions: 47.8 x 39.5 (34.3 x 24.5) cm
Phys. Desc.: Engraved, Uncolored
Language(s): French

Entry (No.): 60
Author: Bellin, Jacques Nicolas (1703–72)
Title: Plan d'Azak où Azof.
Year: 1750
Dimensions: 23.7 x 32.0 (16.4 x 21.4) cm
Phys. Desc.: Engraved, Uncolored
Language(s): French

Entry (No.): 61
Author: Bellin, Jacques Nicolas (1703–72)
Title: Carte Réduite de la Mer Noire, Dressée pour le Service des Vaisseaux du Roy par Ordre de M. de Boynes, Secretaire d'Etat, ayant le Département de la Marine. Par le S. Bellin, Ingénieur de la Marine. 1772.
Year: 1772
Dimensions: 84.0 x 63.5 (81.0 x 53.0) cm
Phys. Desc.: Engraved, Uncolored
Language(s): French

Entry (No.): 62
Author: Berry, William (1639–1718)
Title: The States of the Empire of the Turks in Europe Subdivided according to the extent of its Governments. To the Most Serene and Most Sacred Majesty Charles II. By the Grace of God King of Great Britain, France and Ireland. This Map of Turky [sic] in Europe is Humbly Dedicated and Presented By your Majesties Loyall Subject and Servant William Berry.
Year: ca. 1680–89
Dimensions: 89.8 x 58.2 (86.3 x 56.5)
Phys. Desc.: Engraved, Colored
Language(s): English

Entry (No.): 63
Author: Berry, William (1639–1718)
Title: The Kingdom of Hungary and the States that have been Subject to it, which are now the Northern Parts of Turkey in Europe. To the most Serene and most Sacred Majesty of Charles II. By the grace of God King of Great Britain, France and Ireland. This map of Hungary is Humbly Dedicated and presented by your Majesties Loyall Subject and Servant, William Berry. London Sold by William Berry at the Sign of the Globe between Charing-Cross, and White-Hall.
Year: ca. 1700
Dimensions: 88.0 x 57.5 (86.0 x 56.5) cm
Phys. Desc.: Engraved, Colored
Language(s): English

Entry (No.): 64
Author: Bertius, Petrus (Pierre) 1565–1629 (Bert, Pieter)
Title: Breviarium Totius Orbis Terrarum, una cum serie

Romanorum Imperatorum. autore Petro Bertio.
Year: ca. late 16th–early 17th century
Dimensions: 20.2 x 15.5 cm
Phys. Desc.: Book with Maps, No Cover
Language(s): Latin

Entry (No.): 65
Author: Blankaart (Blanckaert, Blancardus), Nicolas (1629–1703)
Title: Asia Antiqua cum finitimis Africae et Europae Regionibus. Nicolaus Blancardus. Belga, Leidensis, ad lucem aevi veteris delineabat. Nobilissimo Amplissimoque Vivo d. Gerardo Schaep, I.C. Cortenhoevii domino ex legato ad serenissimos Daniae Sueciaeque Reges, magnificaeque reip Amstelae damensis consuli, etc. Nicolaus Blancardus Dedicabat. L.M.Q.
Year: 1670 map; 1684 atlas
Dimensions: 61.6 x 53.5 cm
Phys. Desc.: Engraved
Language(s): Latin

Entry (No.): 66
Author: Blome (Bloome), Richard (1641–1705)
Title: A Mapp of the Estates of the Crowne of Poland, where are the Kingdom of Poland with its Palatinates the Dutches of Russia = noire Cujavia Mozovia Prussia Lithuania Volhynia Podolia +c. designed by Monsieur Sanson Geographer to the French King and Rendred into English by Ric.: Blome, By his Majesties Especiall Command. London. Printed for ric: Blome 1669.
Year: 1669
Dimensions: 44.0 x 36.9 (39.3 x 26.9) cm
Phys. Desc.: Engraved, Colored Outline, Cartouche
Language(s): English

Entry (No.): 67
Author: Blome (Bloome), Richard (1641–1705)
Title: A Mapp of the Estates of the Great Duke of Russia Blanch or Moscovia, Designed by Monsieurr // Sanson... Geographer to the French King. To y[e] R[t] Hon[rbl] S[r] Edw Dering of Surrenden Dering in Kent Bart one of the Lords Comissioners of his Majesty's Treasury, This Mapp is most humbly Dedicated by Richard Blome. /F. Lamb Sculp.
Year: ca. 1682–93
Dimensions: 42.0 x 29.2 (40.2 x 26.4) cm
Phys. Desc.: Engraved, Colored Outline
Language(s): English

Entry (No.): 68
Author: Blome (Bloome), Richard (1641–1705)
Title: A Mapp of y[e] Kingdome of Tartary, by Monsieur Sanson. To the Right Worshipful Sr James Shaen of Shy[ne] in Surrey Knight and Baronet This Map is humbly D.D. by R. Blome.
Year: 1693
Dimensions: 40.0 x 33.5 (39.2 x 25.7) cm
Phys. Desc.: Engraved, Colored
Language(s): English

Entry (No.): 69
Author: Bodenehr, Gabriel (1664–1758)
Title: Prospect von dem Turkischen Berg: Schloß Ussiza welches in der Nacht zwischen dem 1ten und 2ten Octob. 1737 an die Kaysere mit Accord übergangen.
Year: 1700
Dimensions: 29.5 x 18.0 (20.0 x 16.8) cm
Phys. Desc.: Engraved, Uncolored
Language(s): German

Entry (No.): 70
Author: Bodenehr, Johann (Hans) Georg (1631–1704)
Title: Jassi. Die Haupt-Stadt in der Moldau an dem Fluße Pruth....
Year: 1700
Dimensions: 29.5 x 18.0 (19.3 x 11.7) cm
Phys. Desc.: Engraved, Uncolored
Language(s): German

Entry (No.): 71
Author: Bodenehr, Johann (Hans) Georg (1631–1704)
Title: Ungarn samt angräntzenden Ländern Croatien, Dalmatien, Sclavonien, Siebenbürgen, Moldau, Wallacheu, Bulgarien u. Romanien, bis an das Schwartze Meer und Constantinopel.
Year: ca. 1704–37
Dimensions: 29.7 x 18.2 (25.6 x 14.8) cm
Phys. Desc.: Engraved, Colored Outline
Language(s): German
Figure 34

Entry (No.): 72
Author: Bodenehr, Johann (Hans) Georg (1631–1704)
Title: Mongats oder Munkatsch.
Year: 18th c.
Dimensions: 29.5 x 17.6 (21.7 x 13.7)
Phys. Desc.: Engraved, Uncolored
Language(s): German

Entry (No.): 73
Author: Bonné, Rigobert (1727–95)
Title: Partie Occidentale de l'Empire DE RUSSIE, Extrait de l'Atlas Russien et d'autres Cartes Projetée et assujetie aux Observations Astronomiques Posterieures, par M[r]. Bonne, Hydrographe du Roi. A Paris Chez Lattré, ruë St. Jacques, à la Ville de Bordeaux avec Pr. du Roy 1785.
Year: 1785
Dimensions: 38.5 x 51.0 (31.5 x 43.0) cm
Phys. Desc.: Engraved, Colored
Language(s): French
Figure 54

Entry (No.): 74
Author: Bonné, Rigobert (1727–95)
Title: Partie Orientale de l'Empire DE RUSSIE, Extrait de l'Atlas Russien et d'autres Cartes Projettee et assujetie aux Observations astronomiques Posterieures, par M[r]. Bonne Geographe du Roi. A Paris chez Lattre, rue St. Jaques, à la Ville de Bordeaux, avec Pr. du Roy.
Year: 1785
Dimensions: 38.5 x 51.2 (31.0 x 43.0) cm

Phys. Desc.: Engraved, Colored
Language(s): French

Entry (No.): 75
Author: Botero, Giovanni (1540–1617)
Title: MOS // CO // VIA//
Year: 1596
Dimensions: 34.7 x 27.0 (21.0 x 16.0) cm
Phys. Desc.: Engraved
Language(s): Latin

Entry (No.): 76
Author: Bowen, Emanuel (ca. 1693/1694–1767)
Title: The Old Limits between the Russian and Turkish Empires… Eman. Bowen Sculp.
Year: ca. 1744–63
Dimensions: 35.5 x 23.6 (33.0 x 19.8) cm
Phys. Desc.: Engraved, Uncolored
Language(s): English

Entry (No.): 77
Author: Bowen, Emanuel (ca. 1693/1694–1767)
Title: A New and Accurate Map of Poland, Lithuania &c. Divided into its Palatinats. Drawn from the best Authorities, assisted by the most approv'd Modern Maps. The whole being regulated by Astronomical Observations. By Eman[l]. Bowen
Year: 1747
Dimensions: 46.0 x 37.5 (43.0 x 34.3) cm
Phys. Desc.: Engraved, Uncolored
Language(s): English

Entry (No.): 78
Author: Bowen, Emanuel (ca. 1693/1694–1767)
Title: A New and Accurate Map of Turky [sic] in Europe, with the adjacent Countries of Hungary, Little Tartary & c. Collected from the best Authorities. The whole being regulated by Astronomical Observations… By Emanl. Bowen.
Year: 1747
Dimensions: 46.5 x 37.2 (42.5 x 34.8) cm
Phys. Desc.: Engraved, Uncolored
Language(s): English

Entry (No.): 79
Author: Bowen, Emanuel (ca. 1693/1694–1767)
Title: A New and Accurate Map of Moscovy, or Russia in Europe with its Acquisitions. Drawn from Surverys & the best Modern Maps. The whole being Regulated by Astronomical Observations.
Year: 1767
Dimensions: 25.5 x 34.5 (22.2 x 31.8) cm
Phys. Desc.: Engraved, Uncolored
Language(s): English

Entry (No.): 80
Author: Bowen, Emanuel (ca. 1693/1694–1767)
Title: Nieuwe en Naauwkeurige Kaart van't Konigryk Polen en't Groot Hertogdoom Lithauwen in derzelver Palatinaatschappen verdeeld; door Eman. Bowen. Waar by komt, de laaste verdeeling van een groot gedeelte dezer beiden Landshappen, onder de Oostenrykers, Russen en Pruissen in 1771 door W.A. Bachiene.
Year: 1783
Dimensions: 60.0 x 46.5 (41.8 x 34.0) cm
Phys. Desc.: Engraved, Colored
Language(s): Dutch

Entry (No.): 81
Author: Bowen, Thomas (ca. 1732/1733–1790)
Title: Poland, Lithuania and Prussia By Thomas Bowen.
Year: 1752
Dimensions: 19.2 x 17.5 cm
Phys. Desc.: Engraved, Colored
Language(s): English

Entry (No.): 82
Author: Bowles, John (ca. 1701–79) & Bowles, Thomas III (ca. 1712–62)
Title: Bowles's New One-Sheet Map of Europe, Divided into Its Empires, Kingdoms, States, and Subdivisions; laid down from Observations of the most Celebrated Geographers. London: Printed for the Proprietors Bowles and Carver, No. 69 St. Paul's Church Yard.
Year: 1785
Dimensions: 69.0 x 53.5 (56.5 x 48.5) cm
Phys. Desc.: Engraved, Colored
Language(s): English
Figure 55

Entry (No.): 83
Author: Bowles, John (ca. 1701–79) & Bowles, Thomas III (ca. 1712–62)
Title: Bowles's New One-Sheet Map of Poland, and Prussia: With the Roads, +c. ….Bowles + Carver…London.
Year: 1785
Dimensions: 67.8 x 53.1 (52.8 x 45.0) cm
Phys. Desc.: Engraved, Colored
Language(s): English

Entry (No.): 84
Author: Bowles, John (ca. 1701–79) & Bowles, Thomas III (ca. 1712–62)
Title: Bowles's New One-Sheet Map of Poland, and Prussia: With the Roads, +c. …Bowles + Carver…London.
Year: 1785
Dimensions: 68.0 x 54.0 (52.5 x 45.0) cm
Phys. Desc.: Engraved, Colored
Language(s): English

Entry (No.): 85
Author: Braun (Bruin), Georg (Joris) (1541–1622)
Author: Hogenberg, Franciscus (1535–90)
Author: AA, PIETER (PIERRE, PETRUS) VAN DER (1659–1733)
Title: Leopolis. Posnania. Crasno. Premislia. Palanka (Communicant Georgius Houfnaglius) acceptum aliunde A° 1617. Zamos=civm.
Year: ca. 16th c.
Dimensions: 62.8 x 45.7 cm

Phys. Desc.: Copy and Negatives
Language(s): Latin

Entry (No.): 86
Author: Briffaut, Étienne
Title: THEATRE DES OPERATIONS DE LA GUERRE DANS LA TARTARIE DU CUBAN OZAKOW & TARTARIE CRIMÉE, AVEC LES MARCHES & RENDÉS VOUS DE L'ARMÉE DE RUSSIE CONTRE LES TURCS & TARTARES, LE TOUT SUIVANT L'ORIGINAL // DE PETERSBOURG, VENANT DU CAMP DE LA TARTARIE CRIMEÉ MOIS DE JUILLET 736. A VIENNE EN AUTRICHE LE 20. AOUST 1736. CHEZ LE SIEUR BRIFFAUT LIBRAIRE DE L'UNIVERSITEZ. AVEC PRIVILEGE DE S.M.J. &. C.
Year: 1736
Dimensions: 62.5 x 52.5 (44.7 x 40.5) cm
Phys. Desc.: Engraved, Colored
Language(s): French

Entry (No.): 87
Author: Broniowski, Marcin
Title: Tabula, seu Typus Camporum Chersonesi Tauricae, seu Peninsulae.
Year: 1595 (copy)
Dimensions: 46.0 x 40.5 cm
Phys. Desc.: Copy, Uncolored
Language(s): Latin

Entry (No.): 88
Author: Brué, Adrien Hubert (1786–1832)
Title: Carte Générale de L'Empire D'Autriche, du Royau^{me} de Prusse de la Confédération Germanique et du Roy.^{me} de Pologne, Par A.H. Brue.
Year: 1821
Dimensions: 67.8 x 53.9 (50.8 x 36.2) cm
Phys. Desc.: Engraved, Colored
Language(s): French

Entry (No.): 89
Author: Brué, Adrien Hubert (1786–1832)
Title: Carte Générale de la Russie d'Europe, du Royaume de Pologne, des Environs du Caucase et d'une Partie des Pays adjacents, par A. Brué, Géographe du Roi." Paris 1826.
Year: 1826
Dimensions: 54.0 x 68.4 (36.3 x 50.9) cm
Phys. Desc.: Engraved, Colored
Language(s): French

Entry (No.): 90
Author: Carey, Mathew (1760–1839)
Title: The // RUSSIAN EMPIRE, // in // EUROPE AND ASIA.
Year: ca. early 19th c.
Dimensions: 56.0 x 43.0 (40.5 x 29.0) cm
Phys. Desc.: Engraved, Colored
Language(s): English
Figure 60

Entry (No.): 91
Author: Cary, John (the elder) (1755–1835)
Title: A // New Map // of // Poland, // and the // Grand Duchy of Lithuania, // Shewing their Dismemberments and Divisions // between // Austria, Russia and Prussia, // According to the Congress of Vienna // from the Latest Authorities: // By John Cary Engraver. 1819.
Year: 1819
Dimensions: 63.5 x 54.0 (51.5 x 45.5) cm
Phys. Desc.: Engraved, Colored
Language(s): English
Figure 62

Entry (No.): 92
Author: Chaffat, Antonius du (d. 1740)
Title: PROVINCIARUM // TURCICO TARTARICARUM // INTER TANAIM BORYSTHENEM // ET BOGUM SITARUM // QUAS DUOBUS ANNIS VIZ. 1736 ET 1737. // ANNA // AUGUSTISSIMA RUSSORUM AVTOCRATRIX // SUBEGIT // EX PROTOTYPO PETROPOLITANO CONCINNATA // ACCURATISSIMA // DELINEATIO // IN QUA SIMUL OCCUPATIO PENINSULAE // CRIM ET CUBAN NEC NON EXPUGNATIO ASSOW // ET OCZAKOW EXHIBENTUR // STUDIO ANT. DU CHAFFAT, // CIRC. SUEVICI ET CIV.IMP. AUGUST. // CAPIT. ET REI FORTIFIC. PRAEF. // C. P. S. C. M. //
Year: 1738
Dimensions: 75.5 x 62.0 (72.5 x 57.5) cm
Phys. Desc.: Engraved, Colored
Language(s): Latin

Entry (No.): 93
Author: Chaffat, Antonius du (d. 1740)
Title: THEATRUM BELLI // Anno MDCCXXXVII. // A MILITE / AUGUSTAE RUSSORUM IMPERATRICIS // ADVERSUS TURCAS TATTAROSQUE // GESTI // EDITUM EX AUTOGRAPHO PETROPOLITANO // PER ANTONIUM DU CHAFFAT. // CIRC. SUEVICI ET CIV. IMP. AUGUST. // CAPIT. ET REI FORTIFIC. PRAEF. // C. P. S. C. M. //
Year: 1738
Dimensions: 69.0 x 53.0 (64.0 x 47.0) cm
Phys. Desc.: Engraved, Colored
Language(s): Latin

Entry (No.): 94
Author: Chamouin, Jean Baptiste Marie (b. 1768)
Title: Empire D'Autriche.
Year: 1805
Dimensions: 41.0 x 26.0 (30.0 x 22.0) cm
Phys. Desc.: Engraved, Colored Outline
Language(s): French

Entry (No.): 95
Author: Chamouin, Jean Baptiste Marie (b. 1768)
Title: Partie Méridionale de la Russie D'Europe.
Year: 1805
Dimensions: 41.0 x 26.0 (30.0 x 22.0) cm
Phys. Desc.: Engraved, Colored Outline
Language(s): French

Entry (No.): 96

Author: Chamouin, Jean Baptiste Marie (b. 1768)
Title: Partie Septent. de la Russie D'Europe.
Year: 1805
Dimensions: 41.0 x 26.0 (30.0 x 22.0) cm
Phys. Desc.: Engraved, Colored Outline
Language(s): French

Entry (No.): 97
Author: Châtelain, Henri Abraham (1684–1743)
Title: Carte de l'Europe Suivant les Plus Nouvelles Observations de Messieurs de l'Academie des Sciences et des Principaux Geographes avec des Tables et de Nouvelles Instructions Trés Utiles et Trés Curieuses Pour L'Intelligence de l'Histoire.
Year: 1705
Dimensions: 65.0 x 55.5 (58.7 x 47.2) cm
Phys. Desc.: Engraved, Uncolored
Language(s): French

Entry (No.): 98
Author: Châtelain, Henri Abraham (1684–1743)
Title: Carte Ancienne et Moderne des Differents Etats et Païs situez au Long du Danube, pour servir a L'Intelligence de l'Histoire.
Year: 1708
Dimensions: 52.5 x 45.0 (45.0 x 35.0) cm
Phys. Desc.: Engraved, Uncolored
Language(s): French

Entry (No.): 99
Author: Châtelain, Henri Abraham (1684–1743)
Title: Carte de Pologne avec La Chronologie des Rois et des Ducs de Lithuanie, Ainsi Que Des Grands Maitres De L'Ordre Teutonique, les Evéchez et Archevéchez, et les Armes des Provinces, avec une Table des Batailles, et des Villes, Les Plus Considerables de Pologne.
Year: 1714
Dimensions: 63.0 x 52.7 (59.5 x 50.0) cm
Phys. Desc.: Engraved, Uncolored
Language(s): French

Entry (No.): 100
Author: Châtelain, Henri Abraham (1684–1743)
Title: Carte Pour Servir a L'Intelligence de L'Histoire de Gustave Adolphe, de Christine, de Charles Gustave et de Charles XII.
Year: 1714
Dimensions: 55.5 x 48.1 (44.2 x 35.0) cm
Phys. Desc.: Engraved, Uncolored
Language(s): French

Entry (No.): 101
Author: Châtelain, Henri Abraham (1684–1743)
Title: Nouvelle Carte de Moscovie où Sont Representés les Diferents Etats de Sa Maiesté Czarienne en Europe et en Asie et le Chemin d'un de ses Ambassadeurs a Peking Ville Capitale de L'Empereur de la Chine et son Sejour Ordinaire.
Year: 1714
Dimensions: 66.5 x 55.0 (68.5 x 50.5) cm
Phys. Desc.: Engraved, Uncolored
Language(s): French
Figure 40

Entry (No.): 102
Author: Châtelain, Henri Abraham (1684–1743)
Title: Nouvelle Carte des Etats du Grand Duc de Moscovie en Europe.
Year: 1714
Dimensions: 48.8 x 44.5 (46.2 x 36.5) cm
Phys. Desc.: Engraved, Uncolored
Language(s): French

Entry (No.): 103
Author: Châtelain, Henri Abraham (1684–1743)
Title: Nouvelle Carte du Royame de Pologne, Divisée Selon ses Palatinats et Ses Provinces, avec des Remarques Historiques Pour L'Intelligence de Histoire.
Year: 1714
Dimensions: 66.0 x 56.5 (58.2 x 50.5) cm
Phys. Desc.: Engraved, Uncolored
Language(s): French

Entry (No.): 104
Author: Châtelain, Henri Abraham (1684–1743)
Title: Carte Nouvelle de la Grande Tartarie ou de L'Empire du Grand Cham, Dressée sur les Memoires Originaux + sur d'autres observations tres exactes de divers Voyageurs.
Year: ca. 1732–39
Dimensions: 55.5 x 48.0 (50.8 x 40.0) cm
Phys. Desc.: Engraved, Uncolored
Language(s): French

Entry (No.): 105
Author: Châtelain, Henri Abraham (1684–1743)
Title: Genealogie des Anciens Empereurs Tartares, Descendus de Gengiscan.
Year: ca. 1732–39
Dimensions: 56.0 x 48.5 (43.0 x 36.5) cm
Phys. Desc.: Engraved, Uncolored
Language(s): French

Entry (No.): 106
Author: Châtelain, Henri Abraham (1684–1743)
Title: Nouvelle Carte de l'Asie Avec de Tables Alphabetiques Pour Trouver Sans Peine les Etats des Principaux Princes de Cette Partie du Monde, Une Table des Villes Principales et Quelques Remarques Pour L'Intelligence de l'Histoire d'Asie.
Year: 1732
Dimensions: 64.5 x 56.0 (58.0 x 47.5) cm
Phys. Desc.: Engraved, Uncolored
Language(s): French

Entry (No.): 107
Author: Châtelain, Henri Abraham (1684–1743)
Title: Carte Generale des Etats du Czar Empereur de Moscovie, ou L'On Voit ce que ce Prince Possede en Europe et Les Etats que Possede en Asie avec la Plus Part de Leurs Armes et des Tables des Remarques pour Servir a L'Intelligence

de L'Histoire.
Year: 1735
Dimensions: 105.3 x 48.2 (97.3 x 38.5) cm
Phys. Desc.: Engraved, Uncolored
Language(s): French

Entry (No.): 108
Author: Clüver (Cluever), Philipp (1580–1623)
Title: Asia
Year: 1676
Dimensions: 27.5 x 20.0 (25.2 x 18.5) cm
Phys. Desc.: Engraved, Uncolored
Language(s): Latin

Entry (No.): 109
Author: Clüver (Cluever), Philipp (1580–1623)
Title: Nova Europae, Descriptio.
Year: 1676
Dimensions: 28.3 x 20.0 (25.7 x 18.6) cm
Phys. Desc.: Engraved, Uncolored
Language(s): Latin

Entry (No.): 110
Author: Clüver (Cluever), Philipp (1580–1623)
Title: Philippi Cluverii Introductio in Universam Geographiam. Tam Vetererem quam Novam. Libri IV. Amsterdami, Apud Jansonio Waesbergius, Anno MDCLXXVI.
Year: 1676
Dimensions: 20.2 x 15.5 cm
Phys. Desc.: Book Text
Language(s): Latin

Entry (No.): 111
Author: Clüver (Cluever), Philipp (1580–1623)
Title: Russia cum Confiniis.
Year: 1676
Dimensions: 26.5 x 20.0 (19.8 x 14.3) cm
Phys. Desc.: Engraved, Uncolored
Language(s): Latin

Entry (No.): 112
Author: Clüver (Cluever), Philipp (1580–1623)
Title: Sarmatia et Scythia - Russia et Tartaria Europae.
Year: 1711
Dimensions: 32.5 x 25.7 (24.7 x 22.9) cm
Phys. Desc.: Engraved, Uncolored
Language(s): Latin

Entry (No.): 113
Author: Clüver (Cluever), Philipp (1580–1623)
Title: Veteris et Novae Regni Poloniae Magnia Ducatus, Lithuaniae cum suis Palatinatibus ac Confiniis Descriptio.
Year: 1711
Dimensions: 32.4 x 23.5 (24.2 x 20.0) cm
Phys. Desc.: Engraved, Uncolored
Language(s): Latin

Entry (No.): 114
Author: Cochrane, Joseph (publ.)
Author: Pickersgill, Lieutenant Richard
Title: A New Map of the Kingdom of Poland Exhibiting its Ancient Limits, Dismemberments, and Present Boundary as Settled by Act of Congress at Vienna.
Year: after 1815
Dimensions: 60.5 x 51.0 (54.5 x 44.2) cm
Phys. Desc.: Engraved, Colored
Language(s): English

Entry (No.): 115
Author: Coronelli, (Marco) Vincenzo (1650–1718)
Title: Geografia Sagra
Year: ca. 1691–97
Dimensions: 35.5 x 49.5 cm
Phys. Desc.: Text
Language(s): Italian

Entry (No.): 116
Author: Coronelli, (Marco) Vincenzo (1650–1718)
Title: Russia Negra. Ruska ziemia. Palatinato di Luwow. Carpates. Transilvania.
Year: ca. 1691–97
Dimensions: 47.5 x 70.0 (44.0 x 61.0) cm
Phys. Desc.: Engraved, Uncolored
Language(s): Italian
Entry (No.): 117
Author: Coronelli, (Marco) Vincenzo (1650–1718)
Title: Ungheria.
Year: 1691
Dimensions: 48.5 x 70.0 (44.5 x 60.5) cm
Phys. Desc.: Engraved, Uncolored
Language(s): Italian

Entry (No.): 118
Author: Coronelli, (Marco) Vincenzo (1650–1718)
Title: Parte Orientale Dell' Europa, Descritta, e Dedicata Dal P. Cosmografo Coronelli All' Illustrissimo, et Eccellentissimo Signore Giovanni da Mula, Senatore amplissimo Nella Serenissima Republica di Venetia.
Year: 1695
Dimensions: 49.5 x 68.5 (45.5 x 60.5) cm
Phys. Desc.: Engraved, Colored
Language(s): Italian
Figure 24

Entry (No.): 119
Author: Coronelli, (Marco) Vincenzo (1650–1718)
Title: Partie Orientale del Mediterraneo del Padre Cosmografo Coronelli, In Venecia con Privilegio dell' Ecc. Senato.
Year: 1697
Dimensions: 69.5 x 49.3 (60.5 x 44.5) cm
Phys. Desc.: Engraved, Uncolored
Language(s): Italian

Entry (No.): 120
Author: Coronelli, (Marco) Vincenzo (1650–1718)
Title: Ristretto del Mediterraneo, composto Dal P. Maestro Coronelli Cosmografo della Serenissima Republica di Venetia.
Year: 1697

Dimensions: 70.0 x 48.6 (61.0 x 45.0) cm
Phys. Desc.: Engraved, Uncolored
Language(s): Italian

Entry (No.): 121
Author: Covens, Johannes (Jean, Jan) (1697–1774) & Mortier, Cornelis (Corneille) (1699–1783)
Title: Imperii Orientalis et Circumjacentium Regionum sub Constantino Porphyrogenito et ejus Praedecessoribus Descriptio // Auctore Guillelmo Del'Isle e Regiâ Scientiarum Academiâ, ad Novam Editionem Libri Constantini Porphyrogeniti de Administrando Imperio // quae Prodit Studio Anselmi Bandurii Monachi Benedictioni Ragustini Melitensis cum ejusdem Notis et Animaadversionibus. Amsterdami apuel J. Covens et C. Mortier.
Year: ca. late 17th–early 18th c.
Dimensions: 65.5 x 52.8 (62.0 x 44.0) cm
Phys. Desc.: Engraved, Colored Outline
Language(s): Latin

Entry (No.): 122
Author: Covens, Johannes (Jean, Jan) (1697–1774) & Mortier, Cornelis (Corneille) (1699–1783)
Title: Partie Meridionale de Moscovie Dressée Par G. de l'Isle, A Amsterdam chez J. Covens et C. Mortier // Avec privilege.
Year: 1690
Dimensions: 61.6 x 51.5 cm
Phys. Desc.: Engraved, Colored Outline
Language(s): French

Entry (No.): 123
Author: Covens, Johannes (Jean, Jan) (1697–1774) & Mortier, Cornelis (Corneille) (1699–1783)
Title: La Grande // Tartarie // Suivant les // Nouvelles Observations // de // Messrs. de l'Academie Royale // des Sciences, etc. // Augmentes de Nouveau // A Amsterdam // Chez Cóvens et Mortier.
Year: 1730
Dimensions: 39.0 x 26.8 (29.5 x 22.1) cm
Phys. Desc.: Engraved, Colored Outline
Language(s): French

Entry (No.): 124
Author: Covens, Johannes (Jean, Jan) (1697–1774) & Mortier, Cornelis (Corneille) (1699–1783)
Title: Carte // de Tartariae // Dressée sur les Relations // de plusieurs Voyageurs de differentes Nations // et sur quelques Observations // qui ont été faites dans ce païs la // Par Guillaume Del Isle // de l'Academie Royale des Sciences. // A Amsterdam // Chez J. Covens et C. Mortier.
Year: 1733
Dimensions: 64.5 x 52.8 (61.0 x 47.5) cm
Phys. Desc.: Engraved, Colored Outline
Language(s): French

Entry (No.): 125
Author: Covens, Johannes (Jean, Jan) (1697–1774) & Mortier, Cornelis (Corneille) (1699–1783)
Title: Carte D'Asie // dressée pour l'Usage du Roy // Sur les memoires envoyéz par le Czar // a l'Academie Royale des Sciences // Sur ce que les Arabes nous ont laissé de plus exact des pays orientaux // sur un grand nombre de Routiers de terre et de mer // et de Cartes manuscrites detaillées. // Le tout assujeti aux observations de l'Academie // et a Celles des RRPP Jesuites et autres Mathemaciens. // Par Guillaume Delisle Premier Geographe de Sa Majesté // De l'Academie Royale des Sciences. // A Amsterdam // Chez Jean Covens et Corneille Mortier Geographes.
Year: 1733
Dimensions: 65.0 x 52.5 (61.2 x 49.0) cm
Phys. Desc.: Engraved, Colored Outline
Language(s): French

Entry (No.): 126
Author: Covens, Johannes (Jean, Jan) (1697–1774) & Mortier, Cornelis (Corneille) (1699–1783)
Title: Estats De La Couronne de // Pologne // subdivisés Suivant l'estendue des // Palatinats // Dresée sur ce qu'en ont donné Starovolsk, // Beauplan, Hartnoch, et autres Auteurs // Rectifiée par les Observations d'Hevelius etc: // Par Guillaume Del'Isle Geograp: // de l'Academie Royale des Sciences // A Amsterdam // Chez J. Covens et C. Mortier // Avec Privil.
Year: 1733
Dimensions: 64.5 x 53.0 (60.5 x 44.5) cm
Phys. Desc.: Engraved, Colored Outline
Language(s): French

Entry (No.): 127
Author: Covens, Johannes (Jean, Jan) (1697–1774) & Mortier, Cornelis (Corneille) (1699–1783)
Title: La Pologne // Dressée sur ce qu'on ont donné // Starovolsk, Beauplan, Hartnoch, // et autres Auteurs, // Rectifie par les Observations d'Hevelins etc, // Par Guillaume de l'Isle Geographe // del'Academie Royale des Sciences // A Amsterdam // Chez J. Covens et C. Mortier// Geographes.
Year: 1733
Dimensions: 68.6 x 57.5 (61.5 x 49.0) cm
Phys. Desc.: Engraved, Colored Outline
Language(s): French

Entry (No.): 128
Author: Covens, Johannes (Jean, Jan) (1697–1774) & Mortier, Cornelis (Corneille) (1699–1783)
Title: Le Royaume de // Hongrie // et des Pays qui en dependoient auterfois // Dressée sur un grand nombre de memoires // et Cartes manuscrites ou imprimées // Rectifiez par les Observations du Cte Marsilii // et quelches autres // Par Guillaume De L'Isle Geographe // de l'Academie Royale des Sciences // Chez J. Covens et C. Mortier // Geog.[s] // avec Privilege. Nova et Accurata Regni Hungariae Tabula ad usum Serenissimi Burgundiae Ducis.
Year: 1733
Dimensions: 64.5 x 53.0 (57.0 x 46.0) cm
Phys. Desc.: Engraved, Colored Outline
Language(s): French

Entry (No.): 129

Author: Covens, Johannes (Jean, Jan) (1697–1774) & Mortier, Cornelis (Corneille) (1699–1783)
Title: Seconde Partie de La Crimée, La Mer Noire &c. Rectifies par diverses Observations Faites par Guilliaume de l'Isle, // A Amsterdam chez Jean Covens et Corneille Mortier, Geographes.
Year: 1733
Dimensions: 64.5 x 53.0 (61.5 x 48.5) cm
Phys. Desc.: Engraved, Colored Outline
Language(s): French

Entry (No.): 130
Author: Covens, Johannes (Jean, Jan) (1697–1774) & Mortier, Cornelis (Corneille) (1699–1783)
Title: Theatre de la Guerre dans la Petite Tartarie, la Crimée, la Mer Noire, etc. Dressée sur un grand nombre de Cartes Manuscrites et // d'Itineraires, recifiés par diverses observations faites par Guillaume de l'Isle. A Amsterdam chez Jean Covens et Cornielle Mortier, Geographes.
Year: 1733
Dimensions: 65.0 x 52.5 (61.5 x 49.5) cm
Phys. Desc.: Engraved, Colored outline
Language(s): French

Entry (No.): 131
Author: Covens, Johannes (Jean, Jan) (1697–1774) & Mortier, Cornelis (Corneille) (1699–1783)
Title: Theatre de la Guerre dans la Petite Tartarie, la Crimée, la Mer Noire, & c. Dressée sur un grande nombre de Cartes Manuscrites et // et d'Itineraires, rectifiés par diverses Observations faites par Guillaume de l'Isle. A Amsterdam chez Jean Covens + Corneille Mortier, Geographes.
Year: 1739
Dimensions: 65.5 x 54.0 (61.5 x 49.0) cm
Phys. Desc.: Engraved, Colored outline
Language(s): French

Entry (No.): 132
Author: Crépy (Crespy) (18th c. family of publishers)
Title: Royaume de Hongrie // Et Transilvanie, Partie du Royaume de Pologne // Ukraine, Etats du Czar et de la Crimée, // Partie des Etats du Grand // Seigneur // En Europe, Valaquie, Bosnie, Bulgario. // Servie Romanie & c. // Dressée sur les Memoires les plus // Nouveaux par Crepy // Geographe.
Year: 1737
Dimensions: 74.5 x 51.0 (69.1 x 46.0) cm
Phys. Desc.: Engraved, Colored Cartouche
Language(s): French

Entry (No.): 133
Author: Crépy (Crespy) (18th c. family of publishers)
Title: Carte Générale // de l'Empire de Russie // Comprise en Europe. // Pour l'intelligence de la Guerre présente // entre l'Impératrice de toutes les Russies, et // le Turc // d'Apres les Meilleurs Auteurs. // Par le Sr Maichey Ingenieur Géographe // Publiée en 1788. A Paris. chez Crepy rue S[t]. Jacques à S[t]. Pierre pr'es // la rue de la parcheminerie, // On trouvera chez M. Moichey.
Year: 1788
Dimensions: 66.0 x 87.5 (64.0 x 86.0) cm
Phys. Desc.: Engraved, Colored Cartouche
Language(s): French

Entry (No.): 134
Author: Dahlberg, (Count) Erik Jonsson (1625–1703)
Title: Auctior et Correctior Tabula Chorographica // REGNI POLONIAE, // Vicinarumque Regionum, ubi itinera, quae Seren. // Rex Sueciae // CAROLUS GUSTAVUS // una cum Exercitibus suis ab initio usque ad finem belli fecit, // nummeris denotantur, per supr. Castr. Metat. loc. E. I. Dahlberg. // Stationes et loca ubi Sereniss. Rex ipsemet suit Corona indicantur; // reliqua Vero itinera per Eiusdem Generales suere confecta. //
Year: 1696–97
Dimensions: 39.0 x 31.5 (37.5 x 29.0) cm
Phys. Desc.: Engraved
Language(s): Latin
Figure 25

Entry (No.): 135
Author: Dahlberg, (Count) Erik Jonsson (1625–1703)
Title: VRBS ET CASTELLUM BRESTZIE LITEWSKI A S.R.M. SUECIAE DIE 11 // MAY 1657. OBSESSUM, DIE VERO 13. EIUSDEM MENSIS DEDITIONE OCCUPATUM, … // … // PER E. I. DAHLBERGH //
Year: 1696–97
Dimensions: 49.5 x 31.0 (47.0 x 25.5) cm
Phys. Desc.: Engraved
Language(s): Latin

Entry (No.): 136
Author: Danckerts (Danckaerts, Danckers, Dankerts, Danquerts), Cornelius (1664–1717)
Title: Nova // Transilvaniae // Principatus // Tabula. // Novissima descriptia editia // Per // Cornelium Danckerts, Amstelodami // Cum Privilegio.
Year: ca. 1685
Dimensions: 65.0 x 53.0 (58.0 x 49.8) cm
Phys. Desc.: Engraved, Colored
Language(s): Latin

Entry (No.): 137
Author: Danckerts (Danckaerts, Danckers, Dankerts, Danquerts), Cornelius (1664–1717)
Title: Regni // Poloniae // Ducatus // Mazoviae / et Province // Cujaviae // Descriptio Emendata // per // Cornelium Danckerts, // Amsterdami // Cum Privilegio.
Year: ca. 1685
Dimensions: 61.5 x 52.9 cm
Phys. Desc.: Engraved, Colored
Language(s): Latin

Entry (No.): 138
Author: Danckerts (Danckaerts, Danckers, Dankerts, Danquerts), Cornelius (1664–1717)
Title: Novissima et Accuratissima // Totius // Russiae // Vulgo // Muscoviviae // Tabula // a // Justo Dankerts // Amstelodami // cum Privilegio Ordinum Holland: // West-vrisiae.
Year: ca. 1685

Dimensions: 61.5 x 53.0 (57.5 x 49.5) cm
Phys. Desc.: Engraved, Uncolored
Language(s): Latin

Entry (No.): 139
Author: Danckerts (Danckaerts, Danckers, Dankerts, Danquerts), Cornelius (1664–1717)
Title: Exactissima // Asiae // Delineatio. // In Praecipuas Regiones // Caeterasq. Partes divisa. // et denuo in lucem edita // Per // T. Danckerts // Amstelodami cum Privilegio.
Year: ca. 1685
Dimensions: 61.5 x 53.2 (60.0 x 49.5) cm
Phys. Desc.: Engraved, Uncolored
Language(s): Latin
Figure 22

Entry (No.): 140
Author: d'Anville, Jean Baptiste Bourguignon (1697–1782)
Title: Tartariae Sinensis Mappa Geographica ex Tabulis Specialibus R.R.P.P. Iesuitarum nec non Relationibus R.P. Gerbillon per Dom. d'Anville Geographum Parifiensem primum Anno 1732 nunc secundum LL, projectionis Stereographicae in usum translationis Germanicae. Historiae Sinene. Haldianae descripta per Tobiam Mayer, Soc. Cosmogr. Sodal. Curis Homaniorum Heredum Anno 1749 C.P.S.C.M.
Year: 1749
Dimensions: 81.5 x 50.2 (77.5 x 49.2) cm
Phys. Desc.: Engraved, Colored
Language(s): Latin

Entry (No.): 141
Author: d'Anville, Jean Baptiste Bourguignon (1697–1782)
Title: Troisième Partie de la Carte // d'Asie, // Contenant // la Sibérie // et Quelques autres Parties // de la Tartarie // Publiée sous les Auspices de Monseigneur // Louis-Philippe d'Orléans // Duc d'Orleans // Prémier Prince du Sang. // Par le Sr. d'Anville // Sécrétaire de Son Alt. Sereniss. // MDCCLIII // Avec Privilege. [Eastern Siberia]
Year: 1753
Dimensions: 81.6 x 56.5 (53.5 x 51.0) cm
Phys. Desc.: Engraved, Colored
Language(s): French

Entry (No.): 142
Author: d'Anville, Jean Baptiste Bourguignon (1697–1782)
Title: Troisième Partie de la Carte // d'Asie, // Contenant // la Sibérie // et Quelques autres Parties // de la Tartarie // Publiée sous les Auspices de Monseigneur // Louis-Philippe d'Orléans // Duc d'Orleans // Prémier Prince du Sang. // Par le Sr. d'Anville // Sécrétaire de Son Alt. Sereniss. // MDCCLIII // Avec Privilege. [Western Siberia]
Year: 1753
Dimensions: 82.0 x 56.5 (59.0 x 51.0) cm
Phys. Desc.: Engraved, Colored
Language(s): French

Entry (No.): 143
Author: d'Anville, Jean Baptiste Bourguignon (1697–1782)
Title: Partie de l'Empire // de // Russie // Comprise en Europe // Par le S.^{r} d'Anville // de l'Acad.' R.^{le} des Belles-lettres // et de celle des Sc.' de Petersbourg, // Sécret'de S.A.S.M. le D.' d Orleans.
Year: 1759
Dimensions: 38.5 x 49.0 (35.2 x 43.0) cm
Phys. Desc.: Engraved, Uncolored
Language(s): French

Entry (No.): 144
Author: d'Anville, Jean Baptiste Bourguignon (1697–1782)
Title: Troisième Partie de la Carte // d'Europe // Contenant // Le Midi de la Russie, // la Pologne et la Hongrie, // la Turquie // y compris celle d'Asie presqu'entière // Publiee Sous les Aspices de Monseigneur // Louis Philippe d'Orléans // Duc d'Orléans, Premier Prince du Sang. // Par le S.^{r} d'Anville, de l'Académie R.^{le} des Belles-Lettres, et de celle des Sc.'? de Petersbourg. // Sécrétaire de Son Alt.' Seren.' MDCCLX.
Year: 1760
Dimensions: 83.0 x 56.0 (80.5 x 51.5) cm
Phys. Desc.: Engraved, Colored Outline
Language(s): French

Entry (No.): 145
Author: d'Anville, Jean Baptiste Bourguignon (1697–1782)
Title: The // Third Part of // Europe, // Drawn & Engraved // from D'Anvilles // Two Sheet Map // for // J. Harrison N° 115 Newgate Street London." // Feb. 15 1791.
Year: 1791
Dimensions: 54.7 x 79.5 (50.5 x 72.5) cm
Phys. Desc.: Engraved, Colored Outline
Language(s): English
Figure 56

Entry (No.): 146
Author: d'Anville, Jean Baptiste Bourguignon (1697–1782)
Title: The Third Part of Asia Drawn + Engraved from D'Anvilles Two Sheet Map for J. Harrison N^{o} n 5 Newgate Street London Feb. 11th 1791.
Year: 1791
Dimensions: 80.4 x 54.7 (71.5 x 43.0) cm
Phys. Desc.: Engraved, Colored Outline
Language(s): English

Entry (No.): 147
Author: Aefferden (Afferden, Alferden), Francisco de (1653–1709)
Title: ESTATS // DE LA // COURONNE // DE // POLOGNE.
Year: 1692
Dimensions: 20.5 x 18.5 (15.2 x 13.5) cm
Phys. Desc.: Engraved, Colored
Language(s): French

Entry (No.): 148
Author: Fer, Nicolas de (1646–1720)
Title: Kamieniec Podolski // Ville Forte des Estats de Pologne et de la // Haute Podolie avec Titre d'Evéché de la Pa- // latinat, située sur le Sommet d'une Montagne // entourée de la Rivière de Smotrycz, et en en- // velopée de Rochers escarpez, à 48. degrez // 45 minutes le Latitude de 50. deg. 10. min. // de Longitude. Cette Place fut prise par les //

Turcs sur les Polonois [sic] le 29.e Aoust 1672.
Year: 1691
Dimensions: 28.5 x 21.0 cm
Phys. Desc.: Engraved, Uncolored
Language(s): French

Entry (No.): 149
Author: Fer, Nicolas de (1646–1720)
Title: Carte Generale // du // Globe Terrestre // par N. De Fer. Methode pour apprendre facilement la Geographie, contenant Un Abbrege de la Sphere, en ses Continents, Empires, Royaumes, Etats, Republiques, Provinces, etc… Tome Premier. A Paris, Chez Anoine Dezallier, rue St. Jacques, a la Couronne d'Or.
Year: 1703
Dimensions: 17.5 x 16.5 (16.5 x 13.5) cm
Phys. Desc.: Engraved, Colored Outline
Language(s): French

Entry (No.): 150
Author: Fer, Nicolas de (1646–1720)
Title: GRECE OU // PARTIE MERIDIONALE // DE LA TURQUIE EN // EUROPE. // DIVISE EN GRANDS GOUVERNEMENTS // SELON RICAUT ANGLOIS. // PAR N. DE FER. // 1703.
Year: 1703
Dimensions: 17.5 x 16.5 (15.5 x 13.6) cm
Phys. Desc.: Engraved, Colored
Language(s): French

Entry (No.): 151
Author: Fer, Nicolas de (1646–1720)
Title: L'EUROPE // PAR DE FER // 1703.
Year: 1703
Dimensions: 17.5 x 16.5 (15.5 x 14.0) cm
Phys. Desc.: Engraved, Colored
Language(s): French
Figure 30

Entry (No.): 152
Author: Fer, Nicolas de (1646–1720)
Title: RVSSIE // BLANCHE OU // MOSCOVIE // PAR N. DE FER // 1703.
Year: 1703
Dimensions: 17.5 x 16.5 (16.5 x 13.7) cm
Phys. Desc.: Engraved, Colored
Language(s): French

Entry (No.): 153
Author: Fer, Nicolas de (1646–1720)
Title: VKRAINE OU PAYS DES COSAQUES/ /1703/. LE GRAND ROYAUME DE HONGRIE OU // PARTIE SEPTENTRIONALE DE LA TURQUIE EN EUROPE, // DIVISEE PAR GRANDS GOUVERNEMENTS, SUIVANT RICAUT ANGLOIS, // PAR N. DE FER. //
Year: 1703
Dimensions: 17.5 x 16.5 (16.5 x 13.7) cm
Phys. Desc.: Engraved, Colored
Language(s): French

Entry (No.): 154
Author: Fer, Nicolas de (1646–1720)
Title: VKRAINE/ /1703/. ESTATS // DE LA COURONNE // DE POLOGNE // PAR N. DE FER //.
Year: 1703
Dimensions: 17.5 x 16.5 (16.5 x 13.7) cm
Phys. Desc.: Engraved, Colored
Language(s): French

Entry (No.): 155
Author: Fer, Nicolas de (1646–1720)
Title: Profil // de La Fortesse // de Montgast. // Dessiné Sur les lieux par N… Ingenieur de l'Emp // Et mis au jour Par N. De Fer…1705
Year: 1705
Dimensions: 27.0 x 20.0 cm
Phys. Desc.: Engraved, Uncolored
Language(s): French

Entry (No.): 156
Author: Fer, Nicolas de (1646–1720)
Title: Carte Des Estats // de Suede, de Dannemarq, et de Pologne, // sur la Mer Baltique // Pour servir a l'Intilligence [sic] des affaires des Couronnes du Nord // ou se trouvent encore // les Estats de la Couronné de Pologne l'Empire d'Allemagne, // et les XVII. Provinces de Pays Bas. // Avec des Routes de Paris dans tous ces Estats. // Par N. de Fer. Geographe de Monseigneur le Dauphin. // A Paris chez l'Autheur dans l'Isle du Palais sur le Quay de l'Orange a la Sphere Royale // Avec Privilege du Roy 1705.
Year: 1705
Dimensions: 38.2 x 46.5 (35.1 x 43.7) cm
Phys. Desc.: Engraved, Uncolored
Language(s): French

Entry (No.): 157
Author: Fer, Nicolas de (1646–1720)
Title: Plan de la Ville // de Rome / Par de Fer. // A Paris // chez l'Autheur dans l'Isle du Palais sur le Quay de l'Orloge a la Sphere Royale avec Privelege du Roy 1705.
Year: 1705
Dimensions: 39.0 x 27.0 (30.0 x 23.0) cm
Phys. Desc.: Engraved, Uncolored
Language(s): French

Entry (No.): 158
Author: Fer, Nicolas de (1646–1720)
Title: Description de la Bataille de Poltawa Par de Fer.
Year: 1714
Dimensions: 38.7 x 25.8 (23.5 x 17.5) cm
Phys. Desc.: Engraved, Uncolored
Language(s): French

Entry (No.): 159
Author: Fer, Nicolas de (1646–1720)
Title: Description de la Mer Noire. A Paris…chez le Sr. de fer dans l'Isle du Palais sur le Quay de l'Orloge a la Sphere Royale 1714.
Year: 1714
Dimensions: 39.0 x 25.8 (23.3 x 17.2) cm

Phys. Desc.: Engraved, Uncolored
Language(s): French

Entry (No.): 160
Author: Fer, Nicolas de (1646–1720)
Title: La journée de // Poltawa // en Ukraine, // le 8.e juillet 1709. // Entre l'Armée de sa Majesté // Suedoise Charles XII, et celle de // sa Majesté Csarienne Pierre I. // Empereur de la grande Russie, // dit vulgairement Grand Duc de // Moscovie // de Fer. 1714.
Year: 1714
Dimensions: 38.7 x 25.8 (33.0 x 22.7) cm
Phys. Desc.: Engraved, Uncolored
Language(s): French
Figure 36

Entry (No.): 161
Author: Fer, Nicolas de (1646–1720)
Title: Description de Constantinople Par N. de Fer.
Year: ca. 1717
Dimensions: 39.0 x 27.0 (22.0 x 16.3) cm
Phys. Desc.: Engraved, Uncolored
Language(s): French

Entry (No.): 162
Author: Fer, Nicolas de (1646–1720)
Title: Description du Detroit des Dardanelles de Constantinople, par N. de Fer.
Year: ca. 1717
Dimensions: 39.0 x 27.0 (22.5 x 16.2) cm
Phys. Desc.: Engraved, Uncolored
Language(s): French

Entry (No.): 163
Author: Fer, Nicolas de (1646–1720)
Title: Estats de la Couronne // de Pologne // Par de Fer 1717.
Year: 1717
Dimensions: 39.0 x 25.8 (31.5 x 21.9) cm
Phys. Desc.: Engraved, Uncolored
Language(s): French

Entry (No.): 164
Author: Fer, Nicolas de (1646–1720)
Title: Estats de la Couronne // de Pologne // Par de Fer 1717.
Year: 1717
Dimensions: 39.4 x 27.0 (31.5 x 21.9) cm
Phys. Desc.: Engraved, Uncolored
Language(s): French

Entry (No.): 165
Author: Fer, Nicolas de (1646–1720)
Title: La Hongrie // La Transilvanie // La Valaquie // La Moldavie & // Par N. de Fer. Geographe // de sa Majesté Catholique // A Paris // Chez l'Auteur dans l'Isle // du Palais??r le Quay // de l'Orloge a la Sphere // Royale 1717.
Year: 1717
Dimensions: 39.2 x 27.0 (32.0 x 22.0) cm
Phys. Desc.: Engraved, Uncolored
Language(s): French

Entry (No.): 166
Author: Fer, Nicolas de (1646–1720)
Title: La Hongrie // La Transilvanie // La Valaquie // La Moldavie etc. Par // N. de Fer Geographe // de sa Majesté Catolique // A. Paris // Chez l'Auteur dans l'Isle // du Palais sur le Quay // de l'Orloge a la Sphere // Royale 1717.
Year: 1717
Dimensions: 38.8 x 25.8 (32.0 x 22.2) cm
Phys. Desc.: Engraved, Uncolored
Language(s): French

Entry (No.): 167
Author: Fer, Nicolas de (1646–1720)
Title: Veue de la Ville et du Port de Constantinople.
Year: ca. 1717
Dimensions: 39.0 x 27.0 (33.5 x 23.2) cm
Phys. Desc.: Engraved, Uncolored
Language(s): French

Entry (No.): 168
Author: Fer, Nicolas de (1646–1720)
Title: Vue de Dardanelles de Constantinople.
Year: 1717
Dimensions: 39.0 x 27.0 (34.2 x 23.0) cm
Phys. Desc.: Engraved, Uncolored
Language(s): French

Entry (No.): 169
Author: Fer, Nicolas de (1646–1720)
Title: Les Etats du Czar ou Empereur des Russes // en Europe et en Asie, // Avec les Routes Q'uon [sic] tíent ordinairement de Moscow a Pekim. // Dressé sur les Memoires de N. Witzen, Evart Esbrand, le P. Avril et autres // par N. de Fer Geographe de sa Majesté Catolique. // A Paris // Chez J.F. Bernard // gendre du S. de Fer // geographe de sa Majesté Catolique quay // del orlog du Palais // à la Sphere Royale 1722.
Year: 1722
Dimensions: 78.0 x 55.5 (63.5 x 41.4) cm
Phys. Desc.: Engraved, Colored
Language(s): French

Entry (No.): 170
Author: Fer, Nicolas de (1646–1720)
Title: Description de la Moscovie et de la Hongrie Par N. de Fer.
Year: 1725
Dimensions: 38.7 x 25.8 (25.0 x 17.0) cm
Phys. Desc.: Engraved, Uncolored
Language(s): French

Entry (No.): 171
Author: Fer, Nicolas de (1646–1720)
Title: Description de La Pologne par N. de Fer.
Year: 1725
Dimensions: 38.7 x 25.8 (22.0 x 16.5) cm
Phys. Desc.: Engraved, Uncolored
Language(s): French

Entry (No.): 172

Author: Fer, Nicolas de (1646–1720)
Title: Estats du // Grand Duc // de Moscovie // Dressé sur de Nouveaux // Memoire par N. de Fer.
Year: 1725
Dimensions: 38.8 x 25.8 (32.5 x 21.4) cm
Phys. Desc.: Engraved, Uncolored
Language(s): French

Entry (No.): 173
Author: Fer, Nicolas de (1646–1720)
Title: Estats du // Grand Duc // de Moscovie Suivant les dernieré relations par N. de Fer.
Year: 1725
Dimensions: 39.5 x 27.0 (18.3 x 13.5) cm
Phys. Desc.: Engraved, Uncolored
Language(s): French

Entry (No.): 174
Author: Fer, Nicolas de (1646–1720)
Title: La Mer Noire, Autrefois Pont-Euxin, // Appellée par les Tures // Cara-Denghis, Et par les Cosaques Czorno-More // Cette Carte a été tirée d'apres un Manuscript Turc faita Caffa Par de Fer Géographe de sa Majesté Catholique.
Year: ca. early 18th c.
Dimensions: 39.0 x 26.0 (34.0 x 24.0) cm
Phys. Desc.: Engraved, Uncolored
Language(s): French

Entry (No.): 175
Author: Jode, Gerard de (1515–91) and Jode, Cornelius de (1568–1600)
Title: POLONIAE // AMPLISSIMI / REGNI...TYPVS // GEOGRAPHICVS //
Year: 1593
Dimensions: 54.8 x 39.6 (49.2 x 37.8) cm
Phys. Desc.: Engraved, Colored
Language(s): Latin
Figure 8

Entry (No.): 176
Author: Delisle (De L'Isle), Guillaume (1675–1726)
Title: L'EUROPE // DRESSÉE SUR LES OBSERVATIONS DE M.rs // DE L'ACADEMIE ROYALE DES SCIENCES // ET QUELQUES AUTRES; & SUR LES MEMOI- // RES LES PLUS RECENS. // PAR G. DE L'ISLE GEOGRAPHE. // A PARIS, // CHÉZ L'AUTEUR… // 1700. // H. van Loon Sculp.
Year: 1700
Dimensions: 64.5 x 49.7 (58.3 x 46.0) cm
Phys. Desc.: Engraved, Colored Outline
Language(s): French
Figure 29

Entry (No.): 177
Author: Delisle (De L'Isle), Guillaume (1675–1726)
Author: Buache, Philippe (1700–1773)
Title: LA POLOGNE // DRESSÉE SUR CE QU'EN ONT DONNÉ // STAROVOLSK, BEAUPLAN, HARTNOCH, ET AUTRES AUTEUR // RECTIFIÉE PAR LES OBSERVATIONS D'HEVELIUS ETC. // PAR GUILLAUME DE L'ISLE // DE L'ACADEMIE ROYALE DES SCIENCES // ET PREMIER GEOGRAPHE DU ROY // A PARIS // CHEZ L'AUTEUR FUR LE QUAI DE L'HORLOGE AL AIGLE D'OR // AVEC PRIVILEGE DU ROY / / 1703
Year: ca. 1703–45
Dimensions: 70.5 x 52.0 (62.4 x 48.0) cm
Phys. Desc.: Engraved, Colored outline
Language(s): French

Entry (No.): 178
Author: Delisle (De L'Isle), Guillaume (1675–1726)
Title: CARTE // DE MOSCOVIE // DRESSÉE PAR GUILLAUME DE-L'ISLE // DE L'ACADEMIE ROYALE DES SCIENCES // … // A Paris chez l'Auteur… // 1706. // (Inscription in the lower left-hand corner of the 2nd sheet: Gravé par Liebaux le fils.)
Year: 1706
Dimensions: (in two sheets): 65.7 x 49.5 (64.6 x 48.7) cm & 67.0 x 49.8 (64.6 x 48.5) cm
Phys. Desc.: Engraved, Colored Outline
Language(s): French

Entry (No.): 179
Author: Delisle (De L'Isle), Guillaume (1675–1726)
Author: Renard, Louis
Title: CARTE // DE TARTARIE // DRESSÉE SUR LES RELATIONS // DE PLUSIEURS VOYAGEURS DE DIFFERENTES NATIONS // ET SUR QUELQUES OBSERVATIONS // QUI ONT ÉTÉ FAITES DANS CE PAÏS LA // PAR GUILLAUME DE L'ISLE // DE L'ACADEMIE ROYALE DES SCIENCES // A PARIS CHEZ L'AUTEUR FUR LE QUAI DE L'HORLOGE À LA // COUR. DE DIAMAS AVEC PRIVILEGE // 1706. Et s'etrouve a Amsterdam Chez Louis Renard Libraire prez de la Bourse. -N. Guerard...fecit. -des Rosiers fot.
Year: 1706
Dimensions: 67.3 x 51.0 (63.5 x 48.2) cm
Phys. Desc.: Engraved, Colored Outline
Language(s): French

Entry (No.): 180
Author: Delisle (De L'Isle), Guillaume (1675–1726)
Author: Schenck, Pieter, Sr. (1645–1715)
Title: L'EUROPE // DRESSÉE SUR LES OBSERVATIONS DE // M.rs DE L'ACADEMIE ROYALE DES SCIENCES // ET QUELQUES AUTRES, & SUR LES MEMOIRES // LES PLUS RECENS. // PAR G. DE L'ISLE GEOGRAPHE. / A AMSTERDAM // CHÉZ PIERRE SCHENK, // SUR LE VYGENDAM, [AT'?] ATLAS DE SANSON. // AVEC PRIVILEGE. 1708.
Year: 1708
Dimensions: 59.5 x 51.5 (57.0 x 45.0) cm
Phys. Desc.: Engraved, Colored
Language(s): French

Entry (No.): 181
Author: Delisle (De L'Isle), Guillaume (1675–1726)
Author: Buache, Philippe (1700–1773)
Title: CARTE // D'EUROPE // DRESSÉE POUR L'USAGE DU ROY // SUR LES ITINERAIRES ANCIENS ET

MODERNES // ET SUR LES ROUTIERS DE MER // ASSUJETIS AUX OBSERVATIONS ASTRONOMIQUES // PAR G. DELISLE PREMIER GEOGRAPHE DE S. MR. // DE L'ACADEMIE ROYALE DES SCIENCES. // A PARIS // CHEZ L'AUTEUR QUAY De L'HORLOGE // AVEC PRIVILEGE // MAY 1724. // Ph. Buache... Gendre de l'Auteur Avec Privilége du 30 Av. 1745.
Year: ca. 1724–45
Dimensions: 64.0 x 55.7 (60.5 x 49.0) cm
Phys. Desc.: Engraved, Colored
Language(s): French

Entry (No.): 182
Author: Delisle (De L'Isle), Guillaume (1675–1726)
Author: Buache, Philippe (1700–1773)
Title: CARTE PARTICULIERE DE LA HUNGRIE DE LA TRANSYLVANIE DE LA CROATIE ET DE LA SCLAVONIE // DRESSÉE SUR LES OBSERVATIONS DE M. le COMPTE MARSILLI ET SUR PLUSIERS AUTRES MEMOIRES PAR G. DELISLE PREMIER GEOGRAPHE DE L'ACADEMIE R.LE DES SCIENCES A PARIS CHEZ DEZAUCHE SUCCESSUER DES Srs DELISLE ET BUACHE RUE DES NOYERS // AVEC PRIVILEGE // Ph.Buache...Avec privilege de 30 Avril 1745.
Year: 1745
Dimensions: 78.0 x 55.5 (58.5 x 48.0) cm
Phys. Desc.: Engraved, Colored Outline
Language(s): French

Entry (No.): 183
Author: Delisle (De L'Isle), Guillaume (1675–1726)
Author: Buache, Philippe (1700–1773)
Title: CARTE // DE TARTARIE // DRESSÉE SUR LES RELATIONS // DE PLUSIEURS VOYAGEURS DE DIFFERENTES NATIONS // ET SUR QUELQUES OBSERVATIONS // QUI ONT ÉTÉ FAITES DANS CE PAÏS LA // PAR GUILLAUME DE L'ISLE // PREMIER GEOG. DU ROY // DE L'ACADEMIE ROYALE DES SCIENCES // A PARIS CHEZ DEZAUCHE RUE DE NOYERS PRÉS // CELLE DES ANGLOIS // AVEC PRIVILEGE // 1766. // N. Guerard... fecit. - des Rosier fet. Ph.Buache... Gendre de l'auteur. Avec Priv. du Roi.
Year: 1766
Dimensions: 78.4 x 55.5 (63.5 x 48.5) cm
Phys. Desc.: Engraved, Colored Outline
Language(s): French

Entry (No.): 184
Author: Delisle (De L'Isle), Guillaume (1675–1726)
Author: Buache, Philippe (1700–1773)
Title: CARTE // DE MOSCOVIE // DRESSÉE PAR GUILLAUME DE-L'ISLE // PREMIER GÉOGRAPHE DE L'ACADEMIE //...A PARIS CHEZ DEZAUCHE SUCCESSEUR DES SRS. // DE L'ISLE ET BUACHE... // 1780 //
Year: 1780
Dimensions: 77.5 x 55.5 (65.4 x 49.5) cm & 78.0 x 55.5 (65.7 x 49.5) cm
Phys. Desc.: Engraved, Colored Outline
Language(s): French

Entry (No.): 185
Author: Delisle (De L'Isle), Guillaume (1675–1726)
Author: Dezauche, J. A.
Title: THEATRUM // HISTORICUM // AD ANNUM CHRISTI QUADRINGENTESIMU // IN QUO // TU IMPERII ROMANI TU BARBORORUM // CIRCUMINCOLENTIUM STATUS OB OCULOS PONITUR // PARS ORIENTALIS // AUTORE GIULLELMO DELISLE // REGIAE SCIENTARUM ACADEMIAE SICIO // ET PRIMA REGIS GÉOGRAPHO // A PARIS CHEZ DEZAUCHE, SUCCESSEUR // ET POSSESSEUR DU FOND GÉOGRAPHIQUE DES // Srs. DE L'ISLE ET BUACHE, RUE DES NOYERS // CUM PRIVELEGIA REGIS // 1780. // Ph. Buache...
Year: 1780
Dimensions: 78.0 x 55.5 (64.5 x 48.0) cm
Phys. Desc.: Engraved, Colored Outline
Language(s): Latin

Entry (No.): 186
Author: Delisle (De L'Isle), Guillaume (1675–1726)
Author: Buache, Philippe (1700–1773)
Author: Dezauche, J. A.
Title: IMPERII ORIENTALIS ET CIRCUMJACENTIUM REGIONUM SUB CONSTANTINO PORPHYROGENITO ET EIUS PRAEDECESSORIBUS DESCRIPTIO // AUCTORE GUILLELMO DEL'LISLE È REGIÂ SCIENTIARUM ACADEMIÂ, AD NOVAM EDITIONEM LIBRI CONSTANTINI PORPHYROGENITI DE ADMINISTRANDO IMPERIO // QUAE PRODIT STUDIO ANSELMI BANDURII MONACHI BENEDECTINI RAGUSINI MELITENSIS CUM EIUSDEM NOTIS ET ANIMADVERSIONIBUS. A PARIS CHEZ DEZAUCHE, RUE DE NOYERS PRÈS CELLE DES ANGLOIS // Ph.Buache.
Year: 1789
Dimensions: 77.5 x 55.5 (64.0 x 47.7) cm
Phys. Desc.: Engraved, Colored Outline
Language(s): Latin

Entry (No.): 187
Author: Du Val (Du Vall) d'Abbeville, Pierre (1618–1683)
Title: Orbis Vetus // Authore P. du Val Abbarvillaeo // Christianissimi Galliarum Regis // Geographo // A Paris // Chez l'Autheur, Rue des Bons Enfants // au Monde Royal. // Avec Privilege du Roy pour vingt Ans. // 1655.
Year: 1655
Dimensions: 57.0 x 43.0 (51.5 x 35.5) cm
Phys. Desc.: Engraved, Colored Outline
Language(s): French

Entry (No.): 188
Author: Du Val (Du Vall) d'Abbeville, Pierre (1618–1683)
Title: Novissima et Accuratissima // Totius // Hungariae // Tabula // Par P. du Val // Geographe du Roy // Amsterdami // Gedruckt by Justus Dankers.
Year: ca. 1670–87
Dimensions: 61.5 x 53.0 (47.5 x 39.0) cm
Phys. Desc.: Engraved, Colored Outline
Language(s): Latin

Entry (No.): 189
Author: Dubosc, Claude (engr.)
Title: A Map of the // Seat of War in Hungary, // between the Imperialists and the Turks.'
Year: 1736
Dimensions: 51.0 x 40.5 (43.4 x 31.8) cm
Phys. Desc.: Engraved, Uncolored
Language(s): English

Entry (No.): 190
Author: Dury, Andrew (ca. 1715–78)
Title: This Map of the present Seat of War, // between the Russians, Poles, and Turks // is most humbly Inscribed by his // Excellency's most humble and most Obedt Servt, Andrew Dury. To // His Excellency the Count of Czernichew, // Commander in Chief of her Majesty the Empress of all the Russia's Fleet and Gallies // Lieutenant General of her Armies, one of the Lords // of her Admiralty, Ambassador Extraordinary // and Plenipotentiary to his Majesty the King of // Great Britain, // a Knight of the Order of the White Eagle, // and St. Ann, &c. &c. &c. // This map is compiled by A. Dury and P. Bell.
Year: 1770
Dimensions: (in two parts) 129.0 x 53.5 (126.2 x 42.5) cm 129.0 x 23.0 (126.2 x 42.5) cm
Phys. Desc.: Engraved, Colored
Language(s): English

Entry (No.): 191
Author: Dury, Andrew (ca. 1715–78)
Title: Hongrie. Kingdom of Hungary.
Year: 1795
Dimensions: 15.5 x 13.0 (12.0 x 10.2) cm
Phys. Desc.: Engraved, Colored Outline
Language(s): French, English

Entry (No.): 192
Author: Dury, Andrew (ca. 1715–78)
Title: La Pologne. Poland.
Year: 1795
Dimensions: 15.7 x 13.0 (12.0 x 10.2) cm
Phys. Desc.: Engraved, Colored Outline
Language(s): French, English

Entry (No.): 193
Author: Dury, Andrew (ca. 1715–78)
Title: Russie en Asiatique. Russia in Asia.
Year: 1795
Dimensions: 15.5 x 13.0 (12.2 x 13.0) cm
Phys. Desc.: Engraved, Colored Outline
Language(s): French, English

Entry (No.): 194
Author: Dury, Andrew (ca. 1715–78)
Title: Russie Europeenne. Russia in Europe.
Year: 1795
Dimensions: 15.5 x 13.0 (12.0 x 10.2) cm
Phys. Desc.: Engraved, Colored Outline
Language(s): French, English

Entry (No.): 195
Author: Elwe, Jan Barend (Barent) (worked 1785–1809)
Title: Estats del' Empire du Grand Seigneur // des Turcs, // en Europe, en Asie, et en Afrique, // divisé en touts ses // Beglerbeglicz, ou Gouvernements, // où sont aussi remarqués les Estats qui luy sont Tribulaires. // Dressé sur les plus Nouvelles Relations. // A Amsterdam. // Chez Jan Bt. Elwe. // MDCCXCII.
Year: 1792
Dimensions: 64.0 x 55.5 (61.0 x 45.0) cm
Phys. Desc.: Engraved, Uncolored
Language(s): French

Entry (No.): 196
Author: Elwe, Jan Barend (Barent) (worked 1785–1809)
Title: La // Pologne, // Divisee // en // Royaume de Pologne // et les Etats Apartenans // la Pologne Prussienne // Autrichienne & // Russienne // A Amsterdam // Chez J. B. Elwe.
Year: ca. 1796 (reprint)
Dimensions: 64.3 x 53.5 (60.8 x 48.5) cm
Phys. Desc.: Engraved, Colored
Language(s): French

Entry (No.): 197
Author: Euler, Leonhard (1701–83)
Title: Tab: Geogr: // Asiae // ad emendatiora, quae // adhuc prodierunt, exempla // jussu // Acad: Reg: Scient: // et eleg: Citt: Boruss // descripta.
Year: 1760
Dimensions: 48.5 x 37.5 (38.0 x 33.0) cm
Phys. Desc.: Engraved, Colored
Language(s): Latin

Entry (No.): 198
Author: Euler, Leonhard (1701–83)
Title: Tab: Geogr: // Imperii // Russici // ad emendatiora // ad huc edita exempla // jussu // Acad: Reg: Scient:// et eleg Litt Boruss // descripta.
Year: 1760
Dimensions: 50.0 x 39.0 (36.0 x 31.7) cm
Phys. Desc.: Engraved, Colored
Language(s): Latin

Entry (No.): 199
Author: Faden, William (1749–1836)
Title: A Map of // the Kingdom of Poland // and Grand Dutchy of Lithuania // Including Samogitia and Curland, // Divided according to their Dismemberments // with the Kingdom of Prussia // By Willm. Faden, Geographer to the King // MDCCXCIX.
Year: 1799
Dimensions: 79.5 x 57.5 (59.5 x 53.0) cm
Phys. Desc.: Engraved, Colored
Language(s): English
Figure 57

Entry (No.): 200
Author: Faden, William (1749–1836)
Title: The // Russian Dominions // in // Europe // drawn from the latest Maps printed, // by the // Academy of Sciences, St.

Petersburg; // revised and corrected, // with the Post Roads & New Gouvernments, // from the Russian Atlas of 1806: // by // Jasper Nantiat, // London. // Published by W. Vaden, Geographer to His Majesty, and His Royal Highness, the Prince of Wales, Charing Cross, June 4th 1808.
Year: 1808
Dimensions: 96.5 x 59.0 (91.5 x 53.5) cm
Phys. Desc.: Engraved, Colored
Language(s): English

Entry (No.): 201
Author: Felssecker, Adam Jonathan (1683–1729)
Title: Neu=eröfnetes // Kriegs = Theatrvm // in // Ungarn, Ruszland // und der Türckeÿ, // worinnen alle Provintzien, wie sie // an das Türkische Reich angrän - // tzen, zu erzehen // sindt //. Nurnberg bey Adam Jonathan Velsteckers (Velsseckers) Sel Erben A. 1737.
Year: 1737
Dimensions: 58.0 x 45.5 (49.5 x 35.2) cm
Phys. Desc.: Engraved, Colored
Language(s): German

Entry (No.): 202
Author: Gastaldi, Jacopo (Giacomo) (ca. 1500–ca. 1566)
Title: Descriptione de la Mo // scovia per Giacomo // Gastaldo piamontese // Cosmographo in Ve' // netia MDL. [Map of Russia, from Sigismund von Herbertstein's Commentari Della Moscovia, Venice, 1550.]
Year: ca. 1550
Dimensions: 38.0 x 25.5 cm
Phys. Desc.: Copy
Language(s): Italian, Latin
Figure 12

Entry (No.): 203
Author: Gastaldi, Jacopo (Giacomo) (ca. 1500–ca. 1566)
Title: Il Disegno de Geographia Mo // derna del Regno di Polonia, e par // te del Ducado di Moscovia.
Year: 1562
Dimensions: 42.0 x 32.0 cm
Phys. Desc.: Copy and Negative
Language(s): Italian

Entry (No.): 204
Author: Gastaldi, Jacopo (Giacomo) (ca. 1500–ca. 1566)
Title: Il nevo disegno della secondda // parte dil Regno de Polonia. // dell' eccmo. Mp. Giacomo Gastaldo piamontese.// In venetia l'Anno MDLXVIII
Year: 1568
Dimensions: 39.0 x 30.0 cm
Phys. Desc.: Copy
Language(s): Italian

Entry (No.): 205
Author: Gerritsz (Gerritszoon, Gerard, Gerardus, Gheritszoon van Assum), Hessel (1580–1632)
Title: Tabula // Russiae // ex autographo, quod delineandum // curavit Foedor filius Boris desum: // ta; et ad fluvios Dwinam, Zuchanam, aliaque // loca, quantum ex tabulis et notitus ad nos dela: // tis fieri potuit, amplificata: ac // Magno Domino, Tzari Michäeli Foedrowits omnium // Russorum, Autocratori Wolodimeriae, Mos: // coviae et Novogardiae, Tzari Cazaniae, Tzari // Astracaniae, Tzari Siberiae, Domino Plescoviae // Magno Duci Smolenscoviae, Otweriae, Iugoriae, Per: // miae, Wiatkiae, Bulgariae etc: Item Domino // et Magno Duci Novogardiae Inferioris etc: // Domino regionum Iveriae Kartalinie // et Groesiniae Tzari etc: // dedicata ab Hesselo Gerardo MDCXIII.
Year: 1614
Dimensions: 66.6 x 50.5 (54.2 x 42.3) cm
Phys. Desc.: Engraved, Colored
Language(s): Latin

Entry (No.): 206
Author: Gibson, John (sculp.) (worked 1750–92)
Title: A Map of some // Provinces in // Siberia, and the // Borders towards // China, being an Appendage to the // Map of the Russian // Frontiers, as Surveyd // by the Russians // 1747.
Year: 1747
Dimensions: 36.5 x 36.8 (35.0 x 33.5) cm
Phys. Desc.: Engraved, Uncolored
Language(s): English

Entry (No.): 207
Author: Götze (Goetze), Ferdinand
Title: Charte // des // Schwartzen-Meeres // Nach Murdochischer Projection entworfen und nach den // neuesten astronomischen Ortsbestimmulungen berichtiget // auf der Sternwarte Seeberg bey Gotha // gezeichnet von Ferd: Götze, // Weimar // im Verlage des Industrie Comptoirs. // 1800.
Year: 1800
Dimensions: 72.5 x 58.0 (56.5 x 42.0) cm
Phys. Desc.: Engraved, Colored
Language(s): German

Entry (No.): 208
Author: Götze (Goetze), Ferdinand
Title: Charte // von // Ungarn, Siebenbürgen // u.s.l. // Nach den bewährtesten Hulfsmitteln neu entworfen. Und gezeichnet // von // F. Götze. // Weimar // Im Verlage des L. Industrie Comptoirs. 1802.
Year: 1802
Dimensions: 72.5 x 58.0 (61.0 x 42.5) cm
Phys. Desc.: Engraved, Colored
Language(s): German

Entry (No.): 209
Author: Götze (Goetze), Ferdinand
Title: Charte // des // Schwartzen-Meeres // Nach Murdochischer Projection entworfen und nach den // neuesten astronomischen Ortsbestimmulungen berichtiget // auf der Sternwarte Seeberg bey Gotha // gezeichnet von Ferd: Götze, // Weimar // im Verlage des Industrie Comptoirs. // 1804.
Year: 1804
Dimensions: 64.5 x 55.0 (57.0 x 42.5) cm
Phys. Desc.: Engraved, Colored
Language(s): German

Entry (No.): 210
Author: Grata, François
Author: AA, PIETER (PIERRE, PETRUS) VAN DER (1659–1733)
Title: Plan exact des Fortifications du Château de CHOCZIM sur le Niester, avec les Forts et retranchemens de l'Armée des Turcs défaite en 1673 par Jean Sobieski, // Général en Chef des Polonois ensuite Roi de Pologne, sous le nom de Jean III. Dessiné sur les lieux par François Grata, Secretaire de Sa Majesté, et gravé par Rom. de Hooge. // a Leide, Chez Pierre vancer Aa.
Year: 1673
Dimensions: 49.5 x 38.0 (35.0 x 27.5) cm
Phys. Desc.: Engraved
Language(s): French
Figure 21

Entry (No.): 211
Author: Grata, François
Author: AA, PIETER (PIERRE, PETRUS) VAN DER (1659–1733)
Title: LE SIEGE ET DELIVRANCE DU CHÂTEAU TREMBOLA, PAR JEAN SOBIESKI, GENERAL EN CHEF // DES POLONOIS...DESSINÉ SUR LES LIEUX PAR FRANÇOIS GRATA, SECRETAIRE DE SA MAJESTÉ, ET GRAVÉ PAR ROM. DE HOOGE. // A Leide, Chez Pierre vander Aa.
Year: 1675
Dimensions: 49.5 x 38.0 (36.0 x 29.0) cm
Phys. Desc.: Engraved
Language(s): French

Entry (No.): 212
Author: Grodecki (Godreccius), Wacław (Wenceslaus) (ca. 1535–91)
Title: POLONIAE.. FINITIMARUMQUE' // LOCORUM DESCRIP- // TIO. AVCTORE // WENCESLAO.. GODRECCIO // POLONO.
Year: 1572
Dimensions: 54.0 x 40.3 (49.0 x 36.8) cm
Phys. Desc.: Engraved, Colored
Language(s): Latin

Entry (No.): 213
Author: Grodecki (Godreccius), Wacław (Wenceslaus) (ca. 1535–91)
Title: POLONIAE // FINITIMARUQUE // LOCORUM DESCRIP- // TIO. AUOTORE // WECESLAO // GODRECCIO // POLONO.
Year: 1588
Dimensions: 57.0 x 43.2 (49.0 x 36.8) cm
Phys. Desc.: Engraved, Colored
Language(s): Latin

Entry (No.): 214
Author: Grodecki (Godreccius), Wacław (Wenceslaus) (ca. 1535–91)
Title: POLONIAE FINITIMARUM- // QUE LOCORUM DESCRIP- // TIO AUCTORE WENCES- // LAO GODRECCIO POLONO // Henricus Nagel fecit.
Year: 1592
Dimensions: 26.2 x 18.8 (26.0 x 18.2) cm
Phys. Desc.: Engraved
Language(s): Latin

Entry (No.): 215
Author: Grodecki (Godreccius), Wacław (Wenceslaus) (ca. 1535–91)
Title: POLONIA
Year: 1598
Dimensions: 15.8 x 12.0 (11.8 x 7.2) cm
Phys. Desc.: Engraved
Language(s): Latin

Entry (No.): 216
Author: Grodecki (Godreccius), Wacław (Wenceslaus) (ca. 1535–91)
Author: Pograbka (Pograbius), Andrzej, d. 1602
Title: POLONIAE, // LITVANIAEQ. // DESCRIPTIO. // AUCTORE WENCESLAO GODRECCIO; ET COR- // RECTORE ANDREA POGRABIO PILSNENSI.
Year: 1601
Dimensions: 57.3 x 47.4 (48.2 x 37.0) cm
Phys. Desc.: Engraved, Colored
Language(s): Latin
Figure 9

Entry (No.): 217
Author: Grodecki (Godreccius), Wacław (Wenceslaus) (ca. 1535–91)
Title: POLONIA
Year: 1606
Dimensions: 18.5 x 12.0 (12.0 x 8.5) cm
Phys. Desc.: Engraved
Language(s): Latin

Entry (No.): 218
Author: Grodecki (Godreccius), Wacław (Wenceslaus) (ca. 1535–91)
Title: POLONIA
Year: 1616
Dimensions: 20.5 x 12.3 (13.2 x 9.3) cm
Phys. Desc.: Engraved
Language(s): Latin

Entry (No.): 219
Author: Gros, C. (filed under Lavoisne by Krawciw)
Title: MAP OF // POLAND, // PRUSSIA and HUNGARY. // indicating the places // rendered celebrated by // SIEGES and BATTLES // intended for the Elucidation of // LAVOISNE'S // HISTORICAL ATLAS, // by C. GROS // 1813. // London. Published by J. Barfield... 1813.
Year: 1813
Dimensions: 53.0 x 42.5 (29.7 x 29.5) cm
Phys. Desc.: Engraved, Colored
Language(s): English
Figure 61

Entry (No.): 220
Author: Gros, C. (see Lavoisne, C. V.)

Title: MAP OF // RUSSIA // IN // EUROPE // WITH ITS DISMEMBERMENTS FROM POLAND // IN 1773, 1793 AND 1795, INDICATING THE // PLACES RENDERED CELEBRATED BY // BATTLES OR SIEGES. // FOR THE ELUCIDATION OF // LAVOISNES, // HISTORICAL & GENEALOGICAL AND STATISTICAL MAP OF RUSSIA. // London. Published by J. Barfield... [and] J. Walker Sculpt. 1813.
Year: 1813
Dimensions: 53.0 x 42.5 (26.0 x 33.7) cm
Phys. Desc.: Engraved, Colored
Language(s): English

Entry (No.): 221
Author: Güssefeld (Guessefeld), Franz Ludwig (1744–1808)
Title: Charte // von der // Moldau // und // Walachey // nach den Astonomischen Beobachtungen des Russisch // Kayserl. Majors Islenief und denen Charten und Be // schreibung des H. Hauptm. Sulzer, ingl. den Char // ten des ehmal. Russ. Kayserl. Gen. Lieut. v Bauer über die // Moldau u.a.m. entworfen // von F.L. Güssefeld. // Nürnberg // bey den Homaennischen Erben. Anno 1785 // Mit. Rom. Kayserl. Allergn. Freyheit.
Year: 1785
Dimensions: 65.0 x 53.0 (57.0 x 45.0) cm
Phys. Desc.: Engraved, Colored
Language(s): German

Entry (No.): 222
Author: Güssefeld (Guessefeld), Franz Ludwig (1744–1808)
Title: Charte // von // Asien // nach den bewärtesten astronomischen // Beobachtungen, den neuesten Reisen, // und der vorzüglichsten Charten inson - // derheit aber der Geographie des Hrn. // J. C. Gatterers gemals, stereographisch // entworfen von F. L. Güssefeld. Nürnberg // bey Hom. Erben. // Mit. Röm. Kayserl. allergn. Freiheit. 1793.
Year: 1793
Dimensions: 59.5 x 50.5 (57.2 x 47.5) cm
Phys. Desc.: Engraved, Colored
Language(s): German

Entry (No.): 223
Author: Güssefeld (Guessefeld), Franz Ludwig (1744–1808)
Title: General charte // von // Ungarn // mit seinen Nebenländern Croatien, Dalmatien, // Slavonien und Siebenbürgen, ingleichen // Gallizien und der Bukowina // nebst den angrenzenden Provinzen des // Ostmannischen Reichs & cc. // Ney entworfen von V. L. Güssefeld. Nurnberg by den Homannischen Erben // Mit Keyserl. allergnäde. Freiheit.
Year: 1796
Dimensions: 60.5 x 50.5 (56.5 x 46.0) cm
Phys. Desc.: Engraved, Colored
Language(s): German

Entry (No.): 224
Author: Haas (Haase; Has; Hase; Hasius; Hass), Johann Matthias (1684–1742)
Title: Imperii // Russici // et // Tartariae Universal // tam majoris et Asiaticae // quam minoris et Europaeae // Tabula // ex recentissimis et probatissimis // monumentis et relationibis // concinnata // et // ad legitimas projectionum Geo // graphic regulas plane exacta // opera // Ioh. Matthiae Hasii // Math P.P. // juncta Sciagraphia explicationis ejusdem // et descriptionis Russiae et Tataria universae // Geographico - historicae // Impensis Homannianorum Herendum Norimbergae Cum Privil. Sa. Cael. Maj. A. 1739
Year: 1739
Dimensions: 58.0 x 52.0 (54.8 x 47.5) cm
Phys. Desc.: Engraved, Colored
Language(s): Latin

Entry (No.): 225
Author: Haas (Haase; Has; Hase; Hasius; Hass), Johann Matthias (1684–1742)
Title: Imperii // Russici // et // Tartariae Universal // tam majoris et Asiaticae // quam minoris et Europaeae // Tabula // ex recentissimis et probatissimis // monumentis et relationibis // concinnata // et // ad legitimas projectionum Geo // graphic regulas plane exacta // opera // Ioh. Matthiae Hasii // Math P.P. // juncta Sciagraphia explicationis ejusdem // et descriptionis Russiae et Tataria universae // Geographico - historicae // Impensis Homannianorum Herendum Norimbergae. A. 1739. Cum Privil. Sa. Cael. Maj.
Year: 1739
Dimensions: 59.5 x 52.5 (55.0 x 48.0) cm
Phys. Desc.: Engraved, Colored
Language(s): Latin

Entry (No.): 226
Author: Haas (Haase; Has; Hase; Hasius; Hass), Johann Matthias (1684–1742)
Title: Imperii // Russici // et // Tartariae Universal // tam majoris et Asiaticae // quam minoris et Europaeae // Tabula // ex recentissimis et probatissimis // monumentis et relationibis // concinnata // et // ad legitimas projectionum Geo // graphic regulas plane exacta // opera // Ioh. Matthiae Hasii // Math P.P. // juncta Sciagraphia explicationis ejusdem // et descriptionis Russiae et Tataria universae // Geographico - historicae // Impensis Homannianorum Herendum Norimbergae. A. 1739. Cum Privil. Sa. Cael. Maj.
Year: 1739
Dimensions: 63.5 x 54.3 (54.5 x 47.5) cm
Phys. Desc.: Engraved, Colored
Language(s): Latin

Entry (No.): 227
Author: Haas (Haase; Has; Hase; Hasius; Hass), Johann Matthias (1684–1742)
Title: Carte de l'Asiae Minevre ou de la Natolie et du Pont Evxin, tirée des Voyages et des Observations des Anciens et Modernes, et dressée suivant les principes d'une nouvelle Projection par Sr. Has. P.P. imprimée aux depens des Heritiers de Homann. Nuremb. Anno 1743.
Year: 1743
Dimensions: 63.5 x 53.5 (58.2 x 49.0) cm
Phys. Desc.: Engraved, Colored
Language(s): French

Entry (No.): 228

Author: Haas (Haase; Has; Hase; Hasius; Hass), Johann Matthias (1684–1742)
Title: Carte de l'Asie, projette Stereographiquement, tirée des Relations et Observations, contenues dans les voyages modernes, de meme que dans les monumens des anciens, corrigée de tout ce qui sent la fable, et divisée en ses IX Parties methodiques par le Sr. J. M. Hasius, Prof. des Mathem. Recueilliée et desinée sur ses morceaux posthumes par Aug. Gottl. Boehimius, Maitre es Arts. Aux Depens des Heritiers de Homann. l'An 1744. Avec Pr.
Year: 1744
Dimensions: 61.0 x 53.5 (52.8 x 46.5) cm
Phys. Desc.: Engraved, Colored
Language(s): French

Entry (No.): 229
Author: Haas (Haase; Has; Hase; Hasius; Hass), Johann Matthias (1684–1742)
Title: Hungariae // ampliori significatu et veteris vel // Methodicae complexae // Regna Hvngariae Propriae Croatiae // Dalmatiae Bosniae Serviae // Bulgariae Cumaniae // Principatum: Transylvaniae // Despotatus: Walachiae Moldaviae // [exclusis ab eadem alienatis // Galitia et Ludomiriria] in suas Provincias ac partes divisae // et quoad Imperantes ex Avstriacis Tvrcis et Venetis distinctae // [juncta tamen propter comodiatem Romania vel Romelia Tuvrcica // Tabvla ex recentissimis pariter et antiquissimis // relationibus et monumentis concinnata // ac secundum leges Projectionis Stereogra // phicae legitimae descripta a descripta a I.M.Hasio M.P.P. // Curantibus Homannianis Heredibus.
Year: 1744
Dimensions: 61.0 x 54.3 (60.0 x 47.5) cm
Phys. Desc.: Engraved, Colored
Language(s): Latin

Entry (No.): 230
Author: Haraeus (Hareieus, Hareio, Haren, Van der Haer, Verharius), Franciscus (Frans van) (ca. 1555–1631)
Title: Lumen // Historiarum // per // Occidentem // ex conatibus // Fran. Haraei, // Antuerpiae. // Map in coll. B. K. /
Year: 1662
Dimensions: 58.5 x 50.0 (48.0 x 38.0) cm
Phys. Desc.: Engraved, Colored Outline
Language(s): Latin

Entry (No.): 231
Author: Haupt, Gottfried Jacob
Title: Nova et accurata // Tartariae // Europaeae // Seu // Minores // et in specie // Crimeae // Delineatio Geographica, // Cum omnibus // Circa Pontum Euxinum et // Paludem Maeotidem // Jacentibus Provinciis, // exhibentur // a Gottf. Jac. Haupt Calcographo // Aug: VIND..
Year: 1720
Dimensions: 68.0 x 56.5 cm
Phys. Desc.: Engraved, Colored Outline
Language(s): Latin

Entry (No.): 232
Author: Herberstein (Heberstein, Herberstain, Herberstein von Neyberg und Guetenhag), Sigismund von (1486–1566)
Title: De Borysthene flumine, ac cius acco// lis: de natura canum, de Scytharii // moribus, de gemmis, smaragdo, // cyanco, crystallo.
Year: 1538
Dimensions: 29.2 x 24.0 (25.0 x 15.5) cm
Phys. Desc.: Copy and Negative
Language(s): Latin

Entry (No.): 233
Author: Herberstein (Heberstein, Herberstain, Herberstein von Neyberg und Guetenhag), Sigismund von (1486–1566)
Title: [Black Sea]
Year: 1551
Dimensions: 46.1 x 40.5 (37.3 x 26.3) cm
Phys. Desc.: Copy and Negative
Language(s): Latin

Entry (No.): 234
Author: Hirschvogel (Hirschfogel, Hirsvogel, Hirssfogel), Augustin (1503–53)
Title: Moscovia Sigismundi Liberi Baronis in Herberstein, Nieperg, et Gutenhag Anno M.D. XLIX.
Year: 1549
Dimensions: 37.5 x 26.0 cm
Phys. Desc.: Copy and Negative
Language(s): Latin
Figure 6

Entry (No.): 235
Author: Homann, Johann Baptist (Baptista), 1664–1724
Title: Principatus Transylvaniae in suas quasque Nationes earumque Sedes et Regiones cum finitimis vicinorum Statuum Provinciis accurate divisus opera Joh. Baptist. Geographi, e Reg-ae Boru B-ae Societatis Scientiarum memori Norimbergae.
Year: 1699
Dimensions: 59.0 x 50.0 (57.1 x 48.4) cm
Phys. Desc.: Engraved, Colored
Language(s): Latin

Entry (No.): 236
Author: Homann, Johann Baptist (Baptista), 1664–1724
Title: Principatus Transylvaniae in suas quasque Nationes earumque Sedes et Regiones cum finitimis vicinorum Statuum Provinciis accurate divisus opera Joh. Baptista Homanni Sac. Coes. Majest. Geographi, e Reg-ae Boru B-ae Societatis Scientiarum memori Norimbergae.
Year: 1699
Dimensions: 60.5 x 53.5 (57.1 x 48.0) cm
Phys. Desc.: Engraved, Colored
Language(s): Latin

Entry (No.): 237
Author: Homann, Johann Baptist (Baptista), 1664–1724
Title: Accurate Vorstellung der Orientalisch-Kayserliehen Haupt- und Residenz-Stadt CONSTANTINOPEL samt ihrer Gegend und zweyen berühmten Meer-Engen, Bosphoro Thracio, und Hellesponto oder dem freto der Dardanellen herausgegeben von Johann Baptista Homann,

Der. Rom. Kays. Majest. Geographo in Nurnberg.
Year: ca. late 17th-early 18th century
Dimensions: 63.0 x 53.5 (57.5 x 48.5) cm
Phys. Desc.: Engraved, Colored
Language(s): Latin
Figure 41

Entry (No.): 238
Author: Homann, Johann Baptist (Baptista), 1664–1724
Title: Danubii Fluminis (hic ad urbe Belgrado per Mare Nigrum usq. Constantinopolim defluentis exhibiti) Pars Infima, in qua Transylavania, Walachia, Moldavia, Bolgaria, Servia, Romania, et Bessarabia cum vicinis Regionibus ostenduntur à Joh. Bapt. Homanno S. C. M. Geogr. Norimbergae.
Year: 1700
Dimensions: 61.2 x 53.5 (57.1 x 47.0) cm
Phys. Desc.: Engraved, Colored
Language(s): Latin

Entry (No.): 239
Author: Homann, Johann Baptist (Baptista), 1664–1724
Title: Danubii Fluminis (hic ad urbe Belgrado per Mare Nigrum usq. Constrantinopol im defluentis exhibiti) Pars Infima, in qua Transylavania, Wallachia, Moldavia, Bolgaria, Servia, Romania, et Bessarabia cum vicinis Regionibus ostenduntur à Joh. Bapt. Homanno S. C. M. Geogr. Norimbergae.
Year: 1700
Dimensions: 62.2 x 53.0 (58.4 x 47.5) cm
Phys. Desc.: Engraved, Colored
Language(s): Latin

Entry (No.): 240
Author: Homann, Johann Baptist (Baptista), 1664–1724
Title: Generalis Totius // Imperii Moscovitici // Novissima Tabula // Magnam Orbis terrarum partem à Polo Arctica usq. // ad mare Japonicum, et Chinae Septentrionalis // confinia exhibens // cum via Czaricae nuper Legationis ex urbe Mo // scua per universam Tartariam ad magnae Chinae Imperatorem // ex conatibus // Johannis Baptistae Homanni // Norimbergae.
Year: 1710
Dimensions: 60.0 x 51.0 (57.2 x 48.5) cm
Phys. Desc.: Engraved, Colored
Language(s): Latin
Figure 42

Entry (No.): 241
Author: Homann, Johann Baptist (Baptista), 1664–1724
Title: Generalis Totius // Imperii Moscovitici // Novissima Tabula // Magnam Orbis terrarum partem à Polo Arctica usq. // ad mare Japonicum, et Chinae Septentionalis // confinia exhibens // cum via Czarssae nuper Legationis ex urbo Mo // scua per universam Tartariam ad magnae Chinae Imperatorem // ex conatibus // Johannis Baptistae Homanni // Norimbergae.
Year: 1710
Dimensions: 60.5 x 51.5 (57.2 x 48.5) cm
Phys. Desc.: Engraved, Colored
Language(s): Latin

Entry (No.): 242
Author: Homann, Johann Baptist (Baptista), 1664–1724
Title: Imperii // Turcicum // in Europa, Asia et Africa // Regiones // Proprias, Tributarias, Clientelares // sicut et omnes ejusdem // Beglirgebatus seu Praefecturas Generales // exhibens // Sumptibus Io. Baptista Homanni Noribergae.
Year: ca. 1710–30
Dimensions: 61.0 x 64.2 (56.7 x 48.0) cm
Phys. Desc.: Engraved, Colored
Language(s): Latin

Entry (No.): 243
Author: Homann, Johann Baptist (Baptista), 1664–1724
Title: Prospect, Grundris und Gegend der Polnishen vesten Reichs und Handels-Stadt Dantzig und ihrem Werder edirt von Joh. Bapt. Homann S. C. M. Geog. in Nürnberg.
Year: ca. 1710–30
Dimensions: 60.0 x 52.5 (56.5 x 48.2) cm
Phys. Desc.: Engraved, Colored
Language(s): German

Entry (No.): 244
Author: Homann, Johann Baptist (Baptista), 1664–1724
Title: Regni // Poloniae // Magnique Ducat // Lithuaniae // Nova et exacta Tabula // ad mentem Starovolcii descripta // à // Johanne Bapt. Homanno // Sac. Caes. Maj. Geographo Norimbergae.
Year: 1710–30
Dimensions: 63.0 x 53.0 (55.5 x 48.5) cm
Phys. Desc.: Engraved, Colored
Language(s): Latin

Entry (No.): 245
Author: Homann, Johann Baptist (Baptista), 1664–1724
Title: Tabula Geographica // qua pars // Russiae Magnae // Pontus Euxinus // seu // Mare Nigrum // et // Tartaria Minor // cum finitimis Bulgariae, Romaniae // et Natoliae // Provinciis exhibetur // a Joh. Baptista Homanno // Sac. Caes. Majestatis Geographo // Norimbergae. // Cum Privilegio S.C.M.
Year: before 1729
Dimensions: 53.5 x 60.0 (48.4 x 57.7) cm
Phys. Desc.: Engraved, Colored
Language(s): Latin

Entry (No.): 246
Author: Homann, Johann Baptist (Baptista), 1664–1724
Title: Generalis Totius // Imperii Russorum // Novissima Tabula // Magnam Orbis terrarum partem a Polo Arctico usq. // ad mare Iaponicum, et Chinae Septentionalis // confinia exhibens // cum via Czarskae nuper Legationis ex urbo Mo // scua per universam Tartariam ad magnae Chinae Imperatorem // ex conatibus // Johannis Baptistae Homanni // Norimbergae.
Year: 1730
Dimensions: 61.7 x 51.8 (57.0 x 48.5) cm
Phys. Desc.: Engraved, Colored
Language(s): Latin

Entry (No.): 247

Author: Homann, Johann Baptist (Baptista), 1664–1724
Title: Tabula Geographica // qua pars // Russiae Magnae // Pontus Euxinus // seu // Mare Nigrum // et // Tauriae Regnum // cum finitimis Bulgariae, Romaniae et Natoliae // Provinciis exhibētur // a Joh. Baptista Homanno // Sac. Caes. Majestatis Geographo // Norimbergae.
Year: 1730
Dimensions: 54.5 x 64.0 (48.5 x 58.0) cm
Phys. Desc.: Engraved, Colored
Language(s): Latin

Entry (No.): 248
Author: Homann, Johann Baptist (Baptista), 1664–1724
Title: Ukrania // quae et // Terra Cosaccorum // cum vicinis // Walachiae, Moldaviae, Minorisq. Tartariae Provinciis // exhibita // à Joh. Baptista Homanno // Noribergae.
Year: 1730
Dimensions: 60.7 x 51.5 (57.4 x 47.5) cm
Phys. Desc.: Engraved, Colored
Language(s): Latin
Figure 43

Entry (No.): 249
Author: Homännische Erben (Homann Heirs, Heredes Homanniani, Héritiers de Homann) (1730–1813)
Title: Bosphorus Thracicus. // Der Kanal des Schwartzen Meers oder Die Meer - Enge bey Constantinopel // sambt denen // an beiden Ufern desselben gelegenen Stätten, Flecken, Dörfern, Schlössern, Palaesten, Lusthäusern, Wasserleitungen, // geomerisch aufgenommen // durch Johann Baptist von Reben, Kayse. Könige. Ungarl. Ingenieur Hauptmann // herausgegeben durch die Homaenn. Erben zu Nürnberg 1764.
Year: 1764
Dimensions: 53.7 x 89.0 (50.5 x 79.3) cm
Phys. Desc.: Engraved, Colored
Language(s): German

Entry (No.): 250
Author: Homännische Erben (Homann Heirs, Heredes Homanniani, Héritiers de Homann) (1730–1813)
Title: Tabula Geographica // continens // Despotatus Wallachiae // atque Moldaviae // Provinciam Bessarabiae // sub clientala Turcica, // itemque Provinciam Podoliae. // tanquam Regiones, in quibus bellum presens geritur, // ex Hasianis aliisque // novissimus subsidiis // secundum stratum politicum // recentissimum delineata, // in lucem edita // ab // Homannis Heredibus, // 1769. // C.P.S.C.M.
Year: 1769
Dimensions: 61.5 x 54.4 (50.5 x 46.2) cm
Phys. Desc.: Engraved, Colored
Language(s): Latin

Entry (No.): 251
Author: Homännische Erben (Homann Heirs, Heredes Homanniani, Héritiers de Homann) (1730–1813)
Title: Charte // von // Russisch Litauen, // welche die von Polen an Russland // abgetretene Woiewodschaften, // Liefland, Witepsk, Mscislaw, // und einen Theil der Wojewodschaften // Polock und Minsk enthält. Nürnberg, auf Kosten der Homännischen Erben, 1775.
Year: 1775
Dimensions: 49.0 x 58.5 (44.0 x 55.4) cm
Phys. Desc.: Engraved, Colored
Language(s): German

Entry (No.): 252
Author: Homännische Erben (Homann Heirs, Heredes Homanniani, Héritiers de Homann) (1730–1813)
Title: Lubomeriae // et // Galliciae // Regni // Tabula Geographica // Impensis Homannianorum // Haeredum // Cum. Priv. Sac. Caes. // Majestatis // F.L.G.
Year: 1775
Dimensions: 62.0 x 53.0 (58.5 x 44.5) cm
Phys. Desc.: Engraved, Colored
Language(s): Latin
Figure 51

Entry (No.): 253
Author: Homännische Erben (Homann Heirs, Heredes Homanniani, Héritiers de Homann) (1730–1813)
Title: Mappa Geographica // Poloniae // ex novissimis quot quot sunt // mappis specialibus composita, // in partes maiores Austriae. // Russiae, et Borussie. Divisa // et // luci publicae tradita per // Hommannianos Heredes. // Norimbergae. // C.P.S.C.M.
Year: 1800
Dimensions: 58.5 x 49.5 (50.0 x 43.5) cm
Phys. Desc.: Engraved, Colored
Language(s): Latin

Entry (No.): 254
Author: Hondius, Jodocus (Josse de Hondt) (1563–1612)
Title: Tartaria. Jodocus Hondius // lectori slutem // In hac Tartariae descriptione omnem adhibuimus diligentiam ut re: // giones nuper ab Anglis vel a Belgis alijsve detectae suo quaeq. loco habe: // rentur, quamvis mihi ipsi non satisfecerim, quis enim vastum illud et in: // cognitii regnum vere describet-s. at ut possimus dum ut volumus non- // licet usq. dum // melius quoddam.
Year: 1606
Dimensions: 57.5 x 48.4 (49.1 x 34.0) cm
Phys. Desc.: Engraved, Uncolored
Language(s): Latin
Figure 15

Entry (No.): 255
Author: Hondius, Jodocus (Josse de Hondt) (1563–1612)
Title: Asia // recens summa // cura delineata // Auct. Henr. Hodio // 1631.
Year: 1631
Dimensions: 59.0 x 49.0 (50.0 x 37.5) cm
Phys. Desc.: Engraved, Uncolored
Language(s): Latin

Entry (No.): 256
Author: Hondius, Jodocus (Josse de Hondt) (1563–1612)
Title: China // Veteribus // Sinarum Regio // nunc incolis Tame dicta.//
Year: 1649

Dimensions: 59.0 x 48.5 (49.5 x 40.5) cm
Phys. Desc.: Engraved, Colored Outline
Language(s): Latin

Entry (No.): 257
Author: Hondius, Jodocus (Josse de Hondt) (1563–1612)
Title: Tartaria // sive // Magnichami // Imperium.
Year: 1649
Dimensions: 59.3 x 49.0 (50.0 x 38.2) cm
Phys. Desc.: Engraved, Colored Outline
Language(s): Latin

Entry (No.): 258
Author: Houzé de l'Avlnoit, Auguste J. G.
Title: La Russie // La Pologne // La Suède, la Norwège // et le Danemark // à la fin du XIV-e siècle // Dressés sous la Direction // de A. Houzé.
Year: ca. 1849
Dimensions: 33.1 x 26.0 (27.5 x 21.0) cm
Phys. Desc.: Printed, Colored Outline
Language(s): French

Entry (No.): 259
Author: Houzé de l'Avlnoit, Auguste J. G.
Title: La Russie // La Pologne // La Suède, la Norwège // et le Danemark // en 1849 // Dressés sous la Direction // de A. Houzé.
Year: ca. 1849
Dimensions: 33.1 x 26.0 (27.5 x 21.0) cm
Phys. Desc.: Printed, Colored Outline
Language(s): French

Entry (No.): 260
Author: Houzé de l'Avlnoit, Auguste J. G.
Title: La Russie, // La Suède, la Norwège // et le Danemark // à la fin du IX-e siècle // Dressès sous la Direction // de A. Houzé.
Year: ca. 1849
Dimensions: 33.1 x 26.0 (27.5 x 21.0) cm
Phys. Desc.: Printed, Colored Outline
Language(s): French

Entry (No.): 261
Author: Houzé de l'Avlnoit, Auguste J. G.
Title: La Russie, // La Suède, la Norwège // et le Danemark // à la fin du V-e siècle // Dressés sous la Direction // de A. Houzé.
Year: ca. 1849
Dimensions: 33.1 x 26.0 (27.5 x 21.0) cm
Phys. Desc.: Printed, Colored Outline
Language(s): French

Entry (No.): 262
Author: Hutter, Franz Xaver (Francis Xavier)
Title: Charte // des Koenigreichs // Polen // Nach den bewährtesten Hülfsmitteln // und den Theilungs-Tractaten vom // Jahr 1773.u.1793 entworfen" Zu haben bey Joch. Walch in Augsburg. Entworfen u. gest. v. F. X. Hutter.
Year: 1793
Dimensions: 57.5 x 54.5 (58.5 x 46.7) cm
Phys. Desc.: Engraved, Colored Outline
Language(s): German

Entry (No.): 263
Author: Jaillot, Alexis-Hubert (ca. 1632–1712)
Title: La Russie Blanche ou Moscovie Divisée Suivant l'Etendue // des Royuames, Duchés, Principautés, & c:// Presenté a Monseigneur // Le Duc de Bourgogne // Par son Très humble et très Obeissant // Serviteur H. Jaillot a Paris.// se vend a Amsterdam // chez J. Covens & C. Mortier Avec Privilege.
Year: 1692
Dimensions: 64.6 x 53.7 (60.0 x 48.5) cm
Phys. Desc.: Engraved, Colored Outline
Language(s): French

Entry (No.): 264
Author: Janssonius (Jansz. II, Jansz. of Amsterdam), Johannes (Jan, Joannes) (1588–1664)
Title: Turcicum // Imperium // concordia res parvae crescunt // Discordia maximae dilabuntur.
Year: 1649
Dimensions: 32.0 x 49.0 cm (text); 58.0 x 48.5 (51.5 x 41.0) cm (map)
Phys. Desc.: Engraved, Uncolored
Language(s): Latin
Figure 11

Entry (No.): 265
Author: Janvier, Jean Denis [Robert] (mid- to late 18th c.)
Title: Le Royaume // de Hongrie // Divisé // en Haute et Basse Hongrie // Transilvanie Esclavonie et Croatie // Dressé par le S-r. Janvier Géorgraphe // A Paris chez Lattré, rue St. Jacques, à la Ville de Bordeaux. Avec Privil. Du Roy.
Year: 1760
Dimensions: 50.8 x 37.0 (44.0 x 30.5) cm
Phys. Desc.: Engraved, Colored Outline
Language(s): French

Entry (No.): 266
Author: Janvier, Jean Denis [Robert] (mid- to late 18th c.)
Title: Russie D'Europe // avec la partie plus peuplée de celle // d'Asie // Le tout divisé en quatorze Gouvernemens, // Selon le nouvel Atlas de l'Empire Russien. // Par le S-r. Janvier, Géographe. // A Paris. Chez Lattré Graveur rue St. Jacques, au dessus de la Fontaine St. Severin, à la Ville de Bordeaux. Avec. Priv. du Roy.
Year: 1760
Dimensions: 53.5 x 38.2 (44.7 x 30.5) cm
Phys. Desc.: Engraved, Colored Outline
Language(s): French

Entry (No.): 267
Author: Jefferys, Thomas (ca. 1710–71)
Title: Asia. T. Jefferys Scul.p
Year: 1785
Dimensions: 26.0 x 21.0 (22.7 x 17.8) cm
Phys. Desc.: Engraved, Uncolored
Language(s): English

Entry (No.): 268
Author: Jefferys, Thomas (ca. 1710–71)
Title: Europe. T Jeffrys Sc.
Year: 1785
Dimensions: 53.5 x 38.2 (44.7 x 30.5) cm
Phys. Desc.: Engraved, Uncolored
Language(s): English

Entry (No.): 269
Author: Jefferys, Thomas (ca. 1710–71)
Title: Hungary // with // Turkey // in Europe.
Year: 1785
Dimensions: 26.0 x 21.0 (22.0 x 18.0) cm
Phys. Desc.: Engraved, Uncolored
Language(s): English

Entry (No.): 270
Author: Jefferys, Thomas (ca. 1710–71)
Title: Poland // Lithuania // and Prussia. T. Jeffrys Sculp.
Year: 1785
Dimensions: 26.0 x 21.0 (22.0 x 18.0) cm
Phys. Desc.: Engraved, Uncolored
Language(s): English

Entry (No.): 271
Author: Jefferys, Thomas (ca. 1710–71)
Title: Russia or Moscovy in Europe.
Year: 1785
Dimensions: 26.0 x 21.0 (22.0 x 18.0) cm
Phys. Desc.: Engraved, Uncolored
Language(s): English

Entry (No.): 272
Author: Jenkinson, Anthony (ca. 1525–1611)
Title: RVSSIAE, MOSCO- // VIAE ET TARTA- // RIAE DESCRIP- // TIO. AUCTORE // ANTONIO IENKENSONO ANGLO, EDITA // LONDINI ANNO. 1562. & DEDICATA // ILLUSTRISS: D. HERICO SŸDNEO // WALLIĘ PRĘSIDI. //
Year: 1562
Dimensions: 56.8 x 42.0 (44.5 x 35.0) cm
Phys. Desc.: Engraved
Language(s): Latin
Figure 13

Entry (No.): 273
Author: Jenkinson, Anthony (ca. 1525–1611)
Title: Descriptio Rvssiae et Moschoviae.
Year: 1598
Dimensions: 15.8 x 12.0 (10.7 x 7.5) cm
Phys. Desc.: Engraved
Language(s): Latin

Entry (No.): 274
Author: Jenkinson, Anthony (ca. 1525–1611)
Title: Descriptio Rvssiae et Moschoviae
Year: 1606
Dimensions: 19.0 x 12.0 (12.0 x 8.3) cm
Phys. Desc.: Engraved
Language(s): Latin

Entry (No.): 275
Author: Jenkinson, Anthony (ca. 1525–1611)
Title: Descriptio Rvssiae et Moschoviae.
Year: 1616
Dimensions: 20.5 x 12.3 (13.2 x 9.2) cm
Phys. Desc.: Engraved
Language(s): Latin

Entry (No.): 276
Author: Johnston, Alexander Keith I (1804–71)
Title: European Russia by Keith Johnston, V.R.S.E.
Year: after 1832
Dimensions: 44.0 x 57.5 cm
Phys. Desc.: Printed, Colored Outline
Language(s): English

Entry (No.): 277
Author: Johnston, Alexander Keith I (1804–71)
Title: South-West Russia // Showing the extent of // the Kingdom of Poland // previous to its partition in 1772 // by Keith Johnston, F.R.S.E.
Year: after 1832
Dimensions: 57.5 x 44.0 cm
Phys. Desc.: Printed, Colored Outline
Language(s): English
Figure 64

Entry (No.): 278
Author: Judenne, J.
Title: PLAN TOPOGRAPHIQUE // ET PITTORESQUE // DE SÉBASTOPOL, // PAR J. JUDENNE. // 1855 //
Year: 1855
Dimensions: 42.5 x 30.6 (41.0 x 29.3) cm
Phys. Desc.: Lithographed, Colored
Language(s): French

Entry (No.): 279
Author: Keulen (Ceulen), Jan Jansz. (Johannes) van (1635–89)
Title: Niewe Paskaart Van de Geheele // Swarte Zee // Beginnende van // Constantinopolen Tot Asak // Op gestelt door G:v: Keulen // 't Amsterdam Bej Johannes van Keulen // Bock Zeekaart verkooper en Graad-boog-maaker Aande Oost-zÿde vande Niteuwnbrug Inde Gekroonde Lootsman met Previlegie voor [uz?] Iaaren. Ded. To Peter Alexëowits Czaar en Keyser. der Russen, &c. Werd dese Kaart van mÿn Gedienstig Opgedragen.
Year: 1682–86
Dimensions: 100.0 x 62.0 (98.1 x 52.3) cm
Phys. Desc.: Engraved, Uncolored
Language(s): Dutch

Entry (No.): 280
Author: Kiepert, Heinrich C. (1818–99)
Title: Das Europäische // Russland // Revidirt von C. Gräf 1857.
Year: 1857
Dimensions: 58.0 x 71.0 (46.5 x 58.0) cm
Phys. Desc.: Lithographed, Colored Outline
Language(s): German

Figure 73

Entry (No.): 281
Author: Kiepert, Heinrich C. (1818–99)
Title: Ungarn // mit seinen Nebenländern // und // Galizien. Südliches // Dalmatien // im Anschluss an den untern Rand der Hauptkarte. Gezeichnet von H. Kiepert. Revidirt 1857 von C. Gräf.
Year: 1857
Dimensions: 71.0 x 58.0 (64.0 x 52.0) cm
Phys. Desc.: Lithographed, Colored Outline
Language(s): German
Figure 74

Entry (No.): 282
Author: Kinsbergen, Captain Jan Hendrik van (1735–1819)
Title: Carte de la Mer d'Azof et d'une Partie de la Mer Noire // Levée en 1774 par Monsieur Kinsbergen Capitaine de haut-bord de S. M. Impériale de toutes les Russies.
Year: 1772
Dimensions: 50.5 x 32.0 (45.0 x 18.8) cm
Phys. Desc.: Engraved, Uncolored
Language(s): French

Entry (No.): 283
Author: Kinsbergen, Captain Jan Hendrik van (1735–1819)
Title: Carte // de la // Crimée // dressée par // J.H. de Kinsbergen // Contre-Amiral d'Hollande // et de Westfrise, Chevalier de l'ordre de St. George. &c. &c. // donnée par K. Frieseman a Amsterdam 1787.
Year: 1787
Dimensions: 92.0 x 83.0 (88.8 x 79.8) cm
Phys. Desc.: Engraved, Uncolored
Language(s): French

Entry (No.): 284
Author: Kitchin, Thomas (1719–84)
Title: A Map // of the // Eastern Part // of // Poland // by T. Kitchin Geog-r.
Year: 1739
Dimensions: 21.4 x 27.5 (27.8 x 22.5) cm
Phys. Desc.: Engraved, Uncolored
Language(s): English

Entry (No.): 285
Author: Kitchin, Thomas (1719–84)
Title: Poland // showing the claims // of Austria // Russia & // Prussia // By Thomas Kitchin Geo.
Year: after 1772
Dimensions: 35.0 x 22.3 (27.5 x 18.6) cm
Phys. Desc.: Engraved, Colored
Language(s): English

Entry (No.): 286
Author: La Feuille, Jacques (Jacob) de (1668–1719)
Title: Totius Danubii Nova & // Accuratiss. Tabula // Universam Simul Turciam // Europaeam Hungariam // Magnam Germaniae Partem // Totam Pen[é?] Italiam, Atque una Moream // cum Sicilia & Candia Complectens, // Auctore Jacobo de la Feuille // Cum. Priv. Ord Holl. & Westfri.
Year: ca. 1750
Dimensions: 64.4 x 53.5 (60.2 x 51.2) cm
Phys. Desc.: Copy, Colored Outline
Language(s): Latin

Entry (No.): 287
Author: L'Académie de Petersbourg (see also Academia Scientiarum Petropolitana)
Title: Carte Générale de l'Empire de Russie
Year: 1787
Dimensions: 125.0 x 58.0 (199.0 x 55.0) cm
Phys. Desc.: Engraved, Colored Outline
Language(s): French
Figure 48

Entry (No.): 288
Author: Larochette, Ch.
Title: PLAN TOPOGRAPHIQUE // ET PITTORESQUE // DE NICOLAÏEV // KINBURN // OTCHAKOV ET KHERSON // PAR CH. LAROCHETTE // 1856. //
Year: 1856
Dimensions: 43.0 x 30.5 (40.0 x 28.0) cm
Phys. Desc.: Lithographed, Colored
Language(s): French

Entry (No.): 289
Author: Lat, Jan (Joan) de (worked 1734–50)
Title: Klein Tartaryën, // met de // Turksche Provintiën, // grenzende aan de // Zwarte en Azofsche-Zeën; // na de Stellung van der Hr. G. d'L'Isle // Verbeterd volgens // de laatste Waar ne emingen, // Van de Academie // der Weetenshappen // te Petersburg.
Year: 1747
Dimensions: 28.7 x 17.8 cm
Phys. Desc.: Engraved, Colored Outline
Language(s): Dutch

Entry (No.): 290
Author: Lat, Jan (Joan) de (worked 1734–50)
Title: Tzuider-deel // van // Moscovie. // Na de Stelling van // G. de Lisle en Uttgegeven // tot Almelo door J. Keizer en // tot Dventer by J. de Lat.
Year: 1747
Dimensions: 24.0 x 17.6 cm
Phys. Desc.: Engraved, Colored Outline
Language(s): Dutch

Entry (No.): 291
Author: Laurie, Robert (1755–1836) (publisher)
Author: Whittle, James (publisher)
Title: A New Map of // the // Kingdom of Poland // with its Dismembered Provinces // and the Kingdom of Prussia. // London; // Published by Laurie & Whittle, No. 53 Fleet Street, // as the Act directs, 12th May, 1794.
Year: 1794
Dimensions: 73.5 x 53.7 (64.5 x 47.4) cm
Phys. Desc.: Engraved, Colored
Language(s): English

Entry (No.): 292

Author: Laurie, Robert (1755–1836) (publisher)
Author: Whittle, James (publisher)
Title: Kingdom // of // Hungary // Principality of Transilvania, // Sclavonia, Croatia, // with a Part of // Valakia, Bulgaria, // Bosnia and Servia // from the Latest Surveys ascertained by // Astronomical Observations. // London. // Published by Laurie & Whittle, No. 53 Fleet Street, // as the Act directs, 12th May, 1794.
Year: 1794
Dimensions: 71.5 x 51.5 (52.8 x 46.2) cm
Phys. Desc.: Engraved, Colored Outline
Language(s): English

Entry (No.): 293
Author: Laurie, Robert (1755–1836) (publisher)
Author: Whittle, James (publisher)
Title: The Asiatic Part // of the // Russian Empire // From the Maps Published // by the Imperial Academy of St. Petersburg; // with the New Discoveries // of Captain Cook & c.c.
Year: 1794
Dimensions: 131.2 x 53.5 (127.0 x 47.0) cm
Phys. Desc.: Engraved, Colored
Language(s): English

Entry (No.): 294
Author: Laurie, Robert (1755–1836) (publisher)
Author: Whittle, James (publisher)
Title: The European Part // of the // Russian Empire from the maps Published // by the Imperial Academy at St. Petersburg // With the New Provinces // on the Black Sea // London; // Published by Laurie & Whittle, No. 53 Fleet Street, // as the Act directs, 12th May, 1794.
Year: 1794
Dimensions: 131.2 x 53.5 (127.0 x 47.0) cm
Phys. Desc.: Engraved, Colored Outline
Language(s): English

Entry (No.): 295
Author: Laurie, Robert (1755–1836) (publisher)
Author: Whittle, James (publisher)
Title: A New Map of // the Seat of War, // Comprehending // Germany; Poland, // With its Dismemberments, // Prussia; Turkey in Europe, // Italy &c., // From the Maps of Chauchard, Zannoni, &c. Published by Laurie & Whittle, No. 53 Fleet Street London, // as the Act directs. // July 31st 1807.
Year: 1807
Dimensions: 82.0 x 62.5 (80.0 X 59.0) cm
Phys. Desc.: Engraved, Colored
Language(s): English
Figure 58

Entry (No.): 296
Author: Lavoisne, C. V.
Title: GENEALOGICAL, HISTORICAL, AND CHRONOLOGICAL MAP // OF POLAND, FROM THE 9TH CENTURY TO THE DISSOLUTION OF THE KINGDOM; & OF BOHEMIA, & HUNGARY, FROM THE SAME EPOCHA, TO THE REIGN OF MARIA THERESA. // Printed by J. Barfield...
Year: 1814
Dimensions: 53.0 x 42.5 (50.5 x 40.0) cm
Phys. Desc.: Printed, Colored
Language(s): English

Entry (No.): 297
Author: Lavoisne, C. V.
Title: GENEALOGICAL, HISTORICAL, AND CHRONOLOGICAL MAP // OF PRUSSIA, FROM FREDERICK OF NUREMBERG, 1415; AND OF RUSSIA, FROM THE SETTLEMENT OF THE GOVERNMENT AT WLADIMIR, 1157, TO THE YEAR 1813.
Year: 1814
Dimensions: 53.5 x 42.5 (50.0 x 40.0) cm
Phys. Desc.: Printed, Colored
Language(s): English

Entry (No.): 298
Author: Lavoisne, C. V.
Title: GENEALOGICAL, HISTORICAL, AND CHRONOLOGICAL MAP // OF PRUSSIA, FROM FREDERICK OF NUREMBERG, 1415; AND OF RUSSIA, FROM THE SETTLEMENT OF THE GOVERNMENT AT WLADIMIR, TO THE YEAR 1815. Second Philadelphia edition, 1820 - Printed by Thos. H. Palmer, for M. Carey & Son, from the London Edition of 1817, with corrections and additions.
Year: 1820
Dimensions: 56.0 x 44.4 (52.4 x 42.3) cm
Phys. Desc.: Printed, Colored
Language(s): English

Entry (No.): 299
Author: Lavoisne, C. V.
Title: MAP // OF THE // RUSSIAN EMPIRE. // INTENDED FOR THE ELUCIDATION OF // LAVOISNE'S HISTORICAL ATLAS // Improved to 1820. // GEOGRAPHICAL AND STATISTICAL MAP OF RUSSIA. // Second Philadelphia edition, 1820 - Printed by Thos. H. Palmer, for M. Carey & Son, from the London Edition of 1817, with corrections and additions.
Year: 1820
Dimensions: 55.7 x 44.4 (38.5 x 26.0) cm
Phys. Desc.: Engraved, Colored
Language(s): English

Entry (No.): 300
Author: Lavoisne, C. V.
Title: MAP // OF THE // RUSSIAN EMPIRE. // INTENDED FOR THE ELUCIDATION OF // LAVOISNE'S HISTORICAL ATLASS // Improved to 1820. // GEOGRAPHICAL AND STATISTICAL MAP OF RUSSIA. // Second Philadelphia edition, 1829 - Printed by Thos. H. Palmer, for M. Carey & Son, from the London Edition of 1817, with corrections and additions.
Year: 1820
Dimensions: 55.7 x 44.4 (38.5 x 26.0) cm
Phys. Desc.: Engraved, Colored
Language(s): English

Entry (No.): 301
Author: Lazius, Wolfgang (1514–65)
Title: Hungariae Descriptio, Wolfgango Lazio Auct.
Year: 1556
Dimensions: 50.0 x 43.0 (50.0 x 35.5) cm
Phys. Desc.: Engraved, Colored
Language(s): Latin
Figure 7

Entry (No.): 302
Author: Le Rouge (Lerouge), Lieutenant Georges-Louis (ca. 1712–90)
Title: Etats // de Moscovie // dediés // a Monseig-nr le Compte de Maurepas // Ministre et Secretaire d'Etat // par son tres humble et tres // Obeissant Serviteur le Rouge. Ingenieur et Geographe du Roy. // A Paris // Rue des Augustins. 1744.
Year: 1744
Dimensions: 59.0 x 51.0 (54.2 x 47.5) cm
Phys. Desc.: Engraved, Colored Outline
Language(s): French

Entry (No.): 303
Author: Le Rouge (Lerouge), Lieutenant Georges-Louis (ca. 1712–90)
Title: Grand // Duché // de // Lithuanie // Divisé // en 9. // Palatinats.
Year: 1746
Dimensions: 34.5 x 24.0 (37.5 x 20.3) cm
Phys. Desc.: Engraved, Colored Outline
Language(s): French

Entry (No.): 304
Author: Le Rouge (Lerouge), Lieutenant Georges-Louis (ca. 1712–90)
Title: L'Europe.. Suivant les // Nouvelles Observations // A Paris // Par et chez Le. S-r Le Rouge // Ingenieur, Geographe, de // Roy. Rue des Gr-ds // Augustins // 1746.
Year: 1746
Dimensions: 34.5 x 24.0 (28.0 x 21.2) cm
Phys. Desc.: Engraved, Colored Outline
Language(s): French

Entry (No.): 305
Author: Le Sage, Emanuel Auguste Dieudonné (pseudonym of Las Casas, Emmanuel Marie Joseph Auguste Dieudonné, Comte de) (1766–1842)
Title: [Historical texts and Royal Lineage charts]: "de Danemarck,"... "De Russie," ... "De Suede"...
Year: 1805
Dimensions: 70.5 x 54.5 (64.0 x 51.4) cm
Phys. Desc.: Engraved, Colored
Language(s): French

Entry (No.): 306
Author: Le Sage, Emanuel Auguste Dieudonné (pseudonym of Las Casas, Emmanuel Marie Joseph Auguste Dieudonné, Comte de) (1766–1842)
Title: L'Empire Russe en Europe et en Asié // avec les Acquisitions graduelles et caracterisées.
Year: 1805
Dimensions: 70.5 x 54.5 (64.0 x 51.4) cm
Phys. Desc.: Engraved, Colored
Language(s): French

Entry (No.): 307
Author: Le Sage, Emanuel Auguste Dieudonné (pseudonym of Las Casas, Emmanuel Marie Joseph Auguste Dieudonné, Comte de) (1766–1842)
Title: L'Europe Moderne // avec ses Divisions Politiques.
Year: 1805
Dimensions: 70.5 x 54.5 (63.2 x 50.0) cm
Phys. Desc.: Engraved, Colored
Language(s): French
Figure 59

Entry (No.): 308
Author: Le Sage, Emanuel Auguste Dieudonné (pseudonym of Las Casas, Emmanuel Marie Joseph Auguste Dieudonné, Comte de) (1766–1842)
Title: Tableau de la Transmigration // Des Barbares // qui ont envahi l'Empire Romain; montrant la // route qu'ils ont suivi, leur point de départ et ce-/ lui de leur établissement ou de leur destruction.
Year: 1805
Dimensions: 70.5 x 54.5 (62.5 x 47.7) cm
Phys. Desc.: Engraved, Colored
Language(s): French

Entry (No.): 309
Author: Lea, Philip (worked from 1666–1700)
Title: A New Map // of // Antient Europe // To all // Historiographers But more // Especially to the Most Worthy // Clement Mayo G-nt. is humbly // dedicated by Phi. Lea.
Year: 1699–1700
Dimensions: 61.0 x 53.0 (47.2 x 37.8) cm
Phys. Desc.: Engraved, Uncolored
Language(s): English

Entry (No.): 310
Author: Lenglet Dufresnoy, Nicolas (1674–1755)
Title: Le Royaume de Pologne avec ses Confins.
Year: 1736
Dimensions: 22.3 x 16.1 (17.5 x 13.8) cm
Phys. Desc.: Engraved, Colored Outline
Language(s): French

Entry (No.): 311
Author: Lenormandt, Charles François (1837–83)
Title: CARTE DE LA // MER NOIRE // OU PONT-EUXIN // ET SES ENVIRONS // PAR C. LENORMAND // A VERSAILLES 1760. //
Year: 1760
Dimensions: 28.5 x 19.3 (22.0 x 17.0) cm
Phys. Desc.: Manuscript Map, Colored
Language(s): French

Entry (No.): 312
Author: Lotter, Tobias Conrad (1717–77)
Title: Im // perium // Russi- // cum // omnisque Tartaria.

Year: ca. 1744–62
Dimensions: 14.0 x 10.8 (12.3 x 9.5) cm
Phys. Desc.: Engraved, Colored
Language(s): Latin

Entry (No.): 313
Author: Lotter, Tobias Conrad (1717–77)
Title: Pologniae // Regnum, // Ducatusq., // Magnae // Lithuaniae
Year: ca. 1744–62
Dimensions: 14.0 x 10.8 (12.3 x 9.5) cm
Phys. Desc.: Engraved, Colored
Language(s): Latin

Entry (No.): 314
Author: Lotter, Tobias Conrad (1717–77)
Title: Asia // Concimata Secundum // Observationes Academiae // Regalis Scientiarum // et nonnullas alias, et juxta // annotationes recentissimas // Per G. de L'Isle Geograph: // Venalis nunc prostat // Augustae Vindelicorum // apud T. C. Lotter Geogr.
Year: 1750
Dimensions: 62.5 x 53.0 (58.0 x 48.4) cm
Phys. Desc.: Engraved, Colored
Language(s): Latin

Entry (No.): 315
Author: Lotter, Tobias Conrad (1717–77)
Title: Europa // Regnorum, // Provinciarum Regionum, // in iisdemque // Religionis Christianae, // morum artiumq., cultu omnium orbis // terrarum, partium praestantissima, // juxta recentina exemplaria // accuratissime excusa a Tob. Conr. Lotter // Geographo // August. Vindel.
Year: 1750
Dimensions: 63.0 x 53.0 (58.0 x 49.0) cm
Phys. Desc.: Engraved, Colored
Language(s): Latin

Entry (No.): 316
Author: Lotter, Tobias Conrad (1717–77)
Title: Geometrischer Grundris // der // Des Heiligen Römischen Reidis Freyen Stadt // Nürnberg. Verlegt von Tobias Conrad Lotter, Geogr. in Ausburg.
Year: ca. 1750–78
Dimensions: 62.0 x 53.0 (58.5 x 53.0) cm
Phys. Desc.: Engraved, Colored
Language(s): German

Entry (No.): 317
Author: Lotter, Tobias Conrad (1717–77)
Title: Spatiosissimum // Imperium // Russiae Magnae // juxta recentissimas Observationes // Mappa Geographica // accuratissime delineatum // opera et sumptibus // Tobiae Conradi Lotteri, // Geogr. Et Chalcogr. // Augustae Vindel.
Year: 1750
Dimensions: 63.0 x 53.0 (57.0 x 49.5) cm
Phys. Desc.: Engraved, Colored
Language(s): Latin

Entry (No.): 318
Author: Lotter, Tobias Conrad (1717–77)
Title: Mappa // Geographica, // ex novissimis observationibus // repraesentans // Regnum // Poloniae // et // Magnum Ducatum // Lithuaniae, // Cura et sumptibus // Tobiae Conradi Lotter. // Geogr. Aug. Vindel.
Year: 1759
Dimensions: 65.0 x 55.0 (57.6 x 48.0) cm
Phys. Desc.: Engraved, Colored
Language(s): Latin
Figure 49

Entry (No.): 319
Author: Lotter, Tobias Conrad (1717–77)
Title: Carte Geographique // represant // le Théatre de la Guerre // entre les Russes, les Turcs, // et les Polonois Confédéres // C'est à dire: // Les Provinces d'Ukraine, de Nouvelle Servie // de Moldavie, Valachie, Krimée; // le Gouvernment de Woronez // et d'Astracan, // comme aussi la Tartarie de Kuban. // Par Tobie Conrad Lotter, Geogr. // à Augsburg 1769
Year: 1769
Dimensions: 63.0 x 53.0 (49.2 x 48.5) cm
Phys. Desc.: Engraved, Colored
Language(s): French

Entry (No.): 320
Author: Lotter, Tobias Conrad (1717–77)
Title: Carte Geographique // represant // le Théatre de la Guerre // entre les Russes, les Turcs, // et les Polonois Confédéres // C'est à dire: // Les Provinces d'Ukraine, de Nouvelle Servie // de Moldavie, Valachie, Krimée; // le Gouvernment de Woronez // et d'Astracan, // comme aussi la Tartarie de Kuban. // Par Tobie Conrad Lotter, Geogr. // à Augsburg 1769.
Year: 1769
Dimensions: 63.0 x 53.0 (50.0 x 48.5) cm
Phys. Desc.: Engraved, Colored
Language(s): French

Entry (No.): 321
Author: Lotter, Tobias Conrad (1717–77)
Title: Carte Geographique // represant // le Théatre de la Guerre // entre les Russes, les Turcs, // et les Polonois Confédéres // C'est à dire: // Les Provinces d'Ukraine, de Nouvelle Servie // de Moldavie, Valachie, Krimée; // le Gouvernment de Woronez // et d'Astracan, // comme aussi la Tartarie de Kuban. // Par Tobie Conrad Lotter, Geogr. // à Augsburg 1769.
Year: 1769
Dimensions: 63.0 x 53.0 (49.5 x 48.5) cm
Phys. Desc.: Engraved, Colored
Language(s): French

Entry (No.): 322
Author: Lotter, Tobias Conrad (1717–77)
Title: Carte du Gouvernement de Tauride Comprenant la Krimée et les Pays Voisins Par les Freres Lotters a Augsbourg.
Year: ca. 1772–78
Dimensions: 66.0 x 52.0 (63.0 x 43.4) cm

Phys. Desc.: Engraved, Colored
Language(s): French

Entry (No.): 323
Author: Lotter, Tobias Conrad (1717–77)
Title: Carte Géographique // contenant // Le Royaume // de Siberie // divisée en trois departmens, // savoir: // Tobolskago, Ienisseiskago, Irkutskago, // et // queslques autres Parties // de la Tartarie // Publiée par Tobie Conrad Lotter // Geogr: d'Augsburg.
Year: 1772
Dimensions: 107.5 x 53.0 (105.7 x 49.0) cm
Phys. Desc.: Engraved, Colored
Language(s): French

Entry (No.): 324
Author: Lotter, Tobias Conrad (1717–77)
Title: Tabula Synoptica // Totius Fluminis // Danubii // â fontibus usque ad Osta // in qua // omnes regiones, quas perfluit et amines, quos recipit, // imprimis integrum Hungariae Regnum, // Principat Transilvaniae, Moldaviae, Walachiae, // Archiducatus Austriae, Regna et Ducat. Styriae, Carinthiae, Carnioliae, Croatiae, Dalmatiae, Sclavoniae, // Bosniae, Bessarabiae, Bulgariae, Romaniae, cum Magna // parte Turciae, et Tartariae, nec non Mare Adria // ticum et Pontus Euxinus oculis sistuntur, // juxta recentissimas observationes et descriptiones // delineata et aeri incisa // cura et sumptibus // Tobiae Conradi Lotteri, // Geographi, // et Chalc. Augastae Vindel.
Year: 1772
Dimensions: 66.0 x 55.5 (57.4 x 49.3) cm
Phys. Desc.: Engraved, Colored
Language(s): Latin

Entry (No.): 325
Author: Lotter, Tobias Conrad (1717–77)
Title: Tabula Synoptica // Totius Fluminis // Danubii // â fontibus usque ad Osta // in qua // omnes regiones, quas perfluit et amines, quos recipit, // imprimis integrum Hungariae Regnum, // Principat Transilvaniae, Moldaviae, Walachiae, // Archiducatus Austriae, Regna et Ducat. Styriae, Carinthiae, Carnioliae, Croatiae, Dalmatiae, Sclavoniae, // Bosniae, Bessarabiae, Bulgariae, Romaniae, cum Magna // Parte Turciae, et Tartariae, nec non Mare Adria // ticum et Pontus Euxinus oculis sistuntur, // juxta recentissimas observationes et descriptiones // delineata et aeri incisa // cura et sumptibus // Tobiae Conradi Lotteri, // Geographi, // et Chalc. Augastae Vindel.
Year: 1772
Dimensions: 68.0 x 55.5 (57.5 x 49.3) cm
Phys. Desc.: Engraved, Colored
Language(s): Latin

Entry (No.): 326
Author: Lotter, Tobias Conrad (1717–77)
Title: Tabula Synoptica // Totius Fluminis // Danubii // â fontibus usque ad Osta // in qua // omnes regiones, quas perfluit et amines, quos recipit, // imprimis integrum Hungariae Regnum, // Principat Transilvaniae, Moldaviae, Walachiae, // Archiducatus Austriae, Regna et Ducat. Styriae, Carinthiae, Carnioliae, Croatiae, Dalmatiae, Sclavoniae, // Bosniae, Bessarabiae, Bulgariae, Romaniae, cum Magna // parte Turciae, et Tartariae, nec non Mare Adria // ticum et Pontus Euxinus oculis sistuntur, // juxta recentissimas observationes et descriptiones // delineata et aeri incisa // cura et sumptibus // Tobiae Conradi Lotteri, // Geographi, // et Chalc. Augastae Vindel.
Year: 1772
Dimensions: 66.0 x 55.5 (57.8 x 49.3) cm
Phys. Desc.: Engraved, Colored
Language(s): Latin

Entry (No.): 327
Author: Maaskamp, Evert (1769–1834)
Title: Les Provinces du Nord du Roijaume des Paijs - Bas. De Noordeliijke Provinciën van het Koningrijk der Nederlanden. Te Amsterdam, chez bij E. Maakamp.
Year: 1821
Dimensions: 23.5 x 17.3 cm
Phys. Desc.: Engraved, Uncolored
Language(s): French, Dutch

Entry (No.): 328
Author: Maaskamp, Evert (1769–1834)
Title: Nouvelle Carte Generale Itineraire dis Paijs Bas avec Les Routes de Poste, Nieuwe Algemeene Postkaart van det Koningrije der Nederlanden.
Year: 1821
Dimensions: 15.5 x 20.5 cm
Phys. Desc.: Engraved, Uncolored
Language(s): French, Dutch

Entry (No.): 329
Author: Maaskamp, Evert (1769–1834)
Title: Nouvelle Cartes du Roijaume des Paijs Bas indiquant à combien de lieues les principales villes sont situées l'unc de l'autre, à Amsterdam chez E. Maaskamp 1821. Nieuwestedenwijzer der 17 Nederlandsche Provincien, om te weten op hoc red uren gaans der zelter rooraanste Steden ran elkander zyn gelegen.
Year: 1821
Dimensions: 23.0 x 17.5 cm
Phys. Desc.: Engraved, Uncolored
Language(s): French, Dutch

Entry (No.): 330
Author: Mallet, Allain Manesson (1630–1706)
Title: MOLDAVIE // TRANSIL- // VANIE. //
Year: 1683
Dimensions: 14.0 x 21.5 (11.0 x 15.0) cm
Phys. Desc.: Engraved
Language(s): French

Entry (No.): 331
Author: Mallet, Allain Manesson (1630–1706)
Title: MOSCOVIE
Year: 1683
Dimensions: 14.0 x 21.5 (10.5 x 14.5) cm
Phys. Desc.: Engraved
Language(s): French

Entry (No.): 332
Author: Mallet, Allain Manesson (1630–1706)
Title: Natolie.
Year: 1683
Dimensions: 13.5 x 20.5 (9.8 x 14.0) cm
Phys. Desc.: Engraved, Colored Outline
Language(s): French

Entry (No.): 333
Author: Mallet, Allain Manesson (1630–1706)
Title: PETITE.. TARTARIE. //
Year: 1683
Dimensions: 13.7 x 21.5 (9.7 x 14.0) cm
Phys. Desc.: Engraved
Language(s): French

Entry (No.): 334
Author: Mallet, Allain Manesson (1630–1706)
Title: POLOGNE
Year: 1683
Dimensions: 13.5 x 20.7 (10.0 x 14.5) cm
Phys. Desc.: Engraved
Language(s): French

Entry (No.): 335
Author: Mallet, Allain Manesson (1630–1706)
Title: POLOGNE
Year: 1683
Dimensions: 14.0 x 20.7 (11.5 x 15.0) cm
Phys. Desc.: Engraved
Language(s): French

Entry (No.): 336
Author: Mallet, Allain Manesson (1630–1706)
Title: Turquie en Asie.
Year: 1683
Dimensions: 13.5 x 20.5 (10.1 x 14.3) cm
Phys. Desc.: Engraved, Colored Outline
Language(s): French

Entry (No.): 337
Author: Malte-Brun [Brunn], Conrad (1775–1826)
Title: PARTIE // MÉRIDIONALE // DE LA RUSSIE // D'EUROPE. //
Year: 1812
Dimensions: 34.5 x 25.7 (30.0 x 22.0) cm
Phys. Desc.: Engraved, Colored Outline
Language(s): French

Entry (No.): 338
Author: Malte-Brun [Brunn], Conrad (1775–1826)
Title: PARTIE SEPENT-LE. // DE LA RUSSIE // D'EUROPE. //
Year: 1812
Dimensions: 34.5 x 25.7 (30.0 x 21.2) cm
Phys. Desc.: Engraved, Colored Outline
Language(s): French

Entry (No.): 339
Author: Massa, Isaac (1586–1643)
Title: Moscoviae // Pars Avstralis. // auctore // Isaaco Massa.
Year: 1647
Dimensions: 58.0 x 49.0 (50.0 x 38.0) cm
Phys. Desc.: Engraved, Colored
Language(s): Latin
Figure 14

Entry (No.): 340
Author: Massa, Isaac (1586–1643)
Title: Novissima // Russiae // Tabula // Authore Isaaco Massa. Doctrina et humanitate praedito D. // Isaaco Bernart, rerum quaeper Mos- // coviam maxime trahuntur mercatori // peritissimo, hanc Moscoviae tabula // dedicat affinis finis // Hen. Hondius.
Year: 1649
Dimensions: 59.5 x 49.0 (55.0 x 47.0) cm
Phys. Desc.: Engraved, Uncolored
Language(s): Latin

Entry (No.): 341
Author: Massa, Isaac (1586–1643)
Title: Novissima // Russiae // Tabula // Authore Isaaco Massa.
Year: 1658
Dimensions: 61.0 x 50.0 (55.5 x 47.0) cm
Phys. Desc.: Engraved, Uncolored
Language(s): Latin

Entry (No.): 342
Author: Massa, Isaac (1586–1643)
Title: Tabula // Russia // Vulgo // Moscovia.
Year: 1692
Dimensions: 60.5 x 52.5 (55.3 x 44.7) cm
Phys. Desc.: Engraved, Colored
Language(s): Latin

Entry (No.): 343
Author: Mayer, Tobias, Sr. (1723–62) & Mayer, Johann Tobias (1752–1830)
Title: Map // of the // Kingdom of Poland // and // The Grand Dutchy // of // Lithuania // from Tob. Mayer of the Cosmographical // Society of Nuremberg.
Year: before 1772
Dimensions: 72.0 x 53.5 (53.5 x 45.4) cm
Phys. Desc.: Engraved, Colored
Language(s): English

Entry (No.): 344
Author: Mayer, Tobias, Sr. (1723–62) & Mayer, Johann Tobias (1752–1830)
Title: Mappa Geographica // Regni Poloniae // ex novissimus quot quot sunt // mappis specialibus composita // et ad L. L. Stereographica // projectionis revocata a // Tob. Mayero S. C. S. // Luci publicae tradita per Hommanios Heredes. Norimb. A MDCCLXXIII. C.P.S.C.M.
Year: 1773
Dimensions: 61.3 x 54.3 (51.1 x 44.0) cm
Phys. Desc.: Engraved, Colored
Language(s): Latin

Entry (No.): 345

Author: Mercator, Gerard (1512–94)
Title: Lithuania // Miliaria Lithuanica communia.
Year: ca. 1585–95
Dimensions: 34.4 x 23.0 (24.7 x 18.0) cm
Phys. Desc.: Engraved, Uncolored
Language(s): Latin

Entry (No.): 346
Author: Mercator, Gerard (1512–94)
Title: Tavrica // Chersonesvs // Nostra aetate Przecopsca et // Gazara // dicitur.
Year: ca. 1585–95
Dimensions: 27.5 x 20.5 (25.5 x 18.0) cm
Phys. Desc.: Engraved, Uncolored
Language(s): Latin

Entry (No.): 347
Author: Mercator, Gerard (1512–94)
Title: POLONIA // ET // SILESIA // PAR GERARDUM MERCATOREM // ... //
Year: 1613
Dimensions: 57.3 x 48.5 (45.7 x 34.2) cm
Phys. Desc.: Engraved
Language(s): Latin

Entry (No.): 348
Author: Mercator, Gerard (1512–94)
Title: WALACHIA // SERVIA, BVLGA- // RIA, ROMANIA //
Year: 1613
Dimensions: 57.8 x 48.5 (46.8 x 34.5) cm
Phys. Desc.: Engraved
Language(s): Latin

Entry (No.): 349
Author: Mercator, Gerard (1512–94)
Title: HVNGA// RIA
Year: 1616
Dimensions: 56.4 x 47.5 (44.5 x 37.0) cm
Phys. Desc.: Engraved, Colored
Language(s): Latin

Entry (No.): 350
Author: Mercator, Gerard (1512–94)
Title: LITHVA// NIA
Year: 1616
Dimensions: 56.7 x 47.0 (43.7 x 37.0) cm
Phys. Desc.: Engraved, Colored
Language(s): Latin

Entry (No.): 351
Author: Mercator, Gerard (1512–94)
Title: TAVRICA // CHERSONES- // VS. // NOSTRA AETATE // PRZECOPSKA ET // GAZARA // DICTUR //
Year: 1616
Dimensions: 52.0 x 44.0 (39.7 x 31.0) cm
Phys. Desc.: Engraved, Colored
Language(s): Latin

Entry (No.): 352
Author: Mercator, Gerard (1512–94)
Title: LITHV // ANIA
Year: 1628
Dimensions: 24.0 x 17.5 (20.0 x 14.3) cm
Phys. Desc.: Engraved, Colored
Language(s): Latin

Entry (No.): 353
Author: Mercator, Gerard (1512–94)
Title: MOSCO // VIA
Year: 1628
Dimensions: 24.0 x 17.5 (20.0 x 14.0) cm
Phys. Desc.: Engraved, Colored
Language(s): Latin

Entry (No.): 354
Author: Mercator, Gerard (1512–94)
Title: RUSSIA // CUM CONFINIJS //
Year: 1628
Dimensions: 24.0 x 17.4 (19.8 x 14.2) cm
Phys. Desc.: Engraved, Colored
Language(s): Latin

Entry (No.): 355
Author: Mercator, Gerard (1512–94)
Title: TAURICA // CHERSONE- // SUS //
Year: 1628
Dimensions: 24.0 x 17.5 (20.0 x 14.5) cm
Phys. Desc.: Engraved, Colored
Language(s): Latin

Entry (No.): 356
Author: Mercator, Gerard (1512–94)
Title: TRANSYL // VANIA //
Year: 1628
Dimensions: 24.0 x 17.5 (20.0 x 14.0) cm
Phys. Desc.: Engraved, Colored Outline
Language(s): Latin

Entry (No.): 357
Author: Mercator, Gerard (1512–94)
Title: Transylvania, // Sibenburgen.
Year: 1649
Dimensions: 58.5 x 49.5 (43.0 x 34.0) cm
Phys. Desc.: Engraved, Colored Outline
Language(s): Latin

Entry (No.): 358
Author: Mercator, Gerard (1512–94)
Title: TVRCICVM // IMPERIVM //
Year: 1670
Dimensions: 59.5 x 51.0 (55.0 x 45.0) cm
Phys. Desc.: Engraved, Colored
Language(s): Latin
Figure 5

Entry (No.): 359
Author: Merian, Matthäus (Mattheiu, Matheus, Matthew), the elder (1593–1650) & Merian, Matthäus, the younger (1621–87)
Title: TABULA // RUSSIAE // EX AUTOGRAPHO, QUOD

DELI- // NEANDUM CURAVIT FOEDOR // FILIUS TZARIS BORIS DESUMTA // ET AD FLUVIOS DWINAM ZUCH- // ANAW ALIAQUE LOCA QUAN- // DUM EX TABILIS ET NOTITŸS // AD NOS DELATIS [FIEEI?] POTUIT // AMPLIFICACA // ALIAS DICTA // MOSCOVIA //. Inset: ARCHANGELSCKAGOROD.
Year: 1646
Dimensions: 38.2 x 34.5 (34.7 x 28.0) cm
Phys. Desc.: Engraved

Language(s): Latin
Entry (No.): 360
Author: Meyer, Joseph (1796–1856)
Title: DAS // ASIATISCHE // RUSSLAND // JENSEIT DES URAL // 1849. // Entw. u. gez. v. Art. Prem. Ltn. Renner.
Year: 1849
Dimensions: 31.0 x 26.0 (25.5 x 19.5) cm
Phys. Desc.: Lithographed, Colored Outline
Language(s): German

Entry (No.): 361
Author: Meyer, Joseph (1796–1856)
Title: DAS // EUROPÄISCHE RUSSLAND // NEBST DEN ASIATISCHEN // PROVINZEN DIESSEIT DES URAL. // 1849. // Entw. u. gez. v. Art. Prem. Ltn. Renner.
Year: 1849
Dimensions: 26.0 x 31.0 (19.5 x 26.0) cm
Phys. Desc.: Lithographed, Colored Outline
Language(s): German

Entry (No.): 362
Author: Meyer, Joseph (1796–1856)
Title: DIE // REPUBLIK // POLEN // NACH IHREM BESTANDE IM JAHRE 1772 // UND DAS // KÖNIGREICH POLEN // SEIT DEM JAHRE 1815.
Year: 1849–50
Dimensions: 31.0 x 26.0 (25.5 x 18.5) cm
Phys. Desc.: Lithographed, Colored Outline
Language(s): German

Entry (No.): 363
Author: Meyer, Joseph (1796–1856)
Title: KÜSTENLÄNDER // DES // SCHWARZEN // MEERES // 1850. // Entw. u. gez. v. Art. Pr. Ltn. Renner.
Year: 1850
Dimensions: 31.0 x 26.0 (25.0 x 18.5) cm
Phys. Desc.: Lithographed, Colored Outline
Language(s): German

Entry (No.): 364
Author: Moll, Herman (ca. 1654–1732)
Title: To His Most Serene and August Majesty // Peter Alexovitz // Absolute Lord of Russa &c. This Map of Moscovy, Poland, Little Tartary, // and y-e Black Sea &c. is most // Humbly Dedicated by H. Moll // Geographer N.B. This Map has all the Improvements and Corrections of Cap-t. John Perry.
Year: before 1725
Dimensions: 101.5 x 63.5 (97.8 x 61.0) cm
Phys. Desc.: Engraved, Colored Outline
Language(s): English

Entry (No.): 365
Author: Moll, Herman (ca. 1654–1732)
Title: Russia or Moscovy // with its Acquisitions etc. in Sweden. Agreeable to Modern History // By H. Moll Geographer.
Year: 1729
Dimensions: 23.7 x 35.2 (19.3 x 26.1) cm
Phys. Desc.: Engraved, Colored
Language(s): English

Entry (No.): 366
Author: Moll, Herman (ca. 1654–1732)
Title: A General Map of // Turky in Europe, // Hungary &c. // By H. Moll Geographer.
Year: 1736
Dimensions: 30.0 x 24.0 cm
Phys. Desc.: Engraved, Colored Outline
Language(s): English
Figure 45

Entry (No.): 367
Author: Mollo, Tranquillo (1767–1837)
Title: Galizien // nach den neuesten Beobachtungen // Wien bey Tranquillo Mollo.
Year: 1810
Dimensions: 48.0 x 36.5 (45.0 x 32.7) cm
Phys. Desc.: Engraved, Colored Outline
Language(s): German

Entry (No.): 368
Author: Morden, Robert (d. 1703)
Title: TARTARIA // IN // EUROPE // BY ROB-T MORDEN. //
Year: 1688
Dimensions: 15.3 x 20.0 (13.5 x 11.5) cm
Phys. Desc.: Engraved
Language(s): English

Entry (No.): 369
Author: Morse, Jedidiah (1761–1826)
Title: ASIA // FROM THE BEST // AUTHORITIES. //
Year: 1801
Dimensions: 25.0 x 21.3 (22.5 x 19.0) cm
Phys. Desc.: Engraved
Language(s): English

Entry (No.): 370
Author: Morse, Jedidiah (1761–1826)
Title: EUROPE // FROM THE BEST // AUTHORITIES //
Year: 1801
Dimensions: 25.0 x 21.3 (22.3 x 18.0) cm
Phys. Desc.: Engraved
Language(s): English

Entry (No.): 371
Author: Morse, Jedidiah (1761–1826)
Title: PALESTINE // FROM THE BEST // AUTHORITIES //
Year: 1801
Dimensions: 14.4 x 18.8 (13.9 x 18.5) cm

Phys. Desc.: Engraved
Language(s): English

Entry (No.): 372
Author: Morse, Jedidiah (1761–1826)
Title: POLAND // SHEWING THE CLAIMS OF // RUSSIA, PRUSSIA & AUSTRIA // FROM THE BEST AUTHORITIES
Year: 1801
Dimensions: 25.0 x 21.3 (21.5 x 19.0) cm
Phys. Desc.: Engraved
Language(s): English

Entry (No.): 373
Author: Morse, Jedidiah (1761–1826)
Title: TURKEY // IN EUROPE, AND // HUNGARY. //
Year: 1801
Dimensions: 25.0 x 21.3 (21.5 x 18.5) cm
Phys. Desc.: Engraved
Language(s): English

Entry (No.): 374
Author: Münster (Muenster), Sebastian (1488–1552)
Title: POLONIA ET VNGARIA XX -- NOVA TABVLA --
Year: 1545
Dimensions: 38.5 x 32.0 (37.0 x 29.0) cm
Phys. Desc.: Woodcut, Colored
Language(s): Latin
Figure 2

Entry (No.): 375
Author: Münster (Muenster), Sebastian (1488–1552)
Title: Asia wie es jetziger zeit nach den fürnemesten Herzschafften abgetheilet und beschrieben ist.
Year: 1550
Dimensions: 43.0 x 35.0 (36.0 x 31.0) cm
Phys. Desc.: Woodcut, Uncolored
Language(s): German

Entry (No.): 376
Author: Münster (Muenster), Sebastian (1488–1552)
Title: BREUE DESCRITTIONE DEL REGNO DI POLLONIA. // MA LA SEQUENTE TAUOLA E ASSAI PIU DIFFUSA. //
Year: 1558
Dimensions: 21.5 x 31.5 (13.3 x 8.0) cm
Phys. Desc.: Woodcut
Language(s): Italian

Entry (No.): 377
Author: Münster (Muenster), Sebastian (1488–1552)
Title: DEL REGNO & DI TUTTO IL PAESE DELLA POLONIA, & // DELLE PROVINCIE MINORI, CHE GLI SONO ATTORNO. //
Year: 1558
Dimensions: 21.5 x 31.5 (15.1 x 20.0) cm
Phys. Desc.: Woodcut
Language(s): Italian

Entry (No.): 378
Author: Münster (Muenster), Sebastian (1488–1552)
Title: DELLA VNGARIA // TRANSILUANIA PROUNCIA DEL REGNO // ORGHERO. //
Year: 1558
Dimensions: 20.2 x 30.0 (18.5 x 15.0) cm
Phys. Desc.: Woodcut
Language(s): Italian

Entry (No.): 379
Author: Münster (Muenster), Sebastian (1488–1552)
Title: NUOA DESCRIZZIONE DELLA POLONIA, & DEL' VNGARIA.
Year: 1558
Dimensions: 40.0 x 32.0 (34.5 x 25.5) cm
Phys. Desc.: Woodcut
Language(s): Italian

Entry (No.): 380
Author: Münster (Muenster), Sebastian (1488–1552)
Title: MOSCOVIA
Year: 1575
Dimensions: 20.0 x 30.0 (16.0 x 17.5) cm
Phys. Desc.: Woodcut
Language(s): Italian

Entry (No.): 381
Author: Münster (Muenster), Sebastian (1488–1552)
Title: Von den Königreich Polandt. // das in Samatia auch begriffen wirt/ sampt // andern Ländern disem Konigreich zugehörig. // Cap. rlir.
Year: 1578
Dimensions: 20.0 x 31.0 (15.1 x 12.0) cm
Phys. Desc.: Woodcut
Language(s): German

Entry (No.): 382
Author: Münster (Muenster), Sebastian (1488–1552)
Title: Von den Moscowytern. // Moscowyterland Cap. Irrj.
Year: 1578
Dimensions: 20.5 x 31.5 (16.0 x 17.5) cm
Phys. Desc.: Woodcut
Language(s): German

Entry (No.): 383
Author: Münster (Muenster), Sebastian (1488–1552)
Title: LANDTAFEL DES UNGERLANDS / POLANDS / REUSSEN / LITTAW / WALACHEN UND BULGAREN. // SARMATIA EVROPAEA. // DES VNGERLAN- // DES / POLOANDES / REUSSEN / LIT- / TAW / WALACHEY / BULGAREY / etc. BESCHREIBUNG… //
Year: 1591
Dimensions: 41.0 x 34.0 (35.5 x 31.0) cm
Phys. Desc.: Woodcut
Language(s): German

Entry (No.): 384
Author: Münster (Muenster), Sebastian (1488–1552)
Title: SARMATIA ASIE…
Year: 1598
Dimensions: 20.2 x 34.5 (16.0 x 24.0) cm

Phys. Desc.: Woodcut
Language(s): German

Entry (No.): 385
Author: Münster (Muenster), Sebastian (1488–1552)
Title: SARMATIA. // ZUM ERSTEN VOM VNGERLANDT.
Year: 1598
Dimensions: 20.2 x 34.5 (13.5 x 10.0) cm
Phys. Desc.: Woodcut
Language(s): German

Entry (No.): 386
Author: Ortelius, Abraham (1527–98)
Title: Abraham Ortelius Antverpianus, Beneuolis Lectoribus S. D.
Year: 1570
Dimensions: 29.4 x 42.2 cm
Phys. Desc.: Engraved
Language(s): Latin

Entry (No.): 387
Author: Ortelius, Abraham (1527–98)
Title: EUROPAE.
Year: 1572
Dimensions: 54.3 x 40.3 (46.2 x 34.0) cm
Phys. Desc.: Engraved, Colored
Language(s): Latin
Figure 3

Entry (No.): 388
Author: Ortelius, Abraham (1527–98)
Title: PONTVS // EVXINVS. // AEQUOR IASONIO PULSA- // TUM REMIGE PRIMUM... // 1590. // Ex conatibus geographi- // cis Abrahami Ortelij. //
Year: 1590
Dimensions: 58.5 x 46.0 (50.0 x 36.0) cm
Phys. Desc.: Engraved, Colored
Language(s): Latin
Figure 4

Entry (No.): 389
Author: Ortelius, Abraham (1527–98)
Title: EVROPAM // SIVE // CELTICAM // VETEREM, // SIC DESCRIBERE CONABAR // ABRAHAMUS ORTELIUS. //
Year: 1649
Dimensions: 59.0 x 49.0 (47.0 x 35.5) cm
Phys. Desc.: Engraved, Colored
Language(s): Latin

Entry (No.): 390
Author: Ottens, Reinier I (1698–1750)
Author: Frauendorff, Carl (Carolus) von
Title: VERUS // CHERSONESI TAURICAE // SEU CRIMEA // CONSPECTUS // ADJACENTIUM ITEM REGIONUM ITINERISQ; AB EXERCITU RUTHENO // A:° MDCCXXXVI ET MDCCXXXVII // ADUERSUS TATTAROS SUSCEPTIS // EDITUS EX AUTOGRAPHIS CAROLIA FRAUENDORFF // CENTURIONIS IN ACAD: SC: PETROPOLIT: // APUD R. & J. OTTENS // AMSTELAEDAMI. //
Year: 1738
Dimensions: 74.2 x 52.5 (66.0 x 47.4) cm
Phys. Desc.: Engraved, Colored
Language(s): Latin

Entry (No.): 391
Author: Ottens, Reinier I (1698–1750) & Ottens, Josua (Joshua) (1704–65)
Title: Partie Meridionale de Moscovie Dressée par G. de L'Isle Rectifié & Augmente par R. & J. Ottens Geographers a Amsterdam.
Year: 1740
Dimensions: 65.4 x 52.8 (54.0 x 40.4) cm
Phys. Desc.: Engraved, Colored
Language(s): French

Entry (No.): 392
Author: Ottens, Reinier I (1698–1750) & Ottens, Josua (Joshua) (1704–65)
Title: Regni // Poloniae // et Ducatus // Lithuaniae, // Voliniae, Podoliae // Ucrainiae // Prussiae et Curlandiae Descriptio Emendata per R & J. Ottens. Amsterdami Cum Privilegio.
Year: ca. 1740
Dimensions: 64.5 x 53.5 (57.5 x 50.0) cm
Phys. Desc.: Engraved, Colored
Language(s): Latin

Entry (No.): 393
Author: Ottens, Reinier I (1698–1750) & Ottens, Josua (Joshua) (1704–65)
Title: Reipublicae et Status Generalis Poloniae nova Tabula, Comprehendens Majoris et Minoris Poloniae Regni Magni Ducatus Lithuaniae // Ducatus Prussiae, Curlandiae, Samogitiae, Massoviae Volhuniae Podoliae Russiae Ucraniae et de Moscoviae Pars Accuratam Descriptionem Amsterdami apud R & J Ottens in platoa Vulgo dicta, de Niuwe'dyk in signo Mappa Terra.
Year: 1740
Dimensions: 94.5 x 56.0 (88.8 x 51.0) cm
Phys. Desc.: Engraved, Colored
Language(s): Latin

Entry (No.): 394
Author: Petermann, Augustus (August) Heinrich (Henry) (1822–78)
Title: Ost-Europa, N° 6: Süd-Russland & Kaukasien, von A. Peterman.
Year: 1873
Dimensions: 47.0 x 38.3 (40.5 x 33.5) cm
Phys. Desc.: Lithographed, Colored
Language(s): German

Entry (No.): 395
Author: Petermann, Augustus (August) Heinrich (Henry) (1822–78)
Title: Russland // und // Skandinavien // Von A. Peterman.
Year: 1873
Dimensions: 47.0 x 38.3 (40.0 x 33.5) cm

Phys. Desc.: Lithographed, Colored
Language(s): German

Entry (No.): 396
Author: Peutinger, Konrad (1508–47)
Title: Tabulae Pevtingerianae Segmentum V. à Sarmatis Hamaxobiis vsque ad Roxolanas. Segmentum VI. à Sarmatis Rosulanis vsque ad Parnacos.
Year: ca. early 16th c.
Dimensions: 59.5 x 49.3 (50.5 x 39.3) cm
Phys. Desc.: Copy, Engraved, Colored
Language(s): Latin

Entry (No.): 397
Author: Pitt, Moses (1641–97)
Title: Tavrica // Chersonsvs, // Hodie // Przecopsca; // et Gazara dicitur.
Year: 1650
Dimensions: 62.4 x 53.5 (49.3 x 37.8) cm
Phys. Desc.: Engraved, Colored
Language(s): Latin

Entry (No.): 398
Author: Pitt, Moses (1641–97)
Title: The // English // Atlas. // Volume I. // containing a // Description // of the // Places next the North Pole; // as also of // Muscovy, Poland, Sweden, Denmark, // and their several Dependances. Oxford, // Printed at the Theater, for Joh. Jansonius a Waesberge, and Steven Swart, Booksellers in Amsterdam. MDCLXXX.
Year: 1943/1680
Dimensions: 35.0 x 56.0 cm
Phys. Desc.: Printed, Reprint of 1680 atlas
Language(s): English
Figure 27

Entry (No.): 399
Author: Pograbius (Pograbski, Pograbka), Andreas (Andrzej) (d. 1602)
Title: Partis // Sarmatiae Europeae, // Qvae Sigismundo // Avgusto Regi Polo // niae Potentissimo // subiacet // Nova Descriptio. // ....Andreas Pograbivs Pilsnenens... 1569... Venetis...
Year: 1599
Dimensions: 45.8 x 40.3 cm
Phys. Desc.: Copy and Negatives
Language(s): Latin

Entry (No.): 400
Author: Prétot, Étienne André Philippe de (1708–87)
Title: Carte Nouvelle // de l'Empire // de la Russie, // Comprise en Europe, // dirigée par M[r]. Philippe, // Censeur Royal, // et Professeur d'histoire; de l'Acadèmie Royale // des Sciences et belles-lettres // d'Angers.
Year: 1787
Dimensions: 29.3 x 43.0 (24.5 x 37.5) cm
Phys. Desc.: Engraved, Uncolored
Language(s): French

Entry (No.): 401
Author: Ptolemy (Ptolemaeus), Claudius (worked ca. ad 127–80)
Title: The World of the Sixteenth Century according to a Venetian Map of 1562.
Year: 1562
Dimensions: 28.0 x 21.4 cm
Phys. Desc.: Copy
Language(s): English

Entry (No.): 402
Author: Ptolemy (Ptolemaeus), Claudius (worked ca. ad 127–80)
Title: EVR: VIII. TAB: / SARMATIA // EVROPAE // /
Year: 1578
Dimensions: 54.0 x 41.6 (42.5 x 34.0) cm
Phys. Desc.: Copy, Engraved, Colored
Language(s): Latin

Entry (No.): 403
Author: Ptolemy (Ptolemaeus), Claudius (worked ca. ad 127–80)
Title: TAB. II. ASIAE // SARMATIAM // ASIATICAM // REPRAESENTANS // ...
Year: 1730
Dimensions: 53.5 x 45.5 (45.7 x 35.5) cm
Phys. Desc.: Copy, Engraved, Colored
Language(s): Latin
Figure 1

Entry (No.): 404
Author: Radefeld, Major (Hauptmann) Carl Christian Franz (1788–1874)
Title: Europaeisches // Russland // entworfen und gezeichnet vom Hauptm. Radefeld. 1844.
Year: 1844
Dimensions: 37.5 x 44.0 (30.0 x 35.5) cm
Phys. Desc.: Engraved, Colored
Language(s): German
Figure 66

Entry (No.): 405
Author: Radefeld, Major (Hauptmann) Carl Christian Franz (1788–1874)
Title: Asiatisches // Russland // 1845. // Nach den besten Quellen entworf. u. gezechn. v. Hauptm. C. Radefeld.
Year: 1845
Dimensions: 43.5 x 37.5 (35.5 x 29.5) cm
Phys. Desc.: Engraved, Colored
Language(s): German

Entry (No.): 406
Author: Radefeld, Major (Hauptmann) Carl Christian Franz (1788–1874)
Title: Neueste Karte // der // Küstenländer // des // Schwarzen // Meeres. // Entworfen und gezeichnet von // Hauptm. Radefeld. // 1845.
Year: 1845
Dimensions: 44.0 x 37.5 (35.5 x 29.5) cm
Phys. Desc.: Engraved, Colored
Language(s): German

Entry (No.): 407
Author: Radziwiłł, Prince Mikołaj Krzysztof (1549–1616) & Makowski, Tomasz (ca. 1575–1630)
Title: BORYSTHENIS TRACTUS.
Year: 1613
Dimensions: 81.2 x 51.5 (75.0 x 32.2) cm
Phys. Desc.: Engraved, Colored
Language(s): Latin

Entry (No.): 408
Author: Radziwiłł, Prince Mikołaj Krzysztof (1549–1616) & Makowski, Tomasz (ca. 1575–1630)
Title: BORYSTHENIS TRACTUS.
Year: 1613
Dimensions: 80.7 x 46.0 (75.0 x 32.2) cm
Phys. Desc.: Engraved, Uncolored
Language(s): Latin

Entry (No.): 409
Author: Radziwiłł, Prince Mikołaj Krzysztof (1549–1616) & Makowski, Tomasz (ca. 1575–1630)
Title: MAGNI DVCATVS // LITHVANIAE, // CAETERARVMQUE REGIO- // NVM ILLI ADIACENTIVM // EXACTA DESCRIPTIO // ILLSS. mi ac EXCELLS.mi PRINCIPIS ET Dni D. NICOLAI // CHRISTOPHORI RADZIWIL. D.G. OLICAE AC IN NIES-// WIES DUCIS ... OPERA, CURA ET // IMPENSIS FACTA AC IN LUCEM EDITA // Amsterodami // Excudebat GUILHELMUS JANSSONIUS... // Anno 1613. // Sculptum apud Hesselum Gerardum. //
Year: 1613
Dimensions: 80.0 x 77.4 (75.0 x 72.7) cm
Phys. Desc.: Engraved, Colored
Language(s): Latin

Entry (No.): 410
Author: Radziwiłł, Prince Mikołaj Krzysztof (1549–1616) & Makowski, Tomasz (ca. 1575–1630)
Title: MAGNI DVCATVS // LITHVANIAE, // CAETERARVMQUE REGIO- // NVM ILLI ADIACENTIVM // EXACTA DESCRIPTIO // ILLSS. mi ac EXCELLS.mi PRINCIPIS ET Dni D. NICOLAI // CHRISTOPHORI RADZIWIL. D.G. OLICAE AC IN NIES-// WIES DUCIS ... OPERA, CURA ET // IMPENSIS FACTA AC IN LUCEM EDITA // Amsterodami // Excudebat GUILHELMUS JANSSONIUS... // Anno 1613. // Sculptum apud Hesselum Gerardum. //
Year: 1613
Dimensions: 76.0 x 78.0 (73.0 x 74.8) cm
Phys. Desc.: Engraved, Uncolored
Language(s): Latin

Entry (No.): 411
Author: Radziwiłł, Prince Mikołaj Krzysztof (1549–1616) & Makowski, Tomasz (ca. 1575–1630)
Title: MAGNI DVCATVS // LITVANIAE // CAETERARUMQ. REGIONUM ILLI ADIACENTIUM EXACTA DESCRIPTIO // ILLSS.mi ac EXCELLS.mi PRICIPIS ET Dni D. NICOLAI // CHRISTOPHORI RADZIWIL. D.G. OLICAE AC IN NIES-// WIES DUCIS. S. Rom. Imp. Principis in Szylo// wiec ac Mir comitis, et S. Sepulchri Hirosolimitani Militis etc. // OPERA ET CURA IN LUCEM EDITA //
Year: 1647
Dimensions: 58.7 x 49.0 (54.0 x 43.5) cm
Phys. Desc.: Engraved, Colored
Language(s): Latin

Entry (No.): 412
Author: Radziwiłł, Prince Mikołaj Krzysztof (1549–1616) & Makowski, Tomasz (ca. 1575–1630)
Title: MAGNI DVCATVS // LITVANIAE // ET REGIONUM ADIACENTIUM EXACTA DESCRIPTIO // ... NICOLAI CHRISTOPHORI // RADZIWILL... // ...OPERA, CURA ET IMPENSIS OLIM // FACTA, AC NUNC DENUO HAC FORMA EDITA A // I. BLAEU. Ioannes Blaeu // Excudebat. //
Year: 1659–72
Dimensions: 64.0 x 54.5 (52.5 x 44.5) cm
Phys. Desc.: Engraved, Colored
Language(s): Latin
Figure 10

Entry (No.): 413
Author: Radziwiłł, Prince Mikołaj Krzysztof (1549–1616) & Makowski, Tomasz (ca. 1575–1630)
Title: MAGNI DVCATVS // LITVANIAE // CAETERARUMQ. REGIONUM ILLI ADIACENTIUM EXACTA DESCRIP.//.. NICOLAI CHRISTOPHORI RADZIWIL // ... // OPERA ET CURA IN LUCEM EDITA // (BORYSTHENIS TRACTUS...). Sumptibus Janssonio-Waesbergiarum.
Year: 1680
Dimensions: 62.3 x 53.6 (53.7 x 43.5) cm
Phys. Desc.: Engraved, Colored
Language(s): Latin

Entry (No.): 414
Author: Radziwiłł, Prince Mikołaj Krzysztof (1549–1616) & Makowski, Tomasz (ca. 1575–1630)
Title: MAGNI DUCATUS // LITHUANIAE // TABULA, DIVISA TAM IN PALATINATUS, // QUAM IN SUBJACENTES CASTELLANIAS // PER F. DE WITT. AMSTELODAMI... //
Year: 1688
Dimensions: 61.7 x 52.0 (52.3 x 44.3) cm
Phys. Desc.: Engraved, Colored
Language(s): Latin

Entry (No.): 415
Author: Ratelband (Ratelbant), Johannes (1715–93)
Title: CARTE NOUVELLE // DE // TRANSILVANIE // AVEC SES CONFINS. // NIEUVE KAART VAN SEVENBERGEN. //
Year: 1735
Dimensions: 27.0 x 20.2 (26.0 x 18.0) cm
Phys. Desc.: Engraved, Colored
Language(s): French, Dutch

Entry (No.): 416
Author: Ratelband (Ratelbant), Johannes (1715–93)

Title: LE ROYAUME DE POLOGNE AVEC SES CONFINS.
Year: 1735
Dimensions: 27.0 x 16.0 (21.0 x 15.0) cm
Phys. Desc.: Engraved, Colored Outline
Language(s): French

Entry (No.): 417
Author: Ratelband (Ratelbant), Johannes (1715–93)
Title: LE ROYAUME DE POLOGNE AVEC SES CONFINS. GENERALE KAART VAN POLEN.
Year: 1735
Dimensions: 27.0 x 20.3 (21.0 x 15.0) cm
Phys. Desc.: Engraved, Colored
Language(s): French, Dutch

Entry (No.): 418
Author: Reilly, Franz Johann Joseph von (1760–1820)
Title: Schauplatz der Fünf Theile der Welt // Mit bestaendiger Rücksicht auf die besten // Originalwerke in drey Theile zusammengetragen von einer // Gesellschaft Geographen // Nach und zu Büshings grosser Erdbeschreibung // Herausgegeben von F.J.J. von Reilly. Erster Theil. Enthält, I Ungarn und das Osmannishe Reich in Europa, ... Wien 1789. Geschtochen von Ignaz Albrecht.
Year: 1789
Dimensions: 44.0 x 29.0 cm
Phys. Desc.: Engraved, Colored
Language(s): German

Entry (No.): 419
Author: Reilly, Franz Johann Joseph von (1760–1820)
Title: Karte // von dem // Russchen // Reiche // in Europa // Nach der Karte der Gesellschaft der Künste // und Wissenschaften in Petersburg // Neu verseichnet herausgegeben // von // Franz Joh. Jos. von Reilly.
Year: 1796
Dimensions: 88.0 x 62.7 (75.2 x 56.0) cm
Phys. Desc.: Engraved, Colored
Language(s): German

Entry (No.): 420
Author: Reinecke, Johann Matthias Christoph (1770–1818)
Title: CHARTE des ganzen Russishen-Reichs in Europa und Asien. Nach den neuersten und sichersten Ortsbestimmungen entworfen und berichtiget auf der Sternwarte Seeberg bey Gotha. gezeichnet von I.C.M. Reinecke Weimar: im Verlage Industrie Comptoirs 1800.
Year: 1800
Dimensions: 51.2 x 48.8 cm
Phys. Desc.: Engraved, Colored Outline
Language(s): German

Entry (No.): 421
Author: Rizzi-Zannoni, Giovanni Antonio (1736–1814)
Title: CARTE GENERALE // DE LA POLOGNE // avec tons les Etats qui en dependent // dressée sur les meilleurs mémoires et appuyée à plusieurs Observat. Astronom. // par M. =Rizzi Zannoni // de la Societé Cosmog. De Gottingue. // Professeur de Géographie. // A Paris.
Year: ca. 1779–99
Dimensions: 51.0 x 38.6 (45.3 x 30.5) cm
Phys. Desc.: Engraved, Colored Outline
Language(s): French

Entry (No.): 422
Author: Rizzi-Zannoni, Giovanni Antonio (1736–1814)
Title: LI PALATINATI // della // RUSSIA ROSSA PODOLIA // e // WOLHYNIA // Tratta dall' Atlante Polaco // del Sig.r Rizzi Zanoni // VENEZIA 1781 // PRESSO ANTONIO ZATTA // Con Privilegio Dell'Eccmo Senato
Year: 1781
Dimensions: 49.0 x 38.7 (41.2 x 31.5) cm
Phys. Desc.: Engraved, Colored Outline
Language(s): Italian

Entry (No.): 423
Author: Rizzi-Zannoni, Giovanni Antonio (1736–1814)
Title: LI PALATINATI // di // BRACLAW, // e // KIOWIA. // Tratta dall' Atlante Polacco // del Sigr. Rizzi Zanoni. // VENEZIA 1781 // Presso Antonio Zatta // Con Privilegio dell'Eccmo Senato.
Year: 1781
Dimensions: 37.1 x 49.0 (31.5 x 41.0) cm
Phys. Desc.: Engraved, Colored Outline
Language(s): Italian
Figure 52

Entry (No.): 424
Author: Rizzi-Zannoni, Giovanni Antonio (1736–1814)
Title: LI PALATINATI // di // NOWOGRODEK // PODLACHIA // E BRZESK LITTEW // Tratta dall' Atlante Polacco // del Sig.r Rizzi Zanoni // VENEZIA 1781 // Presso Antonio Zatta // Con Privilegio dell'Eccmo Senato.
Year: 1781
Dimensions: 49.0 x 37.1 (40.7 x 31.3) cm
Phys. Desc.: Engraved, Colored Outline
Language(s): Italian

Entry (No.): 425
Author: Rizzi-Zannoni, Giovanni Antonio (1736–1814)
Title: LI PALATINATI // di // NOWOGRODEK // PODLACHIA // E BRZESK LITTEW // Tratta dall' Atlante Polacco del Sig.r Rizzi Zanoni // VENEZIA 1781 // Presso Antonio Zatta // Con Privilegio dell'Eccmo Senato
Year: 1781
Dimensions: 49.0 x 38.7 (40.7 x 31.3) cm
Phys. Desc.: Engraved, Colored Outline
Language(s): Italian

Entry (No.): 426
Author: Rizzi-Zannoni, Giovanni Antonio (1736–1814)
Title: LI PALATINATI di BRACLAW, // e // KIOWIA. // tratta dall' Atlante Polacco // del Sig.r Rizzi Zanoni. // VENEZIA 1781 // Presso Antonio Zatta // Con Privilegio dell'Eccmo Senato.
Year: 1781
Dimensions: 38.6 x 49.0 (31.5 x 41.0) cm
Phys. Desc.: Engraved, Colored Outline
Language(s): Italian

Entry (No.): 427
Author: Rizzi-Zannoni, Giovanni Antonio (1736–1814)
Title: LI PALATINATI // della // RUSSIA ROSSA PODOLIA // e // WOLHYNIA // Tratta dall' Atlante Polacco // del Sig. r Rizzi Zanoni // VENEZIA 1782 // PRESSO ANTONIO ZATTA // Con Privilegio sell'Eccmo Senato
Year: 1782
Dimensions: 49.0 x 37.1 (41.2 x 31.5) cm
Phys. Desc.: Engraved, Colored Outline
Language(s): Italian

Entry (No.): 428
Author: Rizzi-Zannoni, Giovanni Antonio (1736–1814)
Title: PRINCIPATI // di // MOLDAVIA, // E VALLACHIA // Tratti dalle Carte // del' Impero Ottomano // del Sig.r Rizzi Zanoni // VENEZIA 1782 // Presso Antonio Zatta // Con Privilegio dell' Eccmo Senato
Year: 1782
Dimensions: 49.0 x 37.1 (40.3 x 30.8) cm
Phys. Desc.: Engraved, Colored Outline
Language(s): Italian

Entry (No.): 429
Author: Rizzi-Zannoni, Giovanni Antonio (1736–1814)
Title: Polen // in die dermaligen Besitzungen // eingetheilt // Nach der vortreflichen Karte // von Rizzi Lannoni // Herausgegeben in WIEN von Artaria & Comp. // 1807 //// Carte de la // POLOGNE // DIVISÉE // entre les puissances co-partageantes // d'aprés Rizzi-Lannoni // publiée a Vienne par Artaria et Compagnie.
Year: 1807
Dimensions: 95.5 x 83.5 cm
Phys. Desc.: Engraved, Colored
Language(s): German

Entry (No.): 430
Author: Robert de Vaugondy, Didier (1723–86)
Author: Delamarche, Charles François (1740–1817)
Title: PARTIE MÉRIDIONALE // DE LA // RUSSIE EUROPEENNE, // TARTARIE RUSSIENNE, // PETITE TARTARIE. // PAR ROBERT DE VAUGONDY. // REVUES ET CORRIGÉES // PAR C.F. DELAMARCHE // SON SUCCESSEUR. // 1806 //
Year: 1806
Dimensions: 45.0 x 31.0 (29.5 x 24.0) cm
Phys. Desc.: Engraved, Colored Outline
Language(s): French

Entry (No.): 431
Author: Robert de Vaugondy, Didier (1723–86)
Author: Delamarche, Charles François (1740–1817)
Title: ROYAUME DE // POLOGNE // PAR ROBERT DE VAUGONDY. // DIVISÉ ET CORRIGÉ // SELON LES PARTAGES FAITS EN // 1772, 1793 ET 1795, ENTRE LA RUSSIE // LA PRUSSE ET L'AUTRICHE. // PAR C.F. DELAMARCHE GÉOG. //
Year: 1806
Dimensions: 45.0 x 31.0 (28.0 x 24.0) cm
Phys. Desc.: Engraved, Colored Outline
Language(s): French

Entry (No.): 432
Author: Robert de Vaugondy, Didier (1723–86)
Author: Delamarche, Charles François (1740–1817)
Title: Siberie ou Russie Asiatique, Tartarie Chinoise, Pays des Eluts et Isles du Japon. Par Robert de Vaugondy + c.
Year: 1806
Dimensions: 45.0 x 30.6 (30.5 x 24.1) cm
Phys. Desc.: Engraved, Colored Outline
Language(s): French

Entry (No.): 433
Author: Robert de Vaugondy, Gilles (1688–1766)
Title: Partie Orientale // de la Russie Polonoise // où se trouvent // la Basse Volhynie et la // Basse Podolie. // Par l S.r Robert de Vaugondy // fils de M.r Robert Geog. du Roi. // Avec Privilege 1749
Year: 1749
Dimensions: 22.9 x 18.3 (19.1 x 16.4) cm
Phys. Desc.: Engraved, Uncolored
Language(s): French
Figure 32

Entry (No.): 434
Author: Robert de Vaugondy, Gilles (1688–1766)
Title: Carte de // l'Ancien Continent // Selon sa plus grande lonqueur diamètrale // depuis la Pointe de la Tartarie Orientale, // jusqu'au Cap de Bone Esperance. Dressée Sous les yeux de M.r de Buffon // Par le S.r Robert de Vaugondy // Fils de M.r Robert Geogr. ord. du Roi. 1749.
Year: 1749
Dimensions: 18.8 x 22.8 (17.8 x 21.8) cm
Phys. Desc.: Engraved, Uncolored
Language(s): French

Entry (No.): 435
Author: Robert de Vaugondy, Gilles (1688–1766)
Title: Partie Occidentale de // la Russie Polonoise où se trouvent // la Russie Noire, la Haute // Volhynie, et la Haute, Podolie. Par le S.r Robert de Vaugondy // Fils de M.r Robert Geog. du Roi // Avec Privilege 1749.
Year: 1749
Dimensions: 24.3 x 18.3 (23.5 x 16.6) cm
Phys. Desc.: Engraved, Uncolored
Language(s): French

Entry (No.): 436
Author: Robert de Vaugondy, Gilles (1688–1766)
Title: Partie Occidentale // de l'Empire de Russie // où se trouve distinguée // La Russie Européenne // Divisée par ses Gouvernemens ou Provinces de // Dwinskoy, Oloneckoy, Bielozerskoy, Archangel // Nowgorod, Wolockoy, Ustiuskoy, Moscow, Kiowie, // Bielogorod, Woroneskoy, el la Petite Tartarie. // La Partie Asiatique // contient les Provinces de // Berezowskoy, Mangajeiskoy, Wiatka, Solkamskoy, Pelymskoy, // Ufimskoy, Tobolskoy, Surgutskoy, Jeniseiskoy, //Narimskoy, Kuzneckoy, Tomskoy, Krasnojarskoy + c. le Royaume de Kasan, // Le Gouvernment d'Astrachan et les confins du Royaume des Kalmaks Eleuths, // Dressée, d'après les Cartes de l'Atlas Russien, // Par le S.r Robert de Vaugondy, fils de M.r Robert Geographe ordin. du Roi.

Avec Privilege. 1750.
Year: 1750
Dimensions: 80.5 x 55.0 (59.5 x 47.5) cm
Phys. Desc.: Engraved, Colored Outline
Language(s): French

Entry (No.): 437
Author: Robert de Vaugondy, Gilles (1688–1766)
Title: Partie Orientale // de L'Empire de Russie en Asie, // où se trouvent les Provinces de // Jakuckskoy, Nerckzinskoy, Selinginskoy, // Himskoy, Krasnojarskoy, Narimskoy, // Jenisseiskoy, Mangajeiskoy, + c. // et les confins de // La Tartarie Chinoise, // Dressée, d'après les Cartes de l'Atlas Russien // Par le S.r Robert de Vaugondy // fils de M.r Robert, Géographe ord. du Roy. Avec Privilege. 1750.
Year: 1750
Dimensions: 81.2 x 55.0 (59.2 x 46.5) cm
Phys. Desc.: Engraved, Colored Outline
Language(s): French

Entry (No.): 438
Author: Robert de Vaugondy, Gilles (1688–1766)
Title: Le Royaume de // Pologne, // divisé // en ses Duchés et Provinces, et // subdivisé en Palatinats. Par le S.r Robert de Vaugondy // Géographe ordinaire du Roi. Avec Privilege 1752.
Year: 1752
Dimensions: 82.6 x 55.0 (54.0 x 48.0) cm
Phys. Desc.: Engraved, Colored Outline
Language(s): French

Entry (No.): 439
Author: Robert de Vaugondy, Gilles (1688–1766)
Title: Partie Méridionale // de la // Russie Européenne // ou sont distinguées exactement // toutes les Provinces, // d'après le dètail de l'Atlas Russien; // Par le S.r Robert de Vaugondy fils // Géographe ordinaire du Roy // Avec Privilege // 1752.
Year: 1752
Dimensions: 81.7 x 54.9 (59.4 x 48.2) cm
Phys. Desc.: Engraved, Colored Outline
Language(s): French

Entry (No.): 440
Author: Robert de Vaugondy, Gilles (1688–1766)
Title: Le Royaume de Pologne. Par le Sr. Robert de Vaugondy Géog. ord. du Roi de S.M. Pol. Duc de Lorr. et de Bar et de l'Academie royale de Nancy A Paris. Ches le Sr. Delamarche Géographe. Rue du Foin St. Jacques, au Collége de Me. Gervais Avec Privilege 1767
Year: 1767
Dimensions: 68.5 x 55.5 (61.2 x 48.5) cm
Phys. Desc.: Engraved, Colored
Language(s): French
Figure 50

Entry (No.): 441
Author: Robert de Vaugondy, Gilles (1688–1766)
Title: Partie Meridionale // de la // Russie Europeenne, // Tartarie Russienne, // Petite Tartarie. // Par le S.r Robert de Vaugondy. 1778.
Year: 1778
Dimensions: 42.0 x 28.8 (29.4 x 23.7) cm
Phys. Desc.: Engraved, Colored Outline
Language(s): French

Entry (No.): 442
Author: Robert de Vaugondy, Gilles (1688–1766)
Title: Partie Septentrionale // de la // Russie Européenne // Par le S. Robert de Vaugondy // Geog. ordinaire du Roy + c.
Year: 1778
Dimensions: 42.0 x 28.8 (29.3 x 23.7) cm
Phys. Desc.: Engraved, Colored Outline
Language(s): French

Entry (No.): 443
Author: Robert de Vaugondy, Gilles (1688–1766)
Title: Royaume de // Pologne // Par le S. Robert de Vaugondy // Géog. ordinaire du Roy + c.
Year: 1778
Dimensions: 42.0 x 28.8 (28.0 x 23.8) cm
Phys. Desc.: Engraved, Colored Outline
Language(s): French

Entry (No.): 444
Author: Roper, John (1771–ca. 1810)
Title: SARMATIA ANTIQUA
Year: 1801
Dimensions: 28.0 x 22.0 cm
Phys. Desc.: Engraved, Colored
Language(s): English

Entry (No.): 445
Author: Rosaccio (Rosatio), Giuseppe (Gioseppe) (ca. 1530–1620)
Title: SARMATIA - SAMOGITIA - LITVANIA - ROSSIA ROSSA - CODMIA - PODOLIA - LIVONIA - MOSCOVIA - PLESCOVIA - TARTARIA - SIGITIA - TAVRICA & MAR MAGGIORE
Year: 1724
Dimensions: 18.2 x 14.7 (17.0 x 13.0) cm
Phys. Desc.: Copy, Woodcut
Language(s): Italian

Entry (No.): 446
Author: Rossi, Luigi (1764–1824)
Title: Geografia Moderna.
Year: 1820
Dimensions: 47.2 x 30.0 cm
Phys. Desc.: Engraved
Language(s): Italian

Entry (No.): 447
Author: Rossi, Luigi (1764–1824)
Title: La Polonia // colle // Divisioni // del 1772, 1793 e 1795.
Year: 1820–21
Dimensions: 47.5 x 29.1 (20.8 x 17.2) cm
Phys. Desc.: Engraved
Language(s): Italian

Entry (No.): 448
Author: Rossi, Luigi (1764–1824)
Title: Parte Meridionale // Della Russia // ovvero // Moscovia Europea.
Year: 1820–21
Dimensions: 46.5 x 30.0 (21.4 x 17.3) cm
Phys. Desc.: Engraved, Colored Outline
Language(s): Italian

Entry (No.): 449
Author: Ruscelli, Girolamo (ca. 1504–66) (see Ptolemy)
Title: MOSCHOVIA NVOVA TAVOLA.
Year: 1561
Dimensions: 29.0 x 22.0 (26.2 x 19.6) cm
Phys. Desc.: Copy
Language(s): Italian

Entry (No.): 450
Author: Ruscelli, Girolamo (ca. 1504–66) (see Ptolemy)
Title: POLONIA ET HVNGARIA NVOVA TAVOLA.
Year: 1574
Dimensions: 32.0 x 22.5 (26.8 x 19.0) cm
Phys. Desc.: Copy
Language(s): Italian

Entry (No.): 451
Author: Sambucus (Sámbok, Schámboky, Zsámboki, Zsamboky), Professor János (Johannes) (1531–84)
Title: VNGARIAE // LOCA // PRAECIPVA // RECENS // EMENDATA // ATQVE // EDITA PER // IOANNEM // SAMBUCUM // PANNONIVM // IMP. M.s HISTORICUM // 1579.
Year: 1579
Dimensions: 54.5 x 44.0 (50.0 x 34.5) cm
Phys. Desc.: Engraved, Colored
Language(s): Latin

Entry (No.): 452
Author: Sandrart, Jacob (Jakob) von (1630–1708)
Title: Moscoviae // seu // Russiae Magnae // Generalis Tabula // quâ // Lapponia, Norvegia, // Suecia, Dania, Polonia // Maximaeq. partes // Germaniae, Tartariae // Turcici Imperii, // aliaeq. Regiones adjantes // simul ostenduntur de novo correcta et edita // per // Jacobum Sandrart, Chalcographum Norimbergensem.
Year: ca. second half of 17th c.
Dimensions: 39.6 x 30.8 cm
Phys. Desc.: Copy and Negative
Language(s): Latin

Entry (No.): 453
Author: Sanson (d'Abbeville), Guillaume (son of Nicolas) (1633–1703)
Title: Bosphorus Cimmerius // et // Regnum Bosphoranum // Ex Collectionibus Geographicis, // Gulielmi Sanson Nicolai Filii // Regis Christianissimi Geographi.
Year: 1765
Dimensions: 60.0 x 45.5 (50.0 x 40.0) cm
Phys. Desc.: Engraved, Colored Outline
Language(s): Latin

Entry (No.): 454
Author: Sanson (d'Abbeville), Nicolas I (senior) (1600–1667)
Title: La Grande // Tartarie. // Par le S.r Sanson. A Paris. // chez P. Mariette rue St Jacques a l'Esperance. // Avec Privilege pour Vingt Ans. // 1670.
Year: 1670
Dimensions: 33.0 x 24.1 (29.2 x 17.2) cm
Phys. Desc.: Engraved, Colored Outline
Language(s): French

Entry (No.): 455
Author: Sanson (d'Abbeville), Nicolas I (senior) (1600–1667)
Title: La Russie Noire ou Polonoise // qui Comprend les Provinces // de la Russie Noire de Volhynie et de Podolie // divisées en leurs Palatinats // Vulgairement Connües sous le Nom d'Ukraine ou Pays des Cosaques Par les S.rs Sansons 1674.
Year: 1674
Dimensions: 50.7 x 36.8 (45.5 x 34.6) cm
Phys. Desc.: Printed, Uncolored
Language(s): French

Entry (No.): 456
Author: Sanson (d'Abbeville), Nicolas I (senior) (1600–1667)
Author: Jaillot, Alexis-Hubert (ca. 1632–1712)
Title: ESTATS DE //POLOGNE // SUBDIVISÉS SUIVANT L'ESTENDÜE DES // PALATINATS // PAR LE S.r SANSON G. du R. // LES ESTATS DE LA COURONNE DE POLOGNE SUBDIVISÉS EN LEURS PALATINATS QUI COMPRENNENT LE ROYAUME DE POLOGNE, LE GRAND DUCHÉ DE LITHUANIE, // LES DUCHÉS DE PRUSSE, MAZOVIE, SAMOGITIE; LES PROVINCES DE VOLHYNIE, ET DE PODOLIE, OU EST COMPRIS L'UKRAINE OU PAYS DES COSAQUES, SUR LES MEMOIRES LE PLUS NOUVEAUX, 1675. // A Paris Chez H. Jaillot...Auec Privilege du Roy, pour vingt ans 1675.
Year: 1675
Dimensions: 92.5 x 62.5 (88.0 x 55.3) cm
Phys. Desc.: Engraved, Colored Outline
Language(s): French

Entry (No.): 457
Author: Sanson (d'Abbeville), Nicolas I (senior) (1600–1667)
Title: Estats de la Couronne de // Pologne // ou sont les // Royaume de Pologne, // Duchés et Provinces de // Prussie, Cuiavie Mazovie, // Russie Noire + c. // Duches de Lithuanie // Volhynie Podolie + c. // de L'Ukranie + c. Par N. Sanson d'Abbeville // Geographe ordinaire du Roy // A Paris chez l'Autheur avec privil // de Sa Majesté // Renouueté pour Vingt ans // 1679 // aux Galleries du Louvre.
Year: 1679
Dimensions: 60.7 x 45.3 (56.6 x 42.6) cm
Phys. Desc.: Engraved, Colored
Language(s): French

Entry (No.): 458
Author: Sanson (d'Abbeville), Nicolas I (senior) (1600–1667)
Author: Jaillot, Alexis-Hubert (ca. 1632–1712)
Title: ESTATS DE // POLOGNE // Subdivisés suivant l'estendue des // PALATINATS // Par le Sr. Sanson G. du R. // LES

ESTATS DELA COURONNE DE POLOGNE, SUBDIVISÉS EN LEURS PALATINATS, QUI COMPRENNENT LE ROYAUME DE POLOGNE, LE GRAND DUCHÉ DE LITHUANIE, // LES DUCHÉS DE PRUSSE, MAZOVIE, SAMOGITIE, LES PROVINCES DE VOLHYNIE, ET DE PODOLIE, OU EST COMPRIS L'UKRAINE OU PAYS DES COSAQUES... 1692. // A Paris Chez H. Iaillot... 1692.
Year: 1692
Dimensions: 94.2 x 61.2 (86.7 x 53.8) cm
Phys. Desc.: Engraved, Colored
Language(s): French
Figure 26

Entry (No.): 459
Author: Sanson (d'Abbeville), Nicolas I (senior) (1600–1667)
Author: Jaillot, Alexis-Hubert (ca. 1632–1712)
Title: LA RUSSIE BLANCHE OU // MOSCOVIE, // DIVISÉE SUIVANT L'ESTENDUE // DES ROYAUMES, DUCHÉS, PRINCIPAUTÉS, PROVINCES, // ET PEUPLES // QUI SONT PRESENTEMENT SOUS LA DOMINATION // DU CZAR DE LA RUSSIE // COGNU SOUS LE NOM DE // GRAND DUC DE MOSCOVIE // DRESSÉ SUR PLUSIEURS MEMOIRES ET RELATIONS // LES PLUS RÉCENTES // PAR LE Sr SANSON... // PRESENTÉE // A MONSEIGNEUR LE DAUPIN // PAR... // H. IAILLOT. // 1692 // LES ETATS DU CZAAR DE LA RUSSIE BLANCHE OU GRAND DUC DE MOSCOVIE... A Paris Chez H. Iaillot 1692.
Year: 1692
Dimensions: 95.0 x 61.0 (87.5 x 54.2) cm
Phys. Desc.: Engraved, Colored
Language(s): French

Entry (No.): 460
Author: Sanson (d'Abbeville), Nicolas I (senior) (1600–1667)
Author: Jaillot, Alexis-Hubert (ca. 1632–1712)
Title: L'ASIE // DIVISÉE EN SES PRINCIPALES REGIONS, // ET OU SE PEUT VOIR L'ESTENDÜE DES // EMPIRES, MONARCHIES, ROYAUMES ET ESTATS // QUI PARTAGENT PRÉSENTEMENT L'ASIE. // Recueilii de divers Memoires, et sur les Relations les pl. Nouvelles. // PAR LE Sr SANSON, Geographe ordinaire du Roy. // L'ASIE DISTINGUÉE EN SES PRINCIPALES PARTIES, S[Ç]AVOIR LA TURQUIE EN ASIE, L'ARABIE, LA PERSE, L'INDE, LA CHINE, LA TARTARIE, // LES ISLES DU IAPON, DES PHILIPPINES, DES MOLUCQUES, DE LA SONDE, DE CEYLAN, ET DES MALDIVES ou son remarques LES EMPIRES, MONARCHIES, ROYAUMES, et ESTATS, qui s'y trouvent a present, Dressé sur les Relations les pl. Nouvelles, Par le Sr Sanson Geographe ord. du Roy. 1692. A Paris chez H. Iaillot.
Year: 1692
Dimensions: 93.6 x 61.3 (87.6 x 54.8) cm
Phys. Desc.: Engraved, Colored
Language(s): French

Entry (No.): 461
Author: Sanson (d'Abbeville), Nicolas I (senior) (1600–1667)
Title: LE ROYAUME DE // HONGRIE, // ET LES ESTATS QUI EN ONT ESTÉ SUJETS, // ET QUI SONT PRESENTEMENT // LA PARTIE SEPTENTRIONALE // DE LA TURQUIE EN EUROPE // TIRÉ DES MEMOIRES LES PL. NOUUEAUX // PAR LE Sr SANSON... // LE ROYAUME DE HONGRIE, ET LES ESTATS QUI ONT ESTÉ SUJETS ET DEPENDANS DE SA COURONNE, S[Ç]AVOIR LA TRANSILVANIE, MOLDAVIE, VALAQUIE, CROACIE, BOSNIE, DALMACIE, SERVIE, BULGARIE, //... //...DRESSÉ... PAR LES Sr SANSON... // A PARIS, Chez H. Iaillot...1692.
Year: 1692
Dimensions: 94.2 x 61.2 (86.5 x 54.5) cm
Phys. Desc.: Engraved, Colored
Language(s): French

Entry (No.): 462
Author: Sanson (d'Abbeville), Nicolas I (senior) (1600–1667)
Title: Romani Imperii // qua Oriens est // Descriptio Geographica // Auct. N. Sanson Abbavillaeo Christianis Gallram // Regis Math. et Geographo avec Privillege du Roy // 1703.
Year: 1703
Dimensions: 64.2 x 53.3 (56.2 x 39.8) cm
Phys. Desc.: Engraved, Colored
Language(s): Latin

Entry (No.): 463
Author: Sanson (d'Abbeville), Nicolas I (senior) (1600–1667)
Title: Bosphorus Cimmerius // et // Regnum Bosphoranum // Ex Collectionibus Geographicis Sanson // Amsterdami // Apud P. Mortier cum Privilegio.
Year: 1705
Dimensions: 62.0 x 53.0 (48.3 x 39.4) cm
Phys. Desc.: Engraved, Colored
Language(s): Latin

Entry (No.): 464
Author: Sanson (d'Abbeville), Nicolas I (senior) (1600–1667)
Title: Sarmatia // Utraque // Europaea et Asiatica // Autore N. Sanson // Amsterdami // Apud P. Mortier // Cum Privilegio.
Year: 1705
Dimensions: 62.5 x 53.2 (47.6 x 40.4) cm
Phys. Desc.: Engraved, Colored
Language(s): Latin

Entry (No.): 465
Author: Sanson (d'Abbeville), Nicolas I (senior) (1600–1667)
Author: Jaillot, Alexis-Hubert (ca. 1632–1712)
Title: ESTATS DE LA COURONNE DE // POLOGNE // SUBDIVISÉS SUIVANT L'ESTENDÜE DES // PALATINATS // PAR LE Sr SANSON... // DEDIE AU ROY // PAR... // HUBERT IAILLOT // A Paris, Chez H. Iaillot... Avec Privilege du Roy, pour vingt Ans 1708.
Year: 1708
Dimensions: 72.6 x 49.5 (64.5 x 45.5) cm
Phys. Desc.: Engraved, Colored Outline
Language(s): French

Entry (No.): 466
Author: Sanson (d'Abbeville), Nicolas I (senior) (1600–1667)

Title: Ponti // Evxini // Periplus // Auctore N. Sanson. // Amsterdami // Apud J. Covens et C. Mortier.
Year: 1730
Dimensions: 63.3 x 52.5 (59.0 x 47.8) cm
Phys. Desc.: Engraved, Colored
Language(s): Latin

Entry (No.): 467
Author: Sanson (d'Abbeville), Nicolas II (son of Nicolas) (1625–48)
Title: Mer Noire // ou mer Maivre // Par N. Sanson le fils // Geographe du Roy.
Year: 1652
Dimensions: 28.9 x 21.5 (24.0 x 17.5) cm
Phys. Desc.: Engraved, Uncolored
Language(s): French

Entry (No.): 468
Author: Sanson (d'Abbeville), Nicolas II (son of Nicolas) (1625–48)
Title: Estats // de la Couronne de // Pologne // P. N. Sanson le fils G. du R.
Year: 1656–58
Dimensions: 37.0 x 27.0 (25.2 x 19.0) cm
Phys. Desc.: Engraved, Uncolored
Language(s): French

Entry (No.): 469
Author: Sanson (d'Abbeville), Nicolas II (son of Nicolas) (1625–48)
Title: Tartarie Europeenne ou Petit Tartarie où sont Les Tartares du Crim, ou de Perecop; de Nogais, d'Oczakow, et de Budziak. Dressée sur les Relations les plus Recentes, Par le S.'Sanson fils Geographe ord.[re] du Roy. A Paris. Chez. l'Autheur Avec privilege pour Vingt Ans 1665.
Year: 1665
Dimensions: 69.5 x 45.8 (51.0 x 40.0) cm
Phys. Desc.: Engraved, Colored
Language(s): French

Entry (No.): 470
Author: Sanson (d'Abbeville), Nicolas II (son of Nicolas) (1625–48)
Title: Russie // Blanche // ou // Moscovie Par N. Sanson // le Fils Geog. // du Roy.
Year: 1670
Dimensions: 33.2 x 24.2 (24.0 x 18.5) cm
Phys. Desc.: Engraved, Colored
Language(s): French

Entry (No.): 471
Author: Sanson (d'Abbeville), Nicolas II (son of Nicolas) (1625–48)
Author: Visscher, Nicolaes Jansz. II (1649–1702)
Title: TABULA NOVA TOTIUS // REGNI POLONIAE, // IN QUO SUNT // DUCATUS ET PROVINCIAE // PRUSSIA, CUJAVIA, MAZOVIA, // RUSSIA NIGRA, &c. // DUCATUS LITHUANIA, // UKRANIA, &c. IN QUA // VOLHYNIA, ET PODOLIA. / CUM SUIS // PALATINATIBUS AC CONFINIJS. // Authore N. Sansonio Abeevillensi // Geographus Regis Galliae // Amstelodami apud Nicolaum Visscher. //
Year: 1680
Dimensions: 63.0 x 54.0 (55.2 x 42.0) cm
Phys. Desc.: Engraved, Colored
Language(s): Latin

Entry (No.): 472
Author: Sanson (d'Abbeville), Nicolas II (son of Nicolas) (1625–48)
Title: Russie // Blanche // ou // Moscovie // Par N. Sanson // le Fils Geog // du Roy.
Year: 1683
Dimensions: 30.4 x 21.6 (23.7 x 18.3) cm
Phys. Desc.: Engraved, Uncolored
Language(s): French

Entry (No.): 473
Author: Sanson (d'Abbeville), Nicolas II (son of Nicolas) (1625–48)
Title: Estats // de la Couronne de //Pologne // P. N. Sanson le fils G. du R.
Year: 1700
Dimensions: 31.8 x 22.5 (25.5 x 18.8) cm
Phys. Desc.: Engraved, Colored
Language(s): French

Entry (No.): 474
Author: Sanson, Pierre Moullart (Moulard) (d. 1730)
Title: Persarium Imperium // In Viginti Satrapias Vectigales distributum, // ut Eas Ordinaverat // Darius Histapis filius tertius Persarum Rex, // In Scripturis Sacris Assuerus dictus, // Cujus Monarchia, ab India usque Aethiopiam. // In Centrum et viginti Septem divisa Provincias dicitur // in iisdem Scripturis // Haec Tabula Geographica ad Sextam Mundi Aetatem attinet. // Parisiis // Conatibus Petri Moullart = Sanson Regis Christianissimi // Geographi. Anno CIDIDC?XXI. cum Privilegio.
Year: 1721
Dimensions: 64.2 x 52.8 (57.7 x 41.5) cm
Phys. Desc.: Engraved, Colored
Language(s): Latin

Entry (No.): 475
Author: Santini, Paolo (ca. 1729–93, worked 1776–84)
Title: La Crimée, la Nouvelle Russie, les Tatares Nugay et d'Oczakow, les Zaporoviens, et partie de la Circassie á Venice par P. Santini 1777.
Year: 1777
Dimensions: 72.5 x 51.2 (56.0 x 42.0) cm
Phys. Desc.: Engraved, Uncolored
Language(s): French

Entry (No.): 476
Author: Sayer, Robert (1725–94)
Title: The // Kingdom of Poland // and // Great Dutchy of Lithuania // with all their Divisions. // London. Printed for Robert Sayer, Map + Printsetter N° 53 Fleet Street // Published as the Act directs 2d. March 1772.
Year: 1772

Dimensions: 69.5 x 57.5 (65.4 x 47.5) cm
Phys. Desc.: Engraved, Colored Outline
Language(s): English

Entry (No.): 477
Author: Sayer, Robert (1725–94)
Title: A New Map of // the Kingdom of Poland // with // its Dismembered Provinces.
Year: 1787
Dimensions: 73.0 x 54.0 (65.2 x 47.5) cm
Phys. Desc.: Engraved, Colored Outline
Language(s): English

Entry (No.): 478
Author: Schagen (à Schagen), Gerrit (Gerardus a) Lucaszoon van (ca. 1642–ca. 1690)
Title: REGNI // POLONIAE // et Ducatus // LITHVNIAE // VOLINIAE, PODOLIAE // VCRANIAE // PRVSSIAE, LIVONIAE // et CVRLANDIAE descriptio // emendata per G. a Schagen. // Amsterdami.
Year: 1729
Dimensions: 67.5 x 55.5 (55.5 x 48.0) cm
Phys. Desc.: Engraved, Uncolored
Language(s): Latin

Entry (No.): 479
Author: Schenk, Petrus (Pieter), Sr. (1661–1711)
Author: Delisle (De L'Isle), Guillaume (1675–1726)
Title: Ukraine Pays des Cosaques, Russie Moscovite Tartarie Moscovite... Par G. de l'Isle geographe a Amsterdam Chèz Pierre Schenk, avec priv.
Year: ca. 1695–1756
Dimensions: 63.5 x 53.5 (61.2 x 49.0) cm
Phys. Desc.: Engraved, Colored
Language(s): French

Entry (No.): 480
Author: Schenk, Petrus (Pieter), Sr. (1661–1711)
Title: Friderico Augusto // vere Augusto. // Polon. Lithuan. Borus. Pomer. // Regi. Duci. Princip / Saxon. Utr. Duci. s. Imp. elect. // Haec. Imperii. Sui. Regna. // D.D.D. P. Schenkus // Sculpt Regius. // Amsterdami exc: prope forum. // Cum Priv. Ord. Holland. et West Frisiae. 1703. Lithuania, Russia Rubra...Polonia, Russia Alba...
Year: 1705
Dimensions: 61.5 x 53.2 (57.0 x 48.0) cm
Phys. Desc.: Engraved, Colored Outline
Language(s): Latin

Entry (No.): 481
Author: Schenk, Petrus (Pieter), Sr. (1661–1711)
Title: La Russie Blanche ou // Moscovie // Divisée Suivant l'Estendue // des Royaumes, Duchés, // Principautés, Provin^ces // et Peuples, // qui sont Presentement soubs la Domination // Du Czar de la Russie // Cogneusoubs le Nom de Grand Duc de Moscovie. Par le Sr Sanson Geographe Ord.^re du Roy.
Year: 1705
Dimensions: 64.0 x 53.5 (57.0 x 46.2) cm
Phys. Desc.: Engraved, Colored Outline
Language(s): French

Entry (No.): 482
Author: Schenk, Petrus (Pieter), Sr. (1661–1711)
Title: REGNORUM // HUNGARIAE, // DALMATIAE, CROATIAE, SCLA // VONIAE, BOSNIAE, SERVIAE, // ET PRINCIPATUS // TRANSYLVANIAE // Typus accuratior, et plane novus; // Editore Petro Schenck caelatur. // Cum privileg. Regis Polon. et Elect. Sax. et Ord. Holl. et West frisiae. // Amstelaedami 1705. //
Year: 1705
Dimensions: 62.0 x 52.3 (55.0 x 47.5) cm
Phys. Desc.: Engraved, Colored Outline
Language(s): Latin
Figure 35

Entry (No.): 483
Author: Schenken, Petrus (Peter) II (1693–1775)
Title: CARTE DU THEATRE DE LA GUERRE, DANS LA POLOGNE, MOSCOVIE & TURQUIE. PAR PIERRE SCHENCK A AMSTELDAM AVEC PRIVILEGE. 1733. Nunc apud Petrum Schenk Junior.
Year: 1733
Dimensions: 80.0 x 63.3 (72.0 x 59.5) cm
Phys. Desc.: Engraved, Colored
Language(s): French

Entry (No.): 484
Author:: Schlicht, Rudolph
Title: Historisch-statistisch-geographische Darstellung // des // Koenigreichs Polen, 1770, 1773, 1793, 1795, 1807, 1815. Mannheim, Steindruckerey von Rudolph Schlicht, 1831.
Year: 1831
Dimensions: 47.8 x 41.8 cm
Phys. Desc.: Lithographed on Stone, Colored Outline
Language(s): German
Figure 65

Entry (No.): 485
Author: Schlicht, Rudolph
Title: Special-Karte des Königreichs // Polen // Nebst einem Theile // von // Ostpreussen, Lithauen, Wolhynien, Galizien, und Schlesien. Mannheim // 1831 // Steindruckerey von Rudolph Schlicht.
Year: 1831
Dimensions: 56.0 x 42.0 cm
Phys. Desc.: Lithographed on Stone, Colored Outline
Language(s): German

Entry (No.): 486
Author: Schlicht, Rudolph
Title: Vollständige Karte von // Litthauen // Wolhynien, Podolien // und dem Königreiche Polen. // enthabend: // die Schlachten älterer und neuester zeit mit Angabe der commandirenden Feldherrn, // nebst historischen und statistischen Tabellen und Bemerkungen der Poststationen. // Nach den neuesten und zuferlässigsten Hilfsmitteln // entworfen und litographirt von / Rudolph Schlicht // in // Mannheim // im Juni 1831.
Year: 1831

Dimensions: 47.5 x 41.8 cm
Phys. Desc.: Lithographed on Stone, Colored Outline
Language(s): German

Entry (No.): 487
Author: Schmidburg, George Rudolph, Freiherr von
Title: Charte // von // Ost und West // Galizien // nach den neuesten astronomischen // Ortzbestimungen entworfen und, revidirt // auf der Sternwarte Seeberg bey Gotha // gezeichnet von G.R. v. Schmidburg. Weimar // Im Verlage des Industrie Comptoirs // 1801.
Year: 1801
Dimensions: 72.0 x 58.0 (61.5 x 48.0) cm
Phys. Desc.: Engraved, Colored Outline
Language(s): German

Entry (No.): 488
Author: Schmidburg, George Rudolph, Freiherr von
Title: Charte // von // Ost und West // Galizien // nach den neuesten astronomischen // Ortzbestimungen entworfen und, revidirt // auf der Sternwarte Seeberg bey Gotha // gezeichnet von G.R. v. Schmidburg. Weimar im Verlage des Geograph Instituts // Berichtigt nach dem Wiener Frieden vom 14 Octobr. 1810.
Year: 1801
Dimensions: 65.0 x 55.2 (61.5 x 48.5) cm
Phys. Desc.: Engraved, Uncolored
Language(s): German

Entry (No.): 489
Author: Schmidt, [James?]
Title: Crimeae // seu // Chersonesus // Tauricae // item Tatariae Nogayae / Europeae // Tabula Geographica // ex // Autographis Castrametatorum // Exercitus // Augustissimae // Russorum // aliisque novissimis // Concinnata // a // F F Schmidio A. P. adjuncto 1774.
Year: 1774
Dimensions: 65.5 x 52.3 (58.2 x 64.0) cm
Phys. Desc.: Engraved, Colored
Language(s): Latin

Entry (No.): 490
Author: Schreiber, Johann Georg (1676–1750)
Title: Das gantze // Russische Kaeysertum // mit allen seinen // Laendern. // verfertiget von F. George Schreibern // in Leipzig.
Year: 1740
Dimensions: 30.3 x 21.5 (24.7 x 16.2) cm
Phys. Desc.: Engraved, Colored
Language(s): German

Entry (No.): 491
Author: Schreiber, Johann Georg (1676–1750)
Title: Die // Europaeische // oder // Kleine // Tartarey // nebst den // Angraetzenden // Laendern / verfertiget von J. Schreibern in Leipzig.
Year: 1740
Dimensions: 30.3 x 21.5 (23.7 x 15.5) cm
Phys. Desc.: Engraved, Colored
Language(s): German

Entry (No.): 492
Author: Schütz (Schuetz), Carl (1745–1800)
Title: Kriegstheater oder Grenzkarte // Oesterreichs, Russlands, und der Türkey, enthaltend Das Koenigreich Ungarn, Siebenbürgen, Moldau, Bessarabien, Walachey, Bulgarien, Servien, Bosnien, Sclavonien, Croatien, Albanien, Romanien, Macedonien, einen Theil Griechenland, und des Archipelagus, die Dardanellen, das schwarze und asowische Meer, die Krim, Tatarey, Kuban, Circassien &c. &c. Nebst einem Theil von Russland, Polen, Schlesien, Maehren, Oesterreich, Steuermark, &c. Erklaerung der Illumination Roth ist oesterreichisch, Gelb russisch, Grün türkisch, Kolet polnisch, Orangefard venetianisch, zu finden in Wien bey Artaria Campagnie Kunsthändlern am Michaelerplatz. Nach den besten Karten und Handzeichnungen von C. Schütz entworfen 1788.
Year: 1788
Dimensions: 72.0 x 51.5 (70.7 x 41.0) cm
Phys. Desc.: Engraved, Colored
Language(s): German

Entry (No.): 493
Author: Senex, John (ca. 1678–1740)
Title: A // MAP OF // TURKY, ARABIA // AND // PERSIA. // Collected from the latest Travels // & from ye Observations of ye ROY // AL Societys of LONDON & PARIS // By G. DE L'ISLE. Most humbly Inscrib'd to // the Honble Sr. James Stewart // of Goodtrees Bartt.
Year: ca. 1708–25
Dimensions: 59.6 x 53.0 (57.5 x 47.0) cm
Phys. Desc.: Engraved, Colored Outline
Language(s): English

Entry (No.): 494
Author: Senex, John (ca. 1678–1740)
Title: A NEW MAP // OF // EUROPE // From the latest Observations // Inscrib'd to the Honorable Samuel // Molyneux, Esq., Secretary to his R.// H. the Prince.
Year: ca. 1708–25
Dimensions: 50.0 x 53.0 (57.7 x 49.0) cm
Phys. Desc.: Engraved, Colored Outline
Language(s): English
Figure 31

Entry (No.): 495
Author: Senex, John (ca. 1678–1740)
Title: A NEW MAP // OF THE KINGDOM OF // HUNGARY // And of the countries, Provinces &c. // Bordering thereupon; with their respective // BANNATS, COUNTIES, SANJAC SHIPS, // Mountains, Passes, Post Roads, Cities &c. as // also the Places where Battels have been Fought and at what Time: all taken // from the Memorials and Plans of the best Engineers who have serv'd in // those Countries.
Year: ca. 1708–25
Dimensions: 105.8 x 100.0 (67.6 X 65.2) cm
Phys. Desc.: Engraved, Colored
Language(s): English

Entry (No.): 496
Author: Senex, John (ca. 1678–1740)

Title: MOSCOVEY // IN EUROPE // From the latest Observations // Humbly Inscrib'd to // Abel Ketelbey of the // Inner Temple, Esq.... // By John Senex.
Year: ca. 1708–25
Dimensions: 53.1 x 60.2 (49.7 x 59.0) cm
Phys. Desc.: Engraved, Colored Outline
Language(s): English

Entry (No.): 497
Author: Senex, John (ca. 1678–1740)
Title: POLAND // CORRECTED from the Observvations communicated to the // ROYAL SOCIETY at LONDON and the ROYAL ACADEMY // at PARIS. // Is humbly Dedicated To // Sr. James Hallett K. // Citizen of London. By His Obliged Humble Servant // Ion Senex.
Year: ca. 1708–25
Dimensions: 97.0 x 67.5 (95.0 x 65.1) cm
Phys. Desc.: Engraved, Colored Outline
Language(s): English

Entry (No.): 498
Author: Senex, John (ca. 1678–1740)
Title: MOSCOVY // CORRECTED from ye Observations communicat // ed to the Royal Society // of London and Paris // By // John Senex.
Year: 1712
Dimensions: 66.5 x 48.5 (63.5 x 96.0) cm
Phys. Desc.: Engraved, Colored Outline
Language(s): English

Entry (No.): 499
Author: Senex, John (ca. 1678–1740)
Title: POLAND // and other the Countries // belonging to that Crowne // According to the Newest // Observation // 1719
Year: 1719
Dimensions: 58.8 x 53.2 (54.7 x 47.8) cm
Phys. Desc.: Engraved, Colored Outline
Language(s): English

Entry (No.): 500
Author: Senex, John (ca. 1678–1740)
Title: A // New MAP of // ASIA // From the latest Observations // Most Humbly Inscrib'd // to the Most Honorable // George Earl of Washington, &c.
Year: 1721
Dimensions: 58.5 x 53.0 (56.2 x 48.0) cm
Phys. Desc.: Engraved, Colored Outline
Language(s): English

Entry (No.): 501
Author: Seutter, Matthäus (the elder) (1678–1757)
Title: Imperii Moscovici pars Australis in lucem edita per Guillielmum de l'Isle, celebr. Regis Galliae geographum; // nunc excusa et venalis exposita per Mattheum Seutter, S.C.M.G. Augustae Vindelic.
Year: ca. 1720–44
Dimensions: 63.8 x 55.0 (57.8 x 49.7) cm
Phys. Desc.: Engraved, Colored
Language(s): Latin

Entry (No.): 502
Author: Seutter, Matthäus (the elder) (1678–1757)
Title: MAPPA // IMPERII // MOSCOVITICI // pars Septentrionalis, // adornata // per // GUILLIELMUM DE L'ISLE, // Membrum Academie Regalis // Scientarum et celebrr: Reg:// Galliae Geographum; // nunc vero aeri meisa et venalis exposita // à Matth. Seutter S:// Caes. M. Geographe.
Year: ca. 1720–44
Dimensions: 62.0 x 100.0 (58.3 x 97.0) cm
Phys. Desc.: Engraved, Colored
Language(s): Latin

Entry (No.): 503
Author: Seutter, Matthäus (the elder) (1678–1757)
Title: Nova et accurata // TARTARIAE EUROPAE // SEU MINORIS // et in specie // CRIMEAE // DELINEATIO GEOGRAPHICA, // cum omnibus // CIRCA PONTUM EUXINUM ET PALUDEM MAEOTIDEM // JACENTIBUS PROVINCIIS, // manu etsumtibus // MATTHAEI SEUTTERI, SAC. CAES. ET // CATHOL. REG. MAJEST. GEOGR. AUGUSTAE VINDELICOR.
Year: ca. 1720–44
Dimensions: 63.0 x 54.8 (58.0 x 49.3) cm
Phys. Desc.: Engraved, Colored
Language(s): Latin

Entry (No.): 504
Author: Seutter, Matthäus (the elder) (1678–1757)
Title: NOVA MAPPA // GEOGRAPHICA // MARIS ASSOVIENSIS // VEL DI ZABACHE // ET /PALUDIS MAEOTIDIS, // accuratè aeri inci?a et in luce edita // per // MATTHAEUM SEUTTER, S.C.M. GEOGR.
Year: ca. 1720–44
Dimensions: 57.2 x 48.3 cm
Phys. Desc.: Engraved, Colored
Language(s): Latin

Entry (No.): 505
Author: Seutter, Matthäus (the elder) (1678–1757)
Title: POLONIAE // REGNUM // UT ET // MAGNI // DUCATUS LITHUANIAE.// ACCURATISS. DELINEATIONE // REPRAESENTAT.// OPERA ET STUDIO // MATTH. SEUTTERI. // SAC. CAES. MAJ. GEOGR. AUG.
Year: ca. 1720–44
Dimensions: 62.7 x 53.0 (58.5 x 50.0) cm
Phys. Desc.: Engraved, Colored
Language(s): Latin

Entry (No.): 506
Author: Seutter, Matthäus (the elder) (1678–1757)
Title: Theatrum belli // Russorum Victoriia illustratum // sive // Nova et accurata // TURCICARUM ET TARTARICUM // Provinciarum // intra fluvios // Tÿras f. Niester et Tanaim f. Don,// ad Oram Ponti Euxini // et in Peloponeso Taurica // sitarum // Delignatio, // manu et impensis // Matthaei Seutteri // S. Caes. et Reg. Cath. Maj. geogr. // Aug. Vind.
Year: ca. 1720–44
Dimensions: 63.5 x 54.8 (57.5 x 49.4) cm
Phys. Desc.: Engraved, Colored

Language(s): Latin

Entry (No.): 507
Author: Seutter, Matthäus (the elder) (1678–1757)
Title: THEATRUM BELLI // SIVE // NOVISSIMA TABULA // QUA MAXIMA PARS DANUBII ET PRAESERIIM // HUNGARIA // CUM ALIIS ADJACENTIBUS REG-// NIS, NEC NON // GRAECIA MOREA ET ARCHIPELAG // INSULAE // ACCURATE DELINEANTUR // EXCUSA ET VENALIS EXPOSITA // À MATTHAEO SEUTTER // SACRAE CAESARAE, MAJESTATIS, GEOGRAPHO // AUGUSTA VINDELICOR.
Year: ca. 1720–44
Dimensions: 63.3 x 54.8 (58.2 x 49.7) cm
Phys. Desc.: Engraved, Colored
Language(s): Latin

Entry (No.): 508
Author: Seutter, Matthäus (the elder) (1678–1757)
Title: SPATIOSISSIMUM // IMPERIUM // RUSSIAE MAGNAE // juxta recentissimas Observationes // MAPPA GEOGRAPHICA // accuratissime delineatum // opera et sumtibus // MATTHAEI SEUTTERI// S. Caes. et Reg. Cathol. Majest. Geographi// et Chalcogr. Augustae Vindel.
Year: ca. 1740–44
Dimensions: 62.0 x 54.5 (56.9 x 50.0) cm
Phys. Desc.: Engraved, Colored
Language(s): Latin
Figure 44

Entry (No.): 509
Author: Speed, John (1552–1629)
Title: A MAP OF // RUSSIA // Sold by Tho: Basset in Fleet Street/ and Richard Chiswell in St. Pauls // Church yard.
Year: 1626
Dimensions: 50.6 x 39.7 cm
Phys. Desc.: Engraved, Uncolored
Language(s): English

Entry (No.): 510
Author: Speed, John (1552–1629)
Title: The Mape of // Hungari // newly augmented by John Speede // Ano Dom: 1626 // are to be sold in Pops-head alley // by George Humble. Anno 1626.
Year: 1626
Dimensions: 54.3 x 41.9 (51.3 x 39.4) cm
Phys. Desc.: Engraved, Colored
Language(s): English

Entry (No.): 511
Author: Stackhouse, Thomas (1756–1836)
Title: [Russia, Germany, Poland, The Netherlands].
Year: 1798
Dimensions: 53.0 x 42.5 cm
Phys. Desc.: text
Language(s): English

Entry (No.): 512
Author: Stackhouse, Thomas (1756–1836)
Title: Hungary with part of Turkey.
Year: 1798
Dimensions: 42.5 x 53.0 (36.9 x 21.3 and 36.9 x 15.4) cm
Phys. Desc.: Engraved, Colored
Language(s): English

Entry (No.): 513
Author: Stackhouse, Thomas (1756–1836)
Title: Panonia Dacia + c.
Year: 1798
Dimensions: 42.5 x 53.0 (36.5 x 21.1) + (36.5 x 15.2) cm
Phys. Desc.: Engraved, Colored
Language(s): English

Entry (No.): 514
Author: Stackhouse, Thomas (1756–1836)
Title: Poland // with its // Dismembered Provinces // including Prussia.
Year: 1798
Dimensions: 42.5 x 53.0 (41.3 x 38.1) cm
Phys. Desc.: Engraved, Colored
Language(s): English

Entry (No.): 515
Author: Stackhouse, Thomas (1756–1836)
Title: Russia in Europe
Year: 1798
Dimensions: 42.5 x 53.0 (36.2 x 38.5) cm
Phys. Desc.: Engraved, Colored
Language(s): English

Entry (No.): 516
Author: Starowolski (Starovolski), Szymon (Simon, Symon) (1588–1656)
Title: Description // del Reyno // de Polonia, // y otras noticias de su Estado Politico; // Por Simon Starovolsco.
Year: ca. 1657–72
Dimensions: 33.5 x 54.5 cm
Phys. Desc.: Reprint, Text with maps
Language(s): Spanish

Entry (No.): 517
Author: Tasso, Girolamo
Title: L'Impero Austriaco // Distinto ne'sua Domiinii // Venezia 1827. Calcografia Tasso.
Year: 1827
Dimensions: 54.0 x 40.0 (45.9 x 32.4) cm
Phys. Desc.: Engraved, Colored
Language(s): Italian

Entry (No.): 518
Author: Tasso, Girolamo
Title: L'Europa // colle politiche divisioni nel 1827.
Year: 1828
Dimensions: 54.0 x 40.0 (45.4 x 34.1) cm
Phys. Desc.: Engraved, Colored
Language(s): Italian

Entry (No.): 519
Author: Tasso, Girolamo

Title: Transmigrazioni ed Invasioni // de'Barbari // donde provennero, vie da essi per corse, // paesiche occuparono, o dove fuono distrutti. Tavola dell' Atlante Star. Polit. Letter. del Sig.[r] A. Le- Sage Editione di Girolamo Tasso Venecia 1828.
Year: 1828
Dimensions: 54.0 x 40.0 (45.3 x 31.3) cm
Phys. Desc.: Engraved, Colored
Language(s): Italian

Entry (No.): 520
Author: Thompson, John (worked 1814–69)
Title: Austrian // Dominions.
Year: 1817
Dimensions: 72.7 x 52.8 (58.7 x 49.1) cm
Phys. Desc.: Engraved, Colored
Language(s): English

Entry (No.): 521
Author: Thompson, John (worked 1814–69)
Title: European // Russia
Year: 1817
Dimensions: 52.8 x 72.7 (50.0 x 60.1) cm
Phys. Desc.: Engraved, Colored
Language(s): English

Entry (No.): 522
Author: Tirion, Isaac (Isaak) (ca. 1705–ca. 1769)
Title: Nieuwe Kaart // van // Muscovie // of // Russland, na de laatste ondekking // int licht gebracht te Amsterdam // by Isaac Tirion. J. Keyser fecit 1734.
Year: 1734
Dimensions: 38.8 x 32.5 (34.5 x 28.4) cm
Phys. Desc.: Engraved, Uncolored
Language(s): Dutch, English

Entry (No.): 523
Author: Tirion, Isaac (Isaak) (ca. 1705–ca. 1769)
Title: Nieuwe Kaart // van // Muscovie // of // Russland, na de laatste ondekking / int licht gebracht te Amsterdam // by Isaac Tirion. J. Keyser fecit 1734.
Year: 1744
Dimensions: 54.4 x 42.5 (34.5 x 28.4) cm
Phys. Desc.: Engraved, Colored
Language(s): Dutch, English

Entry (No.): 524
Author: Tirion, Isaac (Isaak) (ca. 1705–ca. 1769)
Title: Nieuwe Kaart // van // Tartarie. // Na de alderlaaste ondekking // int licht gebragt. tot Amsterdam, door Isaac Tirion.
Year: 1744
Dimensions: 54.4 x 42.5 (39.8 x 28.1) cm
Phys. Desc.: Engraved, Colored
Language(s): Dutch

Entry (No.): 525
Author: Tirion, Isaac (Isaak) (ca. 1705–ca. 1769)
Title: Nieuwe Kaart // vant // Koninkryk // Hongaryen // en // Zevenbergen // na de laatste ondekkingen. t. Amsterdam by Isaac Tirion.
Year: 1744
Dimensions: 54.1 x 42.5 (33.0 x 28.0) cm
Phys. Desc.: Engraved, Colored
Language(s): Dutch

Entry (No.): 526
Author: Tirion, Isaac (Isaak) (ca. 1705–ca. 1769)
Title: NIEUWE KAART // van het // HOOGHEEMRAADSCHAP // van RYNLAND, als mede van AMSTELLAND // en het // WATERSCHAP VAN WOERDEN //met alle de UITGEVEENDEPLASSEN // DROOGMAKERYEN ENZ. // tot Opheldering van den TEGENWOORDIGEN STAAT // der VEREENIGDE NEDERLANDEN // Urtgegev. te Amst. by Is. Tirion 1745.
Year: 1745
Dimensions: 37.0 x 33.0 cm
Phys. Desc.: Engraved, Uncolored
Language(s): Dutch

Entry (No.): 527
Author: Tirion, Isaac (Isaak) (ca. 1705–ca. 1769)
Title: NIEUWE KAART // van het // ZUID WESTELYKSTE DEEL // VAN HOLLAND, // behelzende de Baljuwschappen // VOORNE, PUTTEN // enz. // te AMSTERDAM by // ISAAK TIRION. // MDCCXLVIII.
Year: 1748
Dimensions: 38.0 x 33.2 cm
Phys. Desc.: Engraved, Uncolored
Language(s): Dutch

Entry (No.): 528
Author: Tirion, Isaac (Isaak) (ca. 1705–ca. 1769)
Title: NIEUWE KAART // van het // BALJUWSCHAF VAN // AMSTELLAND // Ter Ophelderinge van den // TEGENWOORDIGEN STAAT // DER VEREENIGDE NEDERLANDEN. // Uitgegeven te AMSTERDAM // by Isaak Tirion // MDCCXLIX
Year: 1749
Dimensions: 37.0 x 32.5 cm
Phys. Desc.: Engraved, Uncolored
Language(s): Dutch

Entry (No.): 529
Author: Tirion, Isaac (Isaak) (ca. 1705–ca. 1769)
Title: NIEUWE KAART // van het // BALJUWSCHAP VAN // GOOYLAND. // DE LOOSDRECTEN. // MYNDEN, HOLLANDS LOENEN // EN WEESPER KARSPEL. //Ter Opheldering van den // TEGENWOORDIGEN STAAT DER // VEREENIDGE NEDERLANDEN. // Uitgegeven te AMSTERDAM // by ISAAK TIRION. MDCCL.
Year: 1750
Dimensions: 37.0 x 32.5 cm
Phys. Desc.: Engraved, Uncolored
Language(s): Dutch

Entry (No.): 530
Author: Tirion, Isaac (Isaak) (ca. 1705–ca. 1769)
Title: Nuova Carta // del // Regno Di Polonia // Diviso nei suoi Palatinati // Secondo l'ultime osservazioni ed anotazioni //

fata in Amsterdam // per Isak Tirion.
Year: 1750
Dimensions: 46.0 x 36.5 (34.2 x 28.2) cm
Phys. Desc.: Engraved, Uncolored
Language(s): Italian

Entry (No.): 531
Author: Tresscott (Treskot, Trescotti, Trestcott, Trouskot, Truscott), Johann (1721–86)
Title: Mappa Gubernii Orenburgensis Geographica exhibita a Johanne Truskotio 1772.
Year: 1772
Dimensions: 61.5 x 49.0 (57.5 x 45.6) cm
Phys. Desc.: Engraved, Colored
Language(s): Latin

Entry (No.): 532
Author: Tresscott (Treskot, Trescotti, Trestcott, Trouskot, Truscott), Johann (1721–86)
Title: Mappa Gubernii Astrachensis comp. J. Tresscott. 1774.
Year: 1774
Dimensions: 60.5 x 49.5 (56.5 x 46.0) cm
Phys. Desc.: Engraved, Colored
Language(s): Latin

Entry (No.): 533
Author: Valck, Gerard (Gerardus, Gerrit), (1652–1726)
Title: LA MER // MEDITERRANÉE // divisée // en ses Principales Parties ou Mers // par G. Valck // a Amsterdam // sur le Dam avec Privilege
Year: 1720
Dimensions: 87.3 x 53.4 (83.0 x 48.0) cm
Phys. Desc.: Engraved, Colored
Language(s): French

Entry (No.): 534
Author: Visscher (Piscator), Claes Jansz[oon] (1587–1652)
Title: TABULA // RUSSIAE // EX MANDATO FOEDOR BORISSOWITS DELINEATA. // MAGNO STUDIO PLURIMIS IN LOCIS AUCTA. EDITA. // ET // MAGNO DOMINO TZARI, // ET MAGNO DUCI MICHAELI FOEDROWITS... // ... // DEDICATA A // NICOLAO IOANNIS // PISCATORE // 1651. //
Year: 1651
Dimensions: 60.0 x 51.2 (53.5 x 42.5) cm
Phys. Desc.: Engraved
Language(s): Latin
Figure 19

Entry (No.): 535
Author: Visscher (Piscator), Nicolaes I (1618–79)
Title: TABULA // RUSSIAE// ex mandato Foedor Borissowits delineata, // Magno studio plurimis in locis aucta. Edita // et // Mago Domino Tzari... Dedicata a // Nicolao Ioannis Piscatore 1651.
Year: 1651
Dimensions: 60.5 x 51.5 (53.5 x 42.5) cm
Phys. Desc.: Engraved, Uncolored
Language(s): Latin

Entry (No.): 536
Author: Visscher (Piscator), Nicolaes Jansz. II (1649–1702)
Title: EXACTISSIMA TABULA, QUA TAM DANUBII FLUVII PARS INFERIOR, A BELGRADO URBE USQ AD EIUS OSTIA, ET MINORES IN EUM INFLUENTES FLUVII, QUAM REGIONES ADJACENTES, // UT TRANSYLVANIA, VALACHIA, MOLDAVIA, BULGARIA, BESSARABIA ET ROMONIA, ALIIQ UCRANIAE AC PODOLIAE TRACTUS FINITIMI OSTENDUNTUR PER NICOLAUM VISSCHER AMST: BAT: cum Privilegio Ordin. General: Belgii Foederati.
Year: 1680
Dimensions: 63.0 x 54.0 (58.3 x 48.2) cm
Phys. Desc.: Engraved, Colored
Language(s): Latin

Entry (No.): 537
Author: Visscher (Piscator), Nicolaes Jansz. II (1649–1702)
Title: EXACTISSIMA TABULA, QUA TAM DANUBII FLUVII PARS INFERIOR, A BELGRADO URBE USQ AD EIUS OSTIA, ET MINORES IN EUM INFLUENTES FLUVII, QUAM REGIONES ADJACENTES, // UT TRANSYLVANIA, VALACHIA, MOLDAVIA, BULGARIA, BESSARABIA ET ROMONIA, ALIIQ UCRANIAE AC PODOLIAE TRACTUS FINITIMI OSTENDUNTUR PER NICOLAUM VISSCHER AMST: BAT: cum Privilegio Ordin. General: Belgii Foederati.
Year: 1684
Dimensions: 84.0 x 56.0 (79.8 x 43.5) cm
Phys. Desc.: Engraved, Colored
Language(s): Latin

Entry (No.): 538
Author: Visscher (Piscator), Nicolaes Jansz. II (1649–1702)
Title: MOSCOVIAE // SEU // RUSSIAE MAGNAE // GENERALIS TABULA // QUA // LAPPONIA, NORVEGIA, // SUECIA, DANIA, POLONIA, // MAXIMEQUE PARTES // GERMANIAE, TARTARIAE, // TURCICI IMPERII, ALIAEQUE REGIONES ADJACENTES // SIMUL OSTENDUNTUR, / DE NOVO CORRECTA ET EDITA // PER NICOLAUM VISSCHER // AMST: BAT. // cum Privilegio Ordin. General: Belgii Foederati.
Year: 1690
Dimensions: 62.3 x 53.5 (52.5 x 41.5) cm
Phys. Desc.: Engraved, Colored
Language(s): Latin

Entry (No.): 539
Author: Visscher (Piscator), Nicolaes Jansz. II (1649–1702)
Title: EXACTISSIMA TABULA, QUA TAM DANUBII FLUVII PARS MEDIA, AB URBE LENTIAE AD BYDENAM URBEM, ET MINORES IN EUM INFLUENTES FLUVII, QUAM REGIONES ADJACENTES // UT HUNGARIA, SLAVONIA, CROATIA, BOSNIA, DALMATIA, SERVIA, NEC NON GERMANIAE, POLONIAE ET GRECIAE CONFINIA OSTENDUNTUR, PER NICOLAUM VISSCHER AMST: BAT: cum Privilegio Ordin. General: Belgii Foederati.
Year: 1710
Dimensions: 60.2 x 52.0 (55.5 x 48.5) cm

Phys. Desc.: Engraved, Colored
Language(s): Latin

Entry (No.): 540
Author: Visscher (Piscator), Nicolaes Jansz. II (1649–1702)
Title: NOUVELLE CARTE // DE LA MER NOIRE, // ET DU // CANAL // DE // CONSTANTINOPLE, // TRES EXACTE; MISE AU JOUR, // PAR N: VISSCHER AVEC PRIVILEGE. //
Year: 1710
Dimensions: 61.0 x 52.3 (57.2 x 48.0) cm
Phys. Desc.: Engraved
Language(s): French

Entry (No.): 541
Author: Visscher (Piscator), Nicolaes Jansz. II (1649–1702)
Author: Schenk, Petrus (Pieter), Sr. (1661–1711)
Title: NOUVELLE CARTE GEOGRAPHIQUE DU GRAND ROYAUME DE MOSCOVIE REPRESENTANT LA PARTIE MERIDIONALE & C. // PAR NICOLAS VISSCHER a Amsteldam avec Privilege des Etats Generaux. Nunc apud Petrum Schenk Junior. //
Year: 1735
Dimensions: 76.2 x 52.8 (70.5 x 52.0) cm
Phys. Desc.: Engraved, Colored
Language(s): French
Figure 39

Entry (No.): 542
Author: Walker, John & Charles, engr. (worked from 1820)
Title: Austrian Dominions // II // Galizia Eastern Hungary and Transylvania. London, Published by Cha.[s] Knight + Co, 22 Ludgate St., 1845.
Year: 1845
Dimensions: 35.6 x 42.8 (30.5 x 39.0) cm
Phys. Desc.: Engraved, Colored Outline
Language(s): English

Entry (No.): 543
Author: Walker, John & Charles, engr. (worked from 1820)
Title: Russia // in Europe // General Map // Published by Cha.[s] Knight + Co. 22 Ludgate St. London.
Year: 1845
Dimensions: 35.6 x 42.8 (31.5 x 39.5) cm
Phys. Desc.: Engraved, Colored Outline
Language(s): English

Entry (No.): 544
Author: Walker, John & Charles, engr. (worked from 1820)
Title: Russia // in Europe Part IX // and Georgia Publ under the superintendence of the Society for the Diffusion of Useful Knowledge. London, Published by Cha.[s] knight + Co, 22 Ludgate St., 1845.
Year: 1845
Dimensions: 42.8 x 35.6 (39.2 x 31.3) cm
Phys. Desc.: Engraved, Colored Outline
Language(s): English

Entry (No.): 545
Author: Walker, John & Charles, engr. (worked from 1820)
Title: Russia // in Europe Part V // Published under the superintendence of the Society for the // Diffusion of Useful Knowledge.
Year: 1845
Dimensions: 35.6 x 42.8 (32.0 x 39.3) cm
Phys. Desc.: Engraved, Colored Outline
Language(s): English

Entry (No.): 546
Author: Walker, John & Charles, engr. (worked from 1820)
Title: Russia // in Europe Part VI // London, Published by Cha.[s] Knight + Co, 22 Ludgate St., 1845.
Year: 1845
Dimensions: 35.6 x 42.8 (31.3 x 38.7) cm
Phys. Desc.: Engraved, Colored Outline
Language(s): English

Entry (No.): 547
Author: Walker, John & Charles, engr. (worked from 1820)
Title: Russia // in Europe Part VIII // with small index map London, Published by Cha.[s] Knight + co, 22 Ludgate St., 1845.
Year: 1845
Dimensions: 42.8 x 35.6 (39.5 x 35.5) cm
Phys. Desc.: Engraved, Colored Outline
Language(s): English

Entry (No.): 548
Author: Weigel (Weigelius), Christoph (1654–1725)
Title: Ukrania seu Cosacorum. Regio Walachia item Moldavia et Tartaria minor. exc. Christ. Weigelio Norib.
Year: ca. 1670–1725
Dimensions: 47.5 x 37.0 (39.0 x 32.5) cm
Phys. Desc.: Engraved, Colored
Language(s): Latin

Entry (No.): 549
Author: Weigel (Weigelius), Christoph (1654–1725)
Title: Europa // Vetus // per // Christophorum Weigelium. // Norimbergae // cum privilegio Sac. Caesareae Majestatis.
Year: ca. 1720–60
Dimensions: 47.3 x 38.1 (41.3 x 32.8) cm
Phys. Desc.: Engraved, Colored
Language(s): Latin

Entry (No.): 550
Author: Weigel (Weigelius), Christoph (1654–1725)
Title: Imperium // Moscovia // cum regionibus // amplissimis / huc pertinentibus // Exudente // Christoph: Weigelio // Norimbergae.
Year: ca. 1720–45
Dimensions: 44.5 x 36.6 cm
Phys. Desc.: Engraved, Colored
Language(s): Latin

Entry (No.): 551
Author: Weigel (Weigelius), Christoph (1654–1725)
Title: Imperium // Moscovia // cum regionibus // amplissimis / huc pertinentibus // Exudente // Christoph: Weigelio // Norimbergae.

Year: ca. 1720–45
Dimensions: 43.8 x 35.8 (34.0 x 27.5) cm
Phys. Desc.: Engraved, Colored
Language(s): Latin

Entry (No.): 552
Author: Weigel (Weigelius), Christoph (1654–1725)
Title: Charte von Ungarn, Polen, Russland und der Turkey. // Enthaelt Polen, Galicien, die Moldau, Bukowina und Walachey, Bulgarien, Bosnien, Slavonien, Croatien, Dalmatien, Albanien, Griechenland, den Archipelagus, das Schwarze und Asovische Meer, die Krimm und Tartarey, Anadoli; oder Natolien &c. aus den d'Anvillischen und andern Charten, von neuen entworfen G. F. Uz. Ingen Lieut. u. Architect... Nürnberg bey C. Weigel und Schneider 1789.
Year: 1789
Dimensions: 53.0 x 62.0 (51.6 x 58.0) cm
Phys. Desc.: Engraved, Colored
Language(s): German

Entry (No.): 553
Author: Weigel, Johann Christoph (1661–1726)
Title: Regnum // Poloniae // ejusque // confinis.
Year: ca. 1712–60
Dimensions: 22.0 x 16.7 (17.2 x 14.0) cm
Phys. Desc.: Engraved, Colored
Language(s): Latin

Entry (No.): 554
Author: Wells, Edward (1667–1727)
Title: A New Map of EUROPE // according to the Present // General Divisions and Names // of its Countries together with // their Capital Cities Chief Rivers / Mountains &c // Dedicated // To His Highness William // Duke of Gloucester.
Year: 1700
Dimensions: 61.1 x 44.3 (49.6 x 36.7) cm
Phys. Desc.: Engraved, Colored Outline
Language(s): English
Figure 28

Entry (No.): 555
Author: Whittock, Nathaniel (worked from 1828–48)
Title: THE TOWN, FORTS AND HARBOURS OF SEBASTOPOL. // TAKEN FROM THE ILLUSTRATIONS TO THE IMPERIAL SURVEY OF THE BLACK SEA. // DRAWN BY N. WHITTOCK. // London. Published by Lloyd Brothers... August 14th 1854.
Year: 1854
Dimensions: 38.5 x 28.7 (34.0 x 19.2) cm
Phys. Desc.: Lithographed, Colored
Language(s): English

Entry (No.): 556
Author: Wied, Anthon (1500–1558)
Title: [Eastern Europe].
Year: 1532
Dimensions: 31.4 x 30.0 cm
Phys. Desc.: Copy and Negative
Language(s): Latin

Entry (No.): 557
Author: Wied, Anthon (1500–1558)
Title: [Eastern Europe + Tartaria]. Anthonius Wied...lectori S. Moscovia que & Alba Russia n[o?] cotenta Europeae Sarmatie.
Year: 1570
Dimensions: 42.0 x 30.0 cm
Phys. Desc.: Copy and Negative
Language(s): Latin

Entry (No.): 558
Author: Wilkinson, Robert (d. ca. 1825)
Title: Hungary // and // Transilvania. // Drawn from // the Latest Authorities. London, Published Sept.[r] 1st 1794. by R. Wilkinson. No 58 Cornhill.
Year: 1794
Dimensions: 33.5 x 27.5 (28.4 x 22.4) cm
Phys. Desc.: Engraved, Colored
Language(s): English

Entry (No.): 559
Author: Wilkinson, Robert (d. ca. 1825)
Title: Poland // with its // dismembered Provinces, // Drawn from the // Best Authorities. London: Published Jan. 1.st 1794 by R. Wilkinson N° 58 Cornhill.
Year: 1794
Dimensions: 33.5 x 27.5 (26.4 x 21.4) cm
Phys. Desc.: Engraved, Colored
Language(s): English

Entry (No.): 560
Author: Wilkinson, Robert (d. ca. 1825)
Title: Russia // in Europe // with its // Dismemberments // from // Poland // in 1773 1793 and 1795 // London: Published July 1st 1799 by Rob.[t] Wilkinson, N° 58 Cornhill.
Year: 1799
Dimensions: 33.5 x 27.5 (23.5 x 31.2) cm
Phys. Desc.: Engraved, Colored
Language(s): English

Entry (No.): 561
Author: Wit, Frederick de (1629 or 1630–1706)
Title: REGNI // POLONIAE // et Ducatus // LITHVANIAE // VOLINIAE, PODOLIAE // VCRANIAE // PRVSSIAE, LIVONIAE // et CVRLANDIAE descriptio // emendata per F. de Wit. // Amstelodami.
Year: 1654–72
Dimensions: 64.0 x 54.5 (55.0 x 48.0) cm
Phys. Desc.: Engraved, Colored
Language(s): Latin
Figure 20

Entry (No.): 562
Author: Wit, Frederick de (1629 or 1630–1706)
Title: Magnae // Tartariae, // Magni Mogolis Imperii, // Japoniae et Chinae // Nova Descriptio. Gedrucht t'Amsterdam by Frederick de Wit voer aen inde Calverstract // by den Dam inde Witte Pascaart.

Year: 1680
Dimensions: 62.5 x 53.8 (54.5 x 43.3) cm
Phys. Desc.: Engraved, Colored
Language(s): Dutch

Entry (No.): 563
Author: Wit, Frederick de (1629 or 1630–1706)
Title: Nova totius // HUNGARIAE, // TRANSILVANIAE, SERVIAE, ROMANIAE, BULGARIAE, // WALACHIAE, MOLDAVIAE, //SCLAVONIAE, CROATIAE, // BOSNIAE, DALMATIAE // Maximaeq. partis // DANUBII FLUMENIS // Descriptio Emendata a F. de Wit
Year: 1680
Dimensions: 60.8 x 49.0 (57.8 x 44.7) cm
Phys. Desc.: Engraved, Colored
Language(s): Latin

Entry (No.): 564
Author: Wit, Frederick de (1629 or 1630–1706)
Title: Nova et Accurata // totius // EUROPAE // DESCRIPTIO // Authore // FREDERICO DE WIT //...
Year: 1688
Dimensions: 61.4 x 51.9 cm
Phys. Desc.: Engraved, Colored
Language(s): Latin

Entry (No.): 565
Author: Wit, Frederick de (1629 or 1630–1706)
Title: REGNI // HUNGARIAE. // et Regionum, quae ei quondam fuére unitae, ut // TRANSILVANIAE, VALACHIAE, // MOLDAVIAE, SERVIAE, ROMANIAE, // BULGARIAE, BESSARABIAE, // CROATIAE, BOSNIAE, DALMATIAE, // SCLAVONIAE, MORLACHIAE, // RAGUSANAE REIPUBLICAE // Maximaeq Partis // Danubii Fluminis. // Novissima Delineatio. // Cum Previlegio // ex Officina // FREDERICI DE WIT // Amstelodami 1688.
Year: 1688
Dimensions: 91.2 x 52.8 (88.0 x 42.8) cm
Phys. Desc.: Engraved, Colored
Language(s): Latin

Entry (No.): 566
Author: Witsen, Nicolas (1641–1717)
Title: IMPERII // RUSSICI // sive // MOSCOVIAE. // Status Generalis, in ejus // Regna. Ducatus, Provincias, Populosq. Subjacentes Divisus, et ex Tabula // Spectat.mi. Domini N: Witsen, // Urbis Amstelodamensis Consulis etc. // pro Maiori Parte Excerptus Per F. de Witt // Amst. Cum Priv: P.D: Ord: Holl: West:frisiaeq. // ex Officina I. Covens et C. Mortier.
Year: 1700
Dimensions: 64.7 x 56.5 (55.5 x 44.6) cm
Phys. Desc.: Engraved, Colored
Language(s): Latin
Figure 38

Entry (No.): 567
Author: Witsen, Nicolas (1641–1717)
Title: PONTUS EUXINUS of Niewe en Naaukeurige // Paskaart // van de Zwarte Zee, // uytverscheydene stucken van // die gewelten toegefonden,// ontworpen door // N. Witsen, Cons: Amst. // Apud // L. Renard.
Year: 1700
Dimensions: 63.3 x 53.1 (59.1 x 41.6) cm
Phys. Desc.: Engraved, Colored
Language(s): Dutch

Entry (No.): 568
Author: Witsen, Nicolas (1641–1717)
Title: NOUVELLE CARTE // DE // L'EMPIRE de RUSSIE. // Dressée sur les meilleurs // Plans qui aient paru jusqu'à // present, et particulierement // sur celui de Mr. le Bourgue == // maitre Witzen. Levé sur les Lieux et rectifié par Everard Isbrantz Ides.
Year: 1725
Dimensions: 41.6 x 33.5 (36.0 x 28.0) cm
Phys. Desc.: Engraved, Uncolored
Language(s): French

Entry (No.): 569
Author: Wyld, James, Jr. (1812–87)
Title: PLAN OF THE // NORTHERN FORTS // & Attack & Siege Works // SEVASTOPOL // BY // James Wyld. Geographer to the Queen // & H.R.H. Prince Albert // Charing Cross East & 2 Royal Exchange.
Year: 1855
Dimensions: 51.0 x 35.0 (48.0 x 30.50 cm
Phys. Desc.: Printed, Colored Outline
Language(s): English

Entry (No.): 570
Author: Wyld, James, Jr. (1812–87)
Title: The // Crimea // by // James Wyld, Geographer to the Queen & H.R.H. Prince Albert, Charing Cross, East, London. Fifth Edition June 1855.
Year: 1855
Dimensions: 69.0 x 52.0 (64.2 x 47.2) cm
Phys. Desc.: Printed
Language(s): English
Figure 68

Entry (No.): 571
Author: Wyld, James, Sr. (1790–1836)
Title: The Russian Dominions in Europe drawn from the latest Maps printed, by the Academy of Sciences, St. Petersburg; revised and corrected, with the Post Roads, + New Governments, from the Russian Atlas of 1806; by Jasper Nantiat. London. Published by Ja.[s] Wyld Geographer to the Queen.
Year: after 1806
Dimensions: 98.9 x 57.8 (92.5 x 54.0) cm
Phys. Desc.: Engraved, Colored
Language(s): English

Entry (No.): 572
Author: Zatta, Antonio (ca. 1724–1804)
Title: Prospetto della Fortezza d'Oczakow. Di Kilbun. Fiume detto Uzù.
Year: ca. 1779–99
Dimensions: 27.2 x 17.8 cm

Phys. Desc.: Printed, Uncolored
Language(s): Italian

Entry (No.): 573
Author: Zatta, Antonio (ca. 1724–1804)
Title: IL REGNO // di // UNGHERIA / di nuova Projezione // VENEZIA 1781 // Presso Antonio Zatta // Con Privilegio dell'Eccmo Senato.
Year: 1781
Dimensions: 49.0 x 37.1 (40.7 x 31.1) cm
Phys. Desc.: Engraved, Colored Outline
Language(s): Italian

Entry (No.): 574
Author: Zatta, Antonio (ca. 1724–1804)
Title: LA // RUSSIA EUROPEA // divisa // NE SUOI GOVERNI // E PROVINCIE // Tratta dall'Atlante Russo // VENEZIA 1781 Presso Antonio Zatta // Con Privilegio dell'Eccmo Senato.
Year: 1781
Dimensions: 38.6 x 50.0 (31.5 x 40.8) cm
Phys. Desc.: Engraved, Colored Outline
Language(s): Italian

Entry (No.): 575
Author: Zatta, Antonio (ca. 1724–1804) (filed under Rizzi-Zannoni by Krawciw)
Author: Rizzi-Zannoni, Giovanni Antonio (1736–1814)
Title: DUCATI // DI LIVONIA, E DI ESTONIA, // E GOVERNI // DI NOWOGOROD, BIELOGOROD, // E KIOWIA // coi loro Distretti// NELLA RUSSIA EUROPEA // Di nuova Projezione // VENEZIA 1782 // Presso Antonio Zatta // Con Privelegio dell'Eccmo Senato.
Year: 1782
Dimensions: 37.4 x 48.0 (31.4 x 42.2) cm
Phys. Desc.: Engraved, Colored Outline
Language(s): Italian

Entry (No.): 576
Author: Zatta, Antonio (ca. 1724–1804)
Title: LA PICCOLA // TARTARIA // divisa // NE' SUOI TERRITORI // Di nuova Projezione // VENEZIA 1783 // Presso Antonio Zatta // Con Privilegio dell' Eccmo Senato.
Year: 1783
Dimensions: 48.0 x 37.4 (40.4 x 30.8) cm
Phys. Desc.: Engraved, Colored Outline
Language(s): Italian
Figure 53

Entry (No.): 577
Author: Zatta, Antonio (ca. 1724–1804)
Title: Piano // dell' Assedio // della // Fortezza Turchesca // di Oczakow. // Presa d'Assalto da' Rusi // nel mese di Decembre 1788.
Year: 1788
Dimensions: 32.0 x 25.6 (24.2 x 15.1) cm
Phys. Desc.: Engraved, Colored
Language(s): Italian

Entry (No.): 578
Author: Zatta, Antonio (ca. 1724–1804)
Title: Il Regno // di // Polonia // con le Provinie // ora posseduete dalle // Tre confinanti potenze // Prussia Russia // e // Casa d'Austria Presso Antonio Zatta e Figli.
Year: 1793
Dimensions: 40.5 x 27.5 (31.8 x 23.3) cm
Phys. Desc.: Engraved, Colored Outline
Language(s): Italian

Entry (No.): 579
Author: Zündt (Cynthius, Zindt, Zündten, Zunthio, Zynder, Zynndt, Zynnthen), Matthias (ca. 1498–1572)
Title: HVNGARIAE TOTIVS VTI EX COMPVRIBVS ALIORVM GEOGRAPHICIS CHARTIS A MATTHIA ZINTHIO NORICO DELINEATA EST RECENS EDITIO.
Year: 1593
Dimensions: 54.8 x 39.6 (49.0 x 29.5) cm
Phys. Desc.: Reprint, Engraved, Colored
Language(s): Latin

Entry (No.): 580
Author: Zürner (Zuerner), Adam Friedrich (1679–1742)
Title: EUROPAE // IN // TABULA GEOGRAPHICA // DELINEATIO // AD MENTEM NOVISSIMORUM EORUMQ. OPTIMORUM // GEOGRAPHORUM EMENDATA... // OPERA // A. F. Z?RNERI REG: MAJ: POL: ET EL. SAX: // PROVINCIARUM FINIUMQ. COMMISSARII // ET GEOGRAPHI... // EX OFFICINA // PETRI SCHENKII // IN PLATEA VULGO DE VARMOES STRAAT // SUB SIGNO N. VISSCHERS ATHLAS. // Inset: Die Nordischen Schifarten...
Year: 1709
Dimensions: 64.2 x 54.0 (57.3 x 49.5) cm
Phys. Desc.: Engraved, Colored Outline
Language(s): Latin

Maps with No Author Indicated

Entry (No.): 581
Title: Asiae Tavula I.
Year: ca. 16th century
Dimensions: 19.6 x 15.3 (17.0 x 12.4) cm
Phys. Desc.: Engraved, Uncolored
Language(s): Latin

Entry (No.): 582
Title: SARMATIA
Year: ca. 16th century
Dimensions: 31.5 x 22.9 (30.3 x 20.7) cm
Phys. Desc.: Engraved, Colored
Language(s): Latin

Entry (No.): 583
Title: TAVRICA CHERSONESVS
Year: 1633
Dimensions: 19.0 x 28.8 (12.8 x 16.5) cm

Phys. Desc.: Engraved, Uncolored
Language(s): Latin

Entry (No.): 584
Title: GRANDE TARTARIE
Year: 1685
Dimensions: 17.0 x 20.0 (10.6 x 15.2) cm
Phys. Desc.: Engraved, Uncolored
Language(s): French

Entry (No.): 585
Title: NOVA TOTIUS TERRARUM ORBIS GEOGRAPHICA AC HYDROGRAPHICA TABULA //... Bildtnüss des Gantzen Erden Kranses mit allen seinen theilen.
Year: ca. 16th–17th c.
Dimensions: 37.0 x 27.4 (35.7 x 24.8) cm
Phys. Desc.: Engraved (copy)
Language(s): Latin, German

Entry (No.): 586
Title: ASOFF WIE ES VON DEM // TZAAR NACH DER EROBERUNG // IST FORTIFICIERT WORDEN. //
Year: 1710
Dimensions: 29.3 x 18.7 (29.0 x 17.0) cm
Phys. Desc.: Engraved
Language(s): German

Entry (No.): 587
Title: HAUPT KARTE POLEN. XXVI.
Year: 1733
Dimensions: 14.2 x 13.4 (13.0 x 10.8) cm
Phys. Desc.: Engraved, Colored
Language(s): German

PART TWO (BY SUBJECT HEADING)

Entry (No.): 588
Subject: Asia, Siberia (18th, 19th c.)
Author: Giraldon
Title: Sibérie avec L'Amérique Russe
Year: 1805
Dimensions: 41.0 x 26.0 (29.8 x 21.9) cm
Phys. Desc.: Engraved, Colored Outline
Language(s): French

Entry (No.): 589
Subject: Asia, Siberia (18th, 19th c.)
Author: Petermann, Augustus (August) Heinrich (Henry) (1822–78)
Title: Map of the River System of Siberia.
Year: 1882
Dimensions: 33.2 x 29.3 cm
Phys. Desc.: Engraved
Language(s): English

Entry (No.): 590
Subject: Asia, Siberia (18th, 19th c.)
Author: Vivien, L.
Title: Carte Générale de la Russie d'Asie ou Sibérie par L. Vivien Geographe Gravée par Giraldon Bovinet. 1834, Paris.
Year: 1834
Dimensions: 63.0 x 50.8 (56.5 x 35.1) cm
Phys. Desc.: Engraved, Colored Outline
Language(s): French

Entry (No.): 591
Subject: Asia, Siberia (18th, 19th c.)
Author: Weller, Edward (d. 1884)
Title: Russia in Asia.
Year: ca. 1830
Dimensions: 56.5 x 37.8 (49.5 x 34.3) cm
Phys. Desc.: Engraved
Language(s): English

Entry (No.): 592
Subject: Asia, Siberia (18th, 19th c.)
Author: No Author
Title: Europae
Year: ca. 1780
Dimensions: 50.2 x 42.8 (44.5 x 38.6) cm
Phys. Desc.: Engraved, Colored
Language(s): Latin, Italian

Entry (No.): 593
Subject: Austro-Hungary
Author: Cool, Robert N.
Title: Hitler's 'Fortress' Germany.
Year: 1940
Dimensions: 55.5 x 43.0 (38.0 x 34.5) cm
Phys. Desc.: Printed, Uncolored
Language(s): English
Figure 84

Entry (No.): 594
Subject: Austro-Hungary
Author: Cram, George Franklin (1842–1928)
Title: The // Austro-Hungarian // Monarchy // and its Protectorates // Bosnia and Herzegovina.
Year: 1918
Dimensions: 58.0 x 38.0 (50.5 x 33.5) cm
Phys. Desc.: Lithographed, Uncolored
Language(s): English

Entry (No.): 595
Subject: Austro-Hungary
Author: Pasquali, Giovanni Valerio
Title: [Habsburg Länder]
Year: ca. 1830
Dimensions: 27.0 x 39.7 cm
Phys. Desc.: Engraved, Colored
Language(s): Italian

Entry (No.): 596
Subject: Austro-Hungary
Author: Perthes, (Johann Georg) Justus (1749–1816)
Title: Österreichisch. // Ungarische // Monarchie. Bearbeitet von C. Vogel.
Year: 1872
Dimensions: 46.5 x 38.2 (40.2 x 33.5) cm
Phys. Desc.: Engraved, Uncolored
Language(s): German

Entry (No.): 597
Subject: Austro-Hungary
Author: Pinkerton, John (1758–1826)
Author: Neele, [Samuel John] (1758–1824)
Title: Austrian Dominions
Year: 1810
Dimensions: 84.0 x 56.5 (70.5 x 50.2) cm
Phys. Desc.: Engraved, Colored Outline
Language(s): English

Entry (No.): 598
Subject: Austro-Hungary
Author: Renouard, Paul
Title: Carte de L'Allemagne Politique Actuelle à partir du Congrès de Vienne ou Confédération Germanique: refermant occasionellement la Confédération Suisse.
Year: 1823
Dimensions: 68.9 x 51.7 (47.5 x 34.5) cm
Phys. Desc.: Printed, Colored
Language(s): French

Entry (No.): 599
Subject: Austro-Hungary
Author: Renouard, Paul
Title: Carte de L'Allemagne Politique Actuelle à partir du Congrès de Vienne ou Confédération Germanique: renfermant occasionellement la Confédération Suisse.
Year: 1823

Dimensions: 69.0 x 51.8 (66.0 x 49.5) cm
Phys. Desc.: Engraved, Colored
Language(s): French

Entry (No.): 600
Subject: Austro-Hungary
Author: Teleki, Count Pál (1879–1941)
Title: Ethnographical Map // of Hungary // based on // Density of Population // by // Count Paul Teleki // Professor of Geography… According to the census of 1910.
Year: ca. 1919–21
Dimensions: 83.5 x 61.0 (79.5 x 56.5) cm
Phys. Desc.: Printed, Colored
Language(s): English
Figure 79

Entry (No.): 601
Subject: Austro-Hungary
Author: Teleki, Count Pál (1879–1941)
Author: Fodor, Francis
Title: The Treaty of Trianon in the Light of Geography.
Year: 1928
Dimensions: 23.5 x 15.0 cm
Phys. Desc.: book with maps
Language(s): English

Entry (No.): 602
Subject: Austro-Hungary
Author: No Author
Title: Royaume de Hongrie… Autriche… Boheme.
Year: 18th c.
Dimensions: 51.0 x 38.5 (48.1 x 30.0) cm
Phys. Desc.: Engraved, Colored Outline
Language(s): French

Entry (No.): 603
Subject: Austro-Hungary
Author: No Author
Title: La Germania Legislativa.
Year: ca. 1830
Dimensions: 23.2 x 16.0 cm
Phys. Desc.: Engraved
Language(s): Italian

Entry (No.): 604
Subject: Austro-Hungary
Author: No Author
Title: La Germania Ne'Suoi Principali Dominii…
Year: ca. 1830
Dimensions: 23.1 x 15.6 cm
Phys. Desc.: Engraved
Language(s): Italian

Entry (No.): 605
Subject: Austro-Hungary
Author: No Author
Title: Turkey in Europe and Hungary.
Year: 19th c.
Dimensions: 25.5 x 20.0 cm
Phys. Desc.: Printed, Colored Outline
Language(s): English

Entry (No.): 606
Subject: Black Sea
Author: Bugbee, James M.
Title: Map of Russia and Turkey // Showing // Seat of the Present War.
Year: 1877
Dimensions: 53.0 x 35.4 (50.8 x 33.5) cm
Phys. Desc.: Printed, Uncolored
Language(s): English

Entry (No.): 607
Subject: Black Sea
Author: Rolffsen, I. I. [J. J.]
Title: Plan de la Ville d'Otzakow, prise par les Russes l'an 1737.
Year: 1737
Dimensions: 49.5 x 35.0 (34.0 x 26.5) cm
Phys. Desc.: Engraved, Uncolored
Language(s): French

Entry (No.): 608
Subject: Black Sea
Author: Rolffsen I. I. [J. J.]
Title: Theatre de la Guerre // entre // Les Russes et les Turcs. Dans les années: 1736, 1737 & c.
Year: 1737
Dimensions: 54.0 x 44.5 (50.0 x 38.9) cm
Phys. Desc.: Engraved, Uncolored
Language(s): French

Entry (No.): 609
Subject: Black Sea
Author: Routledge, George, publ. (1812–88)
Title: Black Sea.
Year: 1854
Dimensions: 54.0 x 37.0 (52.0 x 33.5)
Phys. Desc.: Engraved on Steel
Language(s): English

Entry (No.): 610
Subject: Black Sea
Author: Rusell, W. H.
Title: Chart of Kherson Bay // and the // Dnieper & Boug Rivers.
Year: ca. 1850s
Dimensions: 48.5 x 36.0 (45.0 x 34.5) cm
Phys. Desc.: Engraved on Steel
Language(s): English

Entry (No.): 611
Subject: Black Sea
Author: Smith, Ed.
Title: A New Map of Ancient Greece, Thrace, Moesia, Illyricum, and the Isles adjoining Dedicated to his Highness William Duke of Clocester.
Year: 1718
Dimensions: 53.8 x 42.5 (48.1 x 35.8) cm
Phys. Desc.: Engraved, Uncolored

Language(s): English

Entry (No.): 612
Subject: Black Sea
Author: Sotzmann [Sotzman], Daniel Friedrich (1754–1840)
Title: Umriss des Schwarzen Meers des Bosphorus, des Propontis, des Hellesponts, des Archipelagus nebst den Jüsten, nach den neuesten Berichtungen von D. F. Sotzmann 1803.
Year: 1803
Dimensions: 36.0 x 24.2 (34.2 x 22.5) cm
Phys. Desc.: Engraved, Colored
Language(s): English

Entry (No.): 613
Subject: Black Sea
Author: No Author
Title: Petite Tatarie dite autrement // Crimée et Precopense.
Year: 1676
Dimensions: 13.5 x 15.0 (12.5 x 10.2) cm
Phys. Desc.: Engraved, Uncolored
Language(s): French

Entry (No.): 614
Subject: Black Sea
Author: No Author
Title: Plan de la Ville d'Asof // Assiegée et prise par l'Armée Russes sous le Marechal de Laseÿ sur les Turcs l'année 1736.
Year: 1736
Dimensions: 55.5 x 44.5 (51.1 x 39.0) cm
Phys. Desc.: Engraved, Uncolored
Language(s): French

Entry (No.): 615
Subject: Black Sea
Author: No Author
Title: Plan de la Ville de Chockzin prise sur les Turcs par les Russes sous les du Marechal Munnich l'An 1739.
Year: 1739
Dimensions: 52.0 x 45.8 (46.5 x 40.5) cm
Phys. Desc.: Engraved, Uncolored
Language(s): French

Entry (No.): 616
Subject: Black Sea
Author: No Author
Title: Bataille de Stavoutschan.
Year: ca. 1850s
Dimensions: 49.5 x 41.2 (42.0 x 34.9) cm
Phys. Desc.: Engraved, Uncolored
Language(s): French

Entry (No.): 617
Subject: Black Sea
Author: No Author
Title: Plan de la Ville d'Otschakow // assiegée par les Turcs & les Tartares, sous le Seraskier Jentsch Ali, Bascha des trois queues // & le Chan de la Crimée qui furent obligés de lever le Siege.
Year: ca. 1850s
Dimensions: 52.8 x 45.2 (50.6 x 38.2) cm
Phys. Desc.: Engraved on Steel
Language(s): French

Entry (No.): 618
Subject: Caspian Sea
Author: Bowen, Emanuel (ca. 1693/1694–1767)
Title: A New and // Accurate Map // of the Caspian Sea, // Laid down // from the Memoirs of // Soskam Sabbas // a Georgian Prince, // assisted by some late Improvements // By Emanl. Bowen.
Year: 1747
Dimensions: 25.2 x 40.4 (22.5 x 32.0) cm
Phys. Desc.: Engraved, Uncolored
Language(s): English

Entry (No.): 619
Subject: Caspian Sea
Author: d'Anville, Jean Baptiste Bourguignon (1697–1782)
Title: Essai d'une Nouvelle Carte de la Mer Caspienne // Par le Sr. d'Anville // de l'Acad. Rle. Des Inscript. Et Belles-Lettres, et de celled des Sciences de Petersbourg. Novembre 1754.
Year: 1754
Dimensions: 40.0 x 56.5 (26.0 x 51.7) cm
Phys. Desc.: Engraved, Uncolored
Language(s): French

Entry (No.): 620
Subject: Caspian Sea
Author: Darmet, J. M.
Title: Carte Générale // du Pays // entre Les Mers // Noire et Caspienne // avec la Désignation // des nouvelles-Frontières // de la Russie et de la Perse // telles qu'elles ont été arrètées // etre les Persans et le Gnl. Major Khatov en 1819.// Présentée // a S.E.Ma. le Bon. de Damas, // Lieutenant-Géneral, Ministre et Secrétaire d'Etat // en Départment des Affaires étrangères, // Par J.M. Darmet.
Year: 1819
Dimensions: 53.0 x 49.6 cm
Phys. Desc.: Engraved, Colored Outline
Language(s): French

Entry (No.): 621
Subject: Caspian Sea
Author: Delisle (De L'Isle), Guillaume (1675–1726)
Title: Carte Marine // de la Mer Caspienne // levée suivant les ordres de S.M. Czarienne, // Par Mr. Carl Vanverden en 1719, 1720, et 1721.// Et Reduite au Meridien de Paris // Par Guillaume De l'Isle Premier Geographe du Roy, // de l'Academier Royale des Sciences. // A Amsterdam // Chez Cóvens et Mortier // Geographes.
Year: 1721
Dimensions: 66.1 x 54.5 (60.5 x 45.8) cm
Phys. Desc.: Engraved, Colored Outline
Language(s): French

Entry (No.): 622
Subject: Caspian Sea

Author: Delisle (De L'Isle), Guillaume (1675–1726)
Title: Carte // des Pays Voisins de la // Mer Caspiene // dressée pour l'usage du Roy // sur la Carte de cete Mer faite par l'ordre du Czar, // sur les Memoires manuscrits // de Soskam-Sabbas Prince de Georgie, // sur ceux de Mrs. Crusins, Zurabek, et Fabritius // Ambassadeurs a la Cour de Perse, / et sur les éclaircissemens tirez d'un grand nombre // des persones intelligentes du païs. // Assujetie aux Observations Astronomiques // Par Guillaume Delisle // Premier Geographe du Roy, // de l'Academie Royale // des Sciences. // 15. Aout // 1723.
Year: 1723
Dimensions: 65.2 x 57.0 (62.0 x 46.0) cm
Phys. Desc.: Engraved, Colored
Language(s): French
Figure 37

Entry (No.): 623
Subject: Caspian Sea
Author: Delisle (De L'Isle), Guillaume (1675–1726)
Title: Carte de // La Mer Caspienne // levée suivant les ordres du Czar // par Mr. Charles van Verden en 1719, 1720, et 1721. // Envoyée a l'Academie Royale des Sciences // Par sa Maiesté Czariéne // et reduite par G. Delisle 1er. Geographe du // Roy de l'Academie Royale des Sciences. 1725.
Year: 1725
Dimensions: 27.8 x 48.5 (23.0 x 44.0) cm
Phys. Desc.: Engraved, Uncolored
Language(s): French

Entry (No.): 624
Subject: Caspian Sea
Author: Delisle (De L'Isle), Guillaume (1675–1726)
Title: Carte // des Pays Voisins de la // Mer Caspiene // dressée pour l'usage du Roy // sur la Carte de cette Mer faite par l'ordre du Czar, // Sur les Memoires manuscrits // de Soskam-Sabbas Prince de Georgie, // sur ceuz de Mrs. Crusins, Zurabek, et Fabritius // Ambassadeurs a la Cour de Perse, / et sur les éclaircissements tirez d'un grand nombre // des persones intelligentes du païs. // Assujetie aux Observations Astronomiques // Par G. Delisle Geographe de l'Academie Roy. des Scien. // A Amsterdam // Chez Jean Cóverns et // Corneille Mortier // Geographes.
Year: ca. 1730–41
Dimensions: 64.5 x 53.0 (60.5 x 45.8) cm
Phys. Desc.: Engraved, Colored Outline
Language(s): French

Entry (No.): 625
Subject: Caspian Sea
Author: Delisle (De L'Isle), Guillaume (1675–1726)
Title: Coste de Perse sur la Mer Caspienne et Partie de Celles de Tartarie.
Year: 1737
Dimensions: 65.6 x 54.3 (60.0 x 42.6) cm
Phys. Desc.: Engraved, Colored Outline
Language(s): French

Entry (No.): 626
Subject: Caspian Sea
Author: Delisle (De L'Isle), Guillaume (1675–1726)
Title: Carte de Perse // dressée pour l'usage du Roy // Par G. Delisle premier Geographe de S.M. de l'Academie Royale des Sciences // A Paris // Chez Dezauche Géographe, successeur des Srs. De Lisle et Phil. Buache. Rue des Noyers.// Avec Privilège du Roi. 1785.
Year: 1785
Dimensions: 78.4 x 55.5 (62.5 x 49.0) cm
Phys. Desc.: Engraved, Colored Outline
Language(s): French

Entry (No.): 627
Subject: Caspian Sea
Author: Homann, Johann Baptist (Baptista) (1664–1724)
Title: Geographica Nova // et Orientis gratiosissima, // duabus tabulis specialissimis contenta // Mare Caspivm,// altera // Kamtzadaliam seu Terram Jedso // curiosè exhib6t. // Editore // Io. Bapt. Homann S.C.M. Geogr. // Norimbergae.
Year: ca. 1710s–20s
Dimensions: 61.0 x 53.6 (57.6 x 48.3) cm
Phys. Desc.: Engraved, Colored
Language(s): Latin

Entry (No.): 628
Subject: Caspian Sea
Author: Lenormandt, Charles François (1837–83)
Title: Mer // Caspiene ou de Tabarestan // Par Charles Lenormand // a Versaille 1760. Turcomanie // Georgie // Commanie // d'apres M. Sanson // par C. Lenormand // Versailles 1760.
Year: 1760
Dimensions: 19.2 x 28.5 (16.2 x 19.2) cm
Phys. Desc.: Engraved, Colored
Language(s): French

Entry (No.): 629
Subject: Caspian Sea
Author: Maas, Abraham
Title: Nova // Maris Caspii // et Regionis // Usbeck // Cum Provincijs adjacentibus // vera Delineatio // In quo itinera Regia et alia // notabiliora accurate denotantur // Per A. Mass. // 1735. // Curatit? Homanianis Heredib?.
Year: 1733
Dimensions: 64.7 x 54.0 (58.6 x 49.0) cm
Phys. Desc.: Engraved, Colored
Language(s): Latin

Entry (No.): 630
Subject: Caspian Sea
Author: Ottens, Reinier I (1698–1750) & Ottens, Joachim (1663–1719)
Title: Nova ac verissima // Maris Caspii // ante hac maximam fere partem nobis incogniti, ac // Regionum adjacentium // Delineatio, // Jussu Invictissimi Principis // Petri Alexii Fil. // Magni Russorum Imperatoris // immenso labore et maximis sumptibus facta. atque // ex Autographo in lucem edita // per Reinerum Ottens Geographum Amstelaedam.
Year: early 18th c.
Dimensions: 65.3 x 54.0 (60.4 x 54.0) cm
Phys. Desc.: Engraved, Colored

Language(s): Latin

Entry (No.): 631
Subject: Caspian Sea
Author: No Author
Title: Route // d'Amsterdam // a // Moscow et de la à Ispahan et Gamron.
Year: ca. 1725
Dimensions: 41.9 x 33.0 cm
Phys. Desc.: Engraved, Uncolored
Language(s): French

Entry (No.): 632
Subject: Cities
Author: Cooke, George (engr.) (worked 1802–16)
Author: Atkinson, [?]
Title: View of St. Petersburg // Engraved by permission from a drawing of Mr. Atkinson's. London. Published by Longman, Hurst, Rees & Orme, Paternoster Row, Febr. 1, 1810.
Year: 1810
Dimensions: 26.6 x 21.0 (25.8 x 14.5) cm
Phys. Desc.: Engraved, Uncolored
Language(s): English

Entry (No.): 633
Subject: Cities
Author: Cooke, George (engr.) (worked 1802–16)
Author: Atkinson, [?]
Title: Constantinople, // from the heights of Eyoub.
Year: 1811
Dimensions: 41.5 x 26.5 (40.0 x 16.8) cm
Phys. Desc.: Engraved, Uncolored
Language(s): English

Entry (No.): 634
Subject: Cities
Author: Fer, Nicolas de (1646–1720)
Title: Plan de la Ville de Constantinople.
Year: 1700
Dimensions: 43.7 x 32.0 (34.8 x 21.0) cm
Phys. Desc.: Engraved, Uncolored
Language(s): French

Entry (No.): 635
Subject: Cities
Author: Rossi (Rubeis), Giovanni Giacomo (Joannes Jacobus) de (1627–91)
Title: NVOVA ET ESSATA PIANTA DEL CONCLAVE FATTO IN SEDE VACANTE DI PAPA INNOCENTO XI. CHE SEDE ANNI XII. MESI X GIORNI XXII. PER L'ELETTIONE DEL NVOVO PONTEFICE, NEL' QVALE ENTRORNO L'EM.[MI] SIG.[RI] CARDINALI A. DI XXIII. AGOSTO MDCLXXXIX.
Year: 1689
Dimensions: 54.4 x 42.0 (52.4 x 39.1) cm
Phys. Desc.: Engraved, Uncolored
Language(s): Italian

Entry (No.): 636
Subject: Cities
Author: Hoefnagel (Hofnagel, Houfnagel, Huefnagel, Hoefnagel), Joris (Georg, Joris) (1542–1600)
Title: Palanca Superioris Hungariae civitas
Year: 1617
Dimensions: 44.5 x 32.8 cm
Phys. Desc.: Engraved, Uncolored
Language(s): Latin

Entry (No.): 637
Subject: Cities
Author: Nolin, Jean-Baptiste II (1686–1762)
Title: La Topografia // Di Roma // Di Gio. Batta Nolli // D'Alla Maggiore // In Gvesta Minor // Tauola // Dal. Medesimo Ridotta.
Year: 1748
Dimensions: 68.3 x 50.7 (67.8 x 45.2) cm
Phys. Desc.: Engraved, Uncolored
Language(s): Italian

Entry (No.): 638
Subject: Cities
Author: Nolin, Jean-Baptiste II (1686–1762)
Title: SILVIO VALENTI // Tit S. Callix. Presb. S.R.E. Card. Camerario // BENEDICTI XIV. P. M. // Administro // VRBIS ICHNOGRAPHIAM // A LEONARDO BUFALINO // Ligneis formis zevulgatam // Servata Proportione contractam // Atq. Aeri incisam // JO. BAPTA NOLLI // Geometra & Architect Novocom. // D.D.D.
Year: 1748
Dimensions: 68.4 x 51.3 (75.5 x 44.5) cm
Phys. Desc.: Engraved, Uncolored
Language(s): Latin

Entry (No.): 639
Subject: Cities
Author: Rolffsen, I. I. [J. J.]
Title: Plan de la Ville de Dantzig // et du Fort de Weichselmunde // assiegés par l'Armée Russe l'an 1734.
Year: ca. 1760–70
Dimensions: 49.1 x 34.1 (41.8 x 28.4) cm
Phys. Desc.: Engraved, Uncolored
Language(s): French

Entry (No.): 640
Subject: Cities
Author: Schleuen, Johann David
Title: Berlin Exudit
Year: 19th c.
Dimensions: 59.3 x 45.0 (56.0 x 40.5) cm
Phys. Desc.: Engraved, Uncolored
Language(s): German

Entry (No.): 641
Subject: Cities
Author: Solinus Gaius Julius (worked 3rd c. ad)
Title: Roma.
Year: 16th c.
Dimensions: 12.9 x 16.3 cm
Phys. Desc.: Copy, Woodcut
Language(s): Latin

Entry (No.): 642
Subject: Cities
Author: Thornton (family), sculp.
Title: A General Perspective View of Constantinople, the Capital of the Turkish Empire, Engraved for Millar's New Complete & Universal System of Geography.
Year: mid-18th c.
Dimensions: 37.0 x 22.7 (28.0 x 17.5) cm
Phys. Desc.: Engraved, Uncolored
Language(s): English

Entry (No.): 643
Subject: Cities
Author: Thornton (family), sculp.
Title: A View of the City of Berlin, Capital of the electorate of Brandenburg, and the residence of the King of Prussia... Engraved for Millar's New Complete Universal System of Geography.
Year: mid-18th c.
Dimensions: 36.5 x 22.9 (28.7 x 19.9) cm
Phys. Desc.: Engraved, Uncolored
Language(s): English

Entry (No.): 644
Subject: Cities
Author: Venutus, Ridolfino Abare
Title: Carta Topografica // della Antichità di Roma // la quale esattamente continene tutte // le vestigia, delle Antiche Fabbriche // secondo il Giro Metodico // usato già dall' Ab. Ridolfino Venuti // e dal medesimo insegnato nella sua Opera Intitolata // Accurata, e Succinta Descrizzione // Topografica delle Antichità di Roma - deli Abare Ridolfino Venuti Cortinese... MDCCLXII.
Year: 1763
Dimensions: 59.8 x 44.5 (53.8 x 38.8) cm
Phys. Desc.: Engraved, Uncolored
Language(s): Italian

Entry (No.): 645
Subject: Cities
Author: No Author
Title: Constantinopolis.
Year: ca. 1493
Dimensions: 30.0 x 43.8 (22.6 x 23.3) cm
Phys. Desc.: Copy, Woodcut
Language(s): Latin

Entry (No.): 646
Subject: Cities
Author: No Author
Title: Roma.
Year: 1493
Dimensions: 61.4 x 43.5 (53.5 x 22.7) cm
Phys. Desc.: Copy, Woodcut
Language(s): Latin

Entry (No.): 647
Subject: Cities
Author: No Author
Title: Constantinopolitanae urbis effigies, quam hodie sub Turcae inhabitione habet.
Year: 15th–16th c.
Dimensions: 40.0 x 26.0 (37.8 x 20.0) cm
Phys. Desc.: Engraved on Wood, Uncolored
Language(s): Latin

Entry (No.): 648
Subject: Cities
Author: No Author
Title: Plan de Kroonstad // et la Passe pour aller a Petersburg.
Year: ca. 1750
Dimensions: 24.0 x 32.0 (17.2 x 21.3) cm
Phys. Desc.: Engraved, Uncolored
Language(s): French

Entry (No.): 649
Subject: Cities
Author: No Author
Title: Plan de la Ville et Port // de St. Petersbourg.
Year: ca. 1750
Dimensions: 24.0 x 32.0 (16.8 x 21.1) cm
Phys. Desc.: Engraved, Uncolored
Language(s): French

Entry (No.): 650
Subject: Cities
Author: No Author
Title: Cronstadt.
Year: 1856
Dimensions: 71.0 x 46.0 (68.0 x 41.0) cm
Phys. Desc.: Lithographed, Uncolored
Language(s): English

Entry (No.): 651
Subject: Cities
Author: No Author
Title: St. Petersburg and Cronstadt.
Year: 1856
Dimensions: 39.0 x 25.5 (35.5 x 16.5) cm
Phys. Desc.: Lithographed, Uncolored
Language(s): English

Entry (No.): 652
Subject: Cities
Author: No Author
Title: La figure de la place, où sut iadiz la cité de Rome, auec ses portes, qvand elle estoit en sa grand' flevr.
Year: 17th c.
Dimensions: 46.7 x 37.3 (22.6 x 21.3) cm
Phys. Desc.: Engraved, Colored
Language(s): French

Entry (No.): 653
Subject: Cities – Crimea
Author: Babkow (Babkov), Ivan Ivanovits (Ivan Ivanovich)
Title: Iuzhnyi Krym.
Year: 1956
Dimensions: 66.3 x 53.0 (48.0 x 37.3) cm
Phys. Desc.: Printed, Colored
Language(s): Russian

Entry (No.): 654
Subject: Cities – Crimea
Author: Desilver, Charles (publ.)
Title: A Complete // Map of the // Crimea // Showing the // Military and Carriage Roads // with Distances // from // Various Points of the Western Coast. // From English & Russian Surveys. // Philadelphia // Published by Charles Desilver 253 Market St. Keen & Lee 148 Lake St. Chicago.
Year: 1856
Dimensions: 68.5 x 44.0 (67.0 x 40.5) cm
Phys. Desc.: Lithographed, Colored
Language(s): English
Figure 70

Entry (No.): 655
Subject: Cities – Crimea
Author: Rapkin, John
Title: The Crimea.
Year: ca. 1850s
Dimensions: 35.0 x 27.0 (32.5 24.0) cm
Phys. Desc.: Lithographed, Colored Outline
Language(s): English

Entry (No.): 656
Subject: Cities – Crimea
Author: Tymm, W. (Timm, V.)
Author: Gross, F.
Author: Miunster, A. (lith.)
Title: Vid´ goroda Simferopolia, s vostochnoi storony, gde izbolil´ byt´ ego Velichestvo gosudar´ imperator´, 27go Oktabria 1855 goda.
Year: 1856
Dimensions: 33.0 x 26.5 cm
Phys. Desc.: Lithographed, Colored
Language(s): Russian

Entry (No.): 657
Subject: Cities – Crimea
Author: Zuev, Nikolai
Title: Karta // Polustrova Tavricheskago Kryma // s chertezhnom okrestnostei Sevastopolia sostavlena N. Zuevym. S. Peterburg 1854.
Year: 1854
Dimensions: 55.9 x 44.5 (51.5 x 40.5) cm
Phys. Desc.: Lithographed, Colored
Language(s): Russian

Entry (No.): 658
Subject: Cities – Crimea
Author: No Author
Title: [No Title] Crimea
Year: ca. 1850s
Dimensions: 24.3 x 14.6 cm
Phys. Desc.: Printed, Colored
Language(s): English

Entry (No.): 659
Subject: Cities – Ukraine
Author: Cazat, [?]
Title: Vue de Kief.
Year: 19th c.
Dimensions: 16.0 x 9.4 cm
Phys. Desc.: Printed
Language(s): French

Entry (No.): 660
Subject: Cities – Ukraine
Author: Gusman, [?]
Title: La Laure de Kief.
Year: 18th–19th c.?
Dimensions: 17.0 x 13.3 cm
Phys. Desc.: Printed
Language(s): French

Entry (No.): 661
Subject: Cities – Ukraine
Author: Habitschek, F.
Author: Rohbock, L.
Title: Käsmark und die Karpaten
Year: 19th c.
Dimensions: 30.5 x 23.6 (18.0 x 12.0) cm
Phys. Desc.: Lithographed, Uncolored
Language(s): German

Entry (No.): 662
Subject: Cities – Ukraine
Author: Kernot, J. H. (sculptor)
Title: Citadelle et Monastère de Kieff.
Year: 19th c.
Dimensions: 27.7 x 17.2 (16.3 x 12.0) cm
Phys. Desc.: Lithographed, Uncolored
Language(s): French

Entry (No.): 663
Subject: Cities – Ukraine
Author: Kurinnyj, Petrus
Author: Povstenko, Alexius
Title: Istorychni pliany Kieva. Tabulae Urbis Kioviae Historica.
Year: 1947
Dimensions: 30.5 x 43.0 cm
Phys. Desc.: Printed, Uncolored
Language(s): English, French, German, Latin, Ukrainian

Entry (No.): 664
Subject: Cities – Ukraine
Author: Orda, Napoleon
Title: Janów nad rzeką Śnidową.
Year: 19th c.
Dimensions: 45.0 x 31.5 (28.8 x 19.0) cm
Phys. Desc.: Lithographed, Colored
Language(s): Polish

Entry (No.): 665
Subject: Cities – Ukraine
Author: Tymm, W. (Timm, V.)
Title: Svatiia Kievskiia Pesheri.
Year: Unknown
Dimensions: 34.7 x 52.1 cm (four inset pieces: 12.4 x 17.0

cm)
Phys. Desc.: Printed
Language(s): Russian

Entry (No.): 666
Subject: Cities – Ukraine
Author: Usetti, Michele Luigi (?)
Title: Kaminietz.
Year: Unknown
Dimensions: 22.2 x 7.0 cm
Phys. Desc.: Engraved, Uncolored
Language(s): Italian

Entry (No.): 667
Subject: Cities – Ukraine
Author: No Author
Title: A Cossak's Return - 1768.
Year: 1768
Dimensions: 26.0 x 18.0 (21.0 x 14.0) cm
Phys. Desc.: Engraved, Uncolored
Language(s): English

Entry (No.): 668
Subject: Cities – Ukraine
Author: No Author
Title: Die Heiligen = Gruft in Kiew.
Year: 19th c.
Dimensions: 26.7 x 19.6 (17.0 x 10.9) cm
Phys. Desc.: Lithographed, Uncolored
Language(s): German

Entry (No.): 669
Subject: Cities – Ukraine
Author: No Author
Title: Kiew.
Year: 19th c.
Dimensions: 21.8 x 17.0 (16.5 x 10.5) cm
Phys. Desc.: Lithographed, Uncolored
Language(s): German
Figure 71

Entry (No.): 670
Subject: Cities – Ukraine
Author: No Author
Title: Lemberg. DCXV
Year: ca. 1850
Dimensions: 28.5 x 19.2 (16.5 x 10.8) cm
Phys. Desc.: Engraved, Uncolored
Language(s): German
Figure 72

Entry (No.): 671
Subject: Cities – Ukraine
Author: No Author
Title: Poltava
Year: ca. 1950
Dimensions: 41.8 x 29.2 cm
Phys. Desc.: Printed
Language(s): English

Entry (No.): 672
Subject: Cities – Ukraine
Author: No Author
Title: The Pechersk Monastery at Kiev.
Year: 20th c.
Dimensions: 14.3 x 11.0 cm
Phys. Desc.: Printed
Language(s): English

Entry (No.): 673
Subject: Cities – Ukraine
Author: No Author
Title: Zaporizhya.
Year: 1950s
Dimensions: 41.8 x 29.2
Phys. Desc.: Printed
Language(s): Ukrainian

Entry (No.): 674
Subject: Cities – Ukraine
Author: No Author
Title: Kyïv.
Year: 1956
Dimensions: 44.4 x 17.9 cm
Phys. Desc.: Printed, Colored
Language(s): Ukrainian

Entry (No.): 675
Subject: Cities – Ukraine
Author: No Author
Title: Skhematychnyi plan Kyieva // naibil′sh tsikavi vyznachni mistsia i pam'iatnyky.
Year: 1958
Dimensions: 55.0 x 40.5 cm
Phys. Desc.: Printed, Colored
Language(s): Ukrainian
Figure 92

Entry (No.): 676
Subject: Cities – Ukraine
Author: No Author
Title: Kyïv: kartoskhema z trasamy mis′koho transportu.
Year: 1967
Dimensions: 78.8 x 61.0 cm
Phys. Desc.: Printed, Colored
Language(s): Ukrainian

Entry (No.): 677
Subject: Cities – Ukraine
Author: No Author
Title: Kyïv: iliustrovana skhema.
Year: 1973
Dimensions: 88.6 x 61.0 cm
Phys. Desc.: Printed, Colored
Language(s): Ukrainian

Entry (No.): 678
Subject: Cities – Ukraine
Author: No Author
Title: No Title [View of a Village Street]

Year: Unknown
Dimensions: 26.0 x 18.0 (17.5 x 11.5) cm
Phys. Desc.: Engraved, Uncolored
Language(s): None

Entry (No.): 679
Subject: Copies of Old Maps
Author: Agnese, Battista (1514–64)
Title: Moskovik
Year: 1525
Dimensions: 34.8 x 26.6 cm
Phys. Desc.: Copy
Language(s): Latin

Entry (No.): 680
Subject: Copies of Old Maps
Author: Barentsz (Barendszoon, Barentszoon, Bernard, Bernardi, Bernardus), Willem (1550–97)
Title: Map Showing Barents' Third Voyage. J.I. Pontani Rerum et Urbis Amstelodamensium Historia. Amst. 1661.
Year: 1882/ca. 1598
Dimensions: 33.4 x 28.3 cm
Phys. Desc.: Reprint
Language(s): Latin, English

Entry (No.): 681
Subject: Copies of Old Maps
Author: Beneventanus, Marco
Title: Tabvla Moderna Polonie Vngarie Boemie Germanie Rvssie Lithuanie. XIX.
Year: 1507
Dimensions: 45.7 x 41.9 cm
Phys. Desc.: Copy
Language(s): Latin

Entry (No.): 682
Subject: Copies of Old Maps
Author: Beneventanus, Marco
Title: Tabvla Moderna Polonie Vngarie Boemie Germanie Rvssie Lithuanie. XIX.
Year: 1507
Dimensions: 42.4 x 32.0 cm
Phys. Desc.: Negative
Language(s): Latin

Entry (No.): 683
Subject: Copies of Old Maps
Author: Doni, Nicolai
Title: Map of North Europe from Nicolai Doni's Edition of Ptolemaei Cosmographia. ULM. 1482.
Year: 1482/ca. 1482
Dimensions: 58.8 x 36.4 cm
Phys. Desc.: Reprint
Language(s): English

Entry (No.): 684
Subject: Copies of Old Maps
Author: Hereford Mappa Mundi
Author: Richard of Haldingham
Title: [Hereford Mappa Mundi]
Year: ca. 1280
Dimensions: 28.3 x 26.1 cm
Phys. Desc.: Copy
Language(s): Latin

Entry (No.): 685
Subject: Copies of Old Maps
Author: Houter, Johan (1498–1549)
Title: [Sarmatia]
Year: ca. 15th–16th c.
Dimensions: 18.6 x 14.6 cm
Phys. Desc.: Negative
Language(s): Latin

Entry (No.): 686
Subject: Copies of Old Maps
Author: Massa, Isaac (1586–1643)
Title: Russian Map of the North Polar Sea from the Beginning of the 17th Century, Published in Holland in 1612 by Isaac Massa.
Year: 1882/1612
Dimensions: 48.7 x 22.3 cm
Phys. Desc.: Reprint
Language(s): English

Entry (No.): 687
Subject: Copies of Old Maps
Author: Reicherstorffer, Georgius de
Title: Moldaviae finitimarusquae regionaum typus. Vienna, 1541.
Year: 1541
Dimensions: 38.2 x 29.3 cm
Phys. Desc.: Copy
Language(s): Latin

Entry (No.): 688
Subject: Copies of Old Maps
Author: Reicherstorffer, Georgius de
Title: Moldaviae finitimarusquae regionaum typus. Vienna, 1541.
Year: 1541
Dimensions: 40.3 x 36.7 cm
Phys. Desc.: Copy
Language(s): Latin

Entry (No.): 689
Subject: Copies of Old Maps
Author: Reicherstorffer, Georgius de
Title: Magni Ducatus Lithvaniae Livoniae et Moscoviae Descriptio.
Year: 16th c.
Dimensions: 45.7 x 41.6 cm
Phys. Desc.: Copy
Language(s): Latin

Entry (No.): 690
Subject: Copies of Old Maps
Author: Reicherstorffer, Georgius de
Title: Magni Ducatus Lithvaniae Livoniae et Moscoviae Descriptio.

Year: 16th c.
Dimensions: 41.4 x 34.5 cm
Phys. Desc.: Negative
Language(s): Latin

Entry (No.): 691
Subject: Copies of Old Maps
Author: Solinus Gaius Julius (worked 3rd c. ad)
Title: Typus Graeciae.
Year: 1538
Dimensions: 37.9 x 29.3 cm
Phys. Desc.: Copy
Language(s): Latin

Entry (No.): 692
Subject: Copies of Old Maps
Author: Solinus Gaius Julius (worked 3rd c. ad)
Title: Typus Graeciae.
Year: 1538
Dimensions: 29.3 x 26.9 cm
Phys. Desc.: Negative
Language(s): Latin

Entry (No.): 693
Subject: Copies of Old Maps
Author: Waldseemüller (Hylacomilus, Hylacomylus, Ilacomilus, Walzenmuller), Martin (Martinus), (ca. 1475–before 1522)
Title: Tabvla Moderna Evr sive Hvngarie Polonie Rvssie Prvssie et Valachie.
Year: 1513
Dimensions: 45.7 x 41.9 cm
Phys. Desc.: Copy
Language(s): Latin

Entry (No.): 694
Subject: Copies of Old Maps
Author: Waldseemüller (Hylacomilus, Hylacomylus, Ilacomilus, Walzenmuller), Martin (Martinus), (ca. 1475–before 1522)
Title: Tabvla Moderna Evr sive Hvngarie Polonie Rvssie Prvssie et Valachie.
Year: 1513
Dimensions: 42.4 x 32.0 cm
Phys. Desc.: Copy
Language(s): Latin

Entry (No.): 695
Subject: Copies of Old Maps
Author: Ziegler, Jakob Melchior (1801–83)
Title: Map of the North, from Jakob Ziegler's Schondia, Strasburg, 1532.
Year: 1882/ca. 1532
Dimensions: 34.6 x 28.0 cm
Phys. Desc.: Reprint
Language(s): English

Entry (No.): 696
Subject: Copies of Old Maps
Author: No Author
Title: Map of North Europe from Olai Magni Historia de Gentium Septentrionalium Varüs Conditionibus, Basel, 1567.
Year: 1882/ca. 1567
Dimensions: 55.2 x 42.3 cm
Phys. Desc.: Reprint
Language(s): English

Entry (No.): 697
Subject: Copies and Negatives
Author: No Author
Title: [Russian Communications Map, ca. 1685.] Bibliothèque du Départ. Hydrographique de la Marine, Paris.
Year: 1685
Dimensions: 117 x 104 cm
Phys. Desc.: Manuscript Map
Language(s): French

Entry (No.): 698
Subject: Copies and Negatives
Author: No Author
Title: Azovs´ke More.
Year: 1935
Dimensions: 61.6 x 46.8 cm
Phys. Desc.: Copy
Language(s): Ukrainian

Entry (No.): 699
Subject: Copies and Negatives
Author: No Author
Title: Azovs´ke More.
Year: 1935
Dimensions: 56.8 x 36.7 cm
Phys. Desc.: Negative
Language(s): Ukrainian

Entry (No.): 700
Subject: Copies and Negatives
Author: No Author
Title: Karta URSR z rozmishchenniam lisiv ta pidpryemstv mebl´ovoï ta derevoobrobnoï promyslovosti za dannymy 1939 roku.
Year: 1939
Dimensions: 62.0 x 46.1 cm
Phys. Desc.: Copy and Negative
Language(s): Ukrainian

Entry (No.): 701
Subject: Copies and Negatives
Author: No Author
Title: Karta URSR z rozmishchenniam lisiv ta pidpryemstv mebl´ovoï ta derevoobrobnoï promyslovosti za dannymy 1939 roku.
Year: 1939
Dimensions: 58.7 x 42.9 cm
Phys. Desc.: Negative
Language(s): Ukrainian

Entry (No.): 702
Subject: Copies and Negatives

Author: No Author
Title: Karta URSR z rozmishchenniam lisiv ta pidpryemstv mebl΄ovoï ta derevoobrobnoï promyslovosti za dannymy 1939 roku.
Year: 1939
Dimensions: 58.6 x 41.2 cm
Phys. Desc.: Negative
Language(s): Ukrainian

Entry (No.): 703
Subject: Copies and Negatives
Author: No Author
Title: Karta URSR z rozmishchenniam lisiv ta pidpryemstv mebl΄ovoï ta derevoobrobnoï promyslovosti za dannymy 1939 roku.
Year: 1939
Dimensions: 62.6 x 41.9 cm
Phys. Desc.: Negative
Language(s): Ukrainian

Entry (No.): 704
Subject: Copies and Negatives
Author: No Author
Title: Wald
Year: 20th c.
Dimensions: 46.2 x 45.9 cm
Phys. Desc.: Copy
Language(s): German

Entry (No.): 705
Subject: Copies and Negatives
Author: No Author
Title: Wald
Year: 20th c.
Dimensions: 45.0 x 39.8 cm
Phys. Desc.: Negative
Language(s): German

Entry (No.): 706
Subject: Europe
Author: Bonné, Rigobert (1727–95)
Title: Carte Générale // de L'Europe. // Par M. Bonne, Ingénr. Hydrograp. // de la Marine. // Nov. 1780.// Avec Priv. du Roi.
Year: 1780
Dimensions: 24.1 x 16.2 cm
Phys. Desc.: Reprint, Uncolored
Language(s): French

Entry (No.): 707
Subject: Europe
Author: Darley, James M.
Author: Grosvenor, Gilbert
Title: Theater of War in Europe // Africa and Western Asia. A. Hoen & Co., Baltimore MD. USA.
Year: 1942
Dimensions: 63.5 x 75.0 cm
Phys. Desc.: Printed, Lithographed
Language(s): English

Entry (No.): 708
Subject: Europe
Author: Darley, James M.
Title: Central Europe // Including the Balkan States... Litho. By A. Hoen & Co., Balt. MD. USA.
Year: 1951
Dimensions: 70.0 x 94.0 cm
Phys. Desc.: Lithographed, Colored
Language(s): English

Entry (No.): 709
Subject: Europe
Author: Gover (Gower), Edward
Title: Europe from the period of the French Revolution, to the Abdication of Napoleon, or the General Peace (1787–1815). By Edw. Gower, Ed. of Univ. Atlas of the Middle Ages, &c., 1853.
Year: 1853
Dimensions: 37.9 x 25.0 (32.8 x 25.0) cm
Phys. Desc.: Engraved, Colored
Language(s): English

Entry (No.): 710
Subject: Europe
Author: Gover (Gower), Edward
Title: Europe from the Xth to the close of the XIIth Century, illustrating the period of the Crusades. London, Varty & Owen.
Year: 1853
Dimensions: 38.5 x 30.0 (32.6 x 25.0) cm
Phys. Desc.: Engraved, Colored
Language(s): English

Entry (No.): 711
Subject: Europe
Author: Gover (Gower), Edward
Title: Europe in the XVIII. Century, and before the Outbreak of the French Revolution. By Edw. Gower, ed., Univ. Atlas of Middle Ages. 1853.
Year: 1853
Dimensions: 38.4 x 30.7 (32.5 x 24.5) cm
Phys. Desc.: Engraved, Colored
Language(s): English

Entry (No.): 712
Subject: Europe
Author: Grosvenor, Gilbert
Title: Europe & the Mediterranean. Washington, A. Hoen & Co. 1938.
Year: 1938
Dimensions: 95.0 x 82.5 cm
Phys. Desc.: Printed, Lithographed, Colored
Language(s): English

Entry (No.): 713
Subject: Europe
Author: Mentelle, Edme (1730–1815)
Title: L'Europe en 1800 Par Mentelle.
Year: 1800
Dimensions: 22.7 x 17.0 cm

Phys. Desc.: Engraved
Language(s): French

Entry (No.): 714
Subject: Europe
Author: No Author
Title: Europe.
Year: 19th c.
Dimensions: 33.8 x 27.0 (28.8 x 24.3) cm
Phys. Desc.: Engraved, Colored
Language(s): English

Entry (No.): 715
Subject: Europe
Author: No Author
Title: Europe
Year: 19th c.
Dimensions: 36.0 x 33.5 cm
Phys. Desc.: Engraved, Colored
Language(s): English

Entry (No.): 716
Subject: Europe
Author: No Author
Title: Krest´ianskoe dvizhenie v Bessarabii 1861–1872.
Year: late 19th c.
Dimensions: 44.0 x 52.0 cm
Phys. Desc.: Uncolored
Language(s): Russian

Entry (No.): 717
Subject: Europe
Author: No Author
Title: Nouvelle Carte de l'Europe Centrale.
Year: 1918
Dimensions: 103.5 x 87.9 cm
Phys. Desc.: Printed
Language(s): French

Entry (No.): 718
Subject: Europe
Author: No Author
Title: Classical Lands of the Meditteranean.
Year: 1948
Dimensions: 81.3 x 55.9 cm
Phys. Desc.: Colored
Language(s): English

Entry (No.): 719
Subject: Poland
Author: Guthrie, William (1708–1800)
Author: Dilly, C.
Title: Poland // Shewing the Claims of Russia, Prussia + Austria // from the best // Authorities. Published as the Act directs January 1st 1785 by C. Dilly + G. Robinson London.
Year: 1785
Dimensions: 46.5 x 37.5 (37.5 x 32.6) cm
Phys. Desc.: Engraved, Colored
Language(s): English

Entry (No.): 720
Subject: Poland
Author: Guthrie, William (1708–1800)
Author: Dilly, C.
Title: Poland // Shewing the Claims of Russia, Prussia + Austria // from the best // Authorities. Published as the Act directs January 1st 1785 by C. Dilly + G. Robinson London.
Year: 1785
Dimensions: 45.0 x 37.0 (37.3 x 32.6) cm
Phys. Desc.: Engraved, Colored
Language(s): English

Entry (No.): 721
Subject: Poland
Author: Kortyko, Stefan
Title: Pierwsza poglądowa Mapa Polski z podziałem na województwa i powiaty z planami miast wojewódzkich oraz z przeglądem przemysłu górniczo-hutniczego. Polish Book Imp. Co. Inc. 38 Union Sq. New York, N.Y. Główny Skład i Ekspedycja Spółka Nakł. "Odrodzenie." Lwów ul. Pilsudskiego 16. 1930.
Year: 1830
Dimensions: 99.7 x 76.6 cm
Phys. Desc.: Book
Language(s): Polish

Entry (No.): 722
Subject: Poland
Author: No Author
Title: A New Map of Poland from the latest Obser.ns.
Year: ca. 1730–50
Dimensions: 20.6 x 19.2 (19.5 x 15.0) cm
Phys. Desc.: Engraved, Uncolored
Language(s): English

Entry (No.): 723
Subject: Poland
Author: No Author
Title: Karte von Polen herausgegeben vom "Völkischen Beobachter."
Year: ca. 1930
Dimensions: 61.5 x 43.2 cm
Phys. Desc.: Printed, Colored
Language(s): German

Entry (No.): 724
Subject: Poland
Author: No Author
Title: Polska Rzeczpospolita Ludowa Mapa administracyjna. Państwowe Przedsiębiorstwa Wydawnictw Kartograficznych. Warszawa 1957.
Year: 1957
Dimensions: 89.2 x 75.7 cm
Phys. Desc.: Printed
Language(s): Polish

Entry (No.): 725
Subject: Russia (18th, 19th, 20th c.)
Author: Gray & Son
Title: Russia // in Europe, // South Part.

Year: ca. 1820s
Dimensions: 26.5 x 21.2 (24.5 x 20.0) cm
Phys. Desc.: Engraved, Uncolored
Language(s): English

Entry (No.): 726
Subject: Russia (18th, 19th, 20th c.)
Author: Illman, Thomas
Title: Russia in Europe. Entered according to act of Congress by Tho.s. Illman in the Clerks Office of the district court for the Southern District of New York.
Year: 1834
Dimensions: 31.0 x 37.8 (27.5 x 31.2) cm
Phys. Desc.: Engraved, Colored
Language(s): English

Entry (No.): 727
Subject: Russia (18th, 19th, 20th c.)
Author: Juigne, R.
Title: The Russian Empire, with its gradual acquisitions traced + explained.
Year: 1818
Dimensions: 72.2 x 50.0 (67.5 x 46.6) cm
Phys. Desc.: Engraved, Colored
Language(s): English

Entry (No.): 728
Subject: Russia (18th, 19th, 20th c.)
Author: Mikhailov, I. N. (Mykhajlow, Y. N.)
Title: Karta Sibirii Turkestana s ukazaniem polozheniiam granits Rossiiskoi Imperii. Moskva 1912 g.
Year: 1912
Dimensions: 39.9 x 29.5 cm
Phys. Desc.: Lithographed, Colored
Language(s): Russian

Entry (No.): 729
Subject: Russia (18th, 19th, 20th c.)
Author: Mitchell, Samuel Augustus Senior (1792–1868)
Title: Russia in Europe, // Sweden // and // Norway.
Year: 1860
Dimensions: 38.5 x 31.3 (34.5 x 27.0) cm
Phys. Desc.: Lithographed, Colored
Language(s): English

Entry (No.): 730
Subject: Russia (18th, 19th, 20th c.)
Author: Nolin, Jean-Baptiste II (1686–1762)
Title: Carte Générale de L'Empire des Russes et de la Tartarie Universelle. Suivant les Nouvelles Observations par J.B. Nolin Géographe du Roi.
Year: 1720
Dimensions: 32.0 x 22.8 (265.7 x 20.5) cm
Phys. Desc.: Engraved, Colored
Language(s): French

Entry (No.): 731
Subject: Russia (18th, 19th, 20th c.)
Author: Nolin, Jean-Baptiste II (1686–1762)
Title: La Russie D'Europe. Divisée par gouvernemens. Assujetties aux Nouvelles Observations.
Year: 1720
Dimensions: 32.0 x 22.8 (26.1 x 20.0) cm
Phys. Desc.: Engraved, Colored Outline
Language(s): French

Entry (No.): 732
Subject: Russia (18th, 19th, 20th c.)
Author: Probst, Johann Michael the younger (1757–1809)
Title: Nouvelle Carte Generale d'Empire de Russie et de la Tartarie grande et petite en Europe et Asie. Nova Mappa generalis Totius Orbis amplissimae Russorum Monarchiae Opera et Sumptibus Johan: Michael Probst Geogr: Augustae Vind: Ao MDCCXXCIII.
Year: 1783
Dimensions: 65.6 x 54.5 (53.3 x 46.7) cm
Phys. Desc.: Engraved, Colored
Language(s): French

Entry (No.): 733
Subject: Russia (18th, 19th, 20th c.)
Author: Protas, F. N.
Title: Fizicheskaia karta Evropeiskoi chasti Soiuza SSR.
Year: 1947
Dimensions: 24.8 x 31.4 cm
Phys. Desc.: Printed, Colored
Language(s): Russian

Entry (No.): 734
Subject: Russia (18th, 19th, 20th c.)
Author: Protas, F. N.
Title: Fizicheskaia karta Soiuza SSR.
Year: 1947
Dimensions: 42.0 x 26.0 cm
Phys. Desc.: Printed, Colored
Language(s): Russian

Entry (No.): 735
Subject: Russia (18th, 19th, 20th c.)
Author: Protas, F. N.
Title: Politisko-administrativnaia karta Soiuza SSR.
Year: 1947
Dimensions: 43.1 x 28.8 cm
Phys. Desc.: Printed, Colored
Language(s): Russian

Entry (No.): 736
Subject: Russia (18th, 19th, 20th c.)
Author: Renouard, Paul
Title: Tableau Géographique Statistique &c. de L'Empire Russe Actuel en Europe et en Asie avec ses Acquisitions graduelles et caractérisées. Gravé par Moisy. Ecrit par Bénard.
Year: 1828
Dimensions: 69.0 x 51.6 (52.1 x 31.0) cm
Phys. Desc.: Printed, Colored
Language(s): French

Entry (No.): 737
Subject: Russia (18th, 19th, 20th c.)
Author: Stieler, Adolf (1775–1836)

Title: Des Russichen Reichs: Europäischer Theil.
Year: ca. 1816–31
Dimensions: 35.2 x 43.0 (29.0 x 37.3) cm
Phys. Desc.: Engraved, Colored Outline
Language(s): German

Entry (No.): 738
Subject: Russia (18th, 19th, 20th c.)
Author: Swanston, George H.
Title: Russian Empire according to Kiepert + Others.
Year: 1856
Dimensions: 58.8 x 46.0 (50.5 x 41.0) cm
Phys. Desc.: Engraved, Colored
Language(s): English

Entry (No.): 739
Subject: Russia (18th, 19th, 20th c.)
Author: Sytin, I. D.
Title: Etnograficheskaia karta Rossiiskoi Imperii.
Year: 1912
Dimensions: 29.3 x 18.4 cm
Phys. Desc.: Lithographed, Colored
Language(s): Russian
Figure 76

Entry (No.): 740
Subject: Russia (18th, 19th, 20th c.)
Author: No Author
Title: Carte Nouvelle de Tout L'Empire de la Grande Russie: Dans L'Estat ou il s'est trouvé à la mort de Pierre le Grand – dressée sur les observations toutes nouvelles et dediée a l'Immortelle Memoire de ce Grand Monarque. Avec Privilege, à Amsterdam.
Year: 1725
Dimensions: 76.8 x 53.5 (66.3 x 47.5) cm
Phys. Desc.: Engraved, Colored
Language(s): French

Entry (No.): 741
Subject: Russia (18th, 19th, 20th c.)
Author: No Author
Title: Chart of the Course of the Don + the Volga through the Region of the Kosaks.
Year: 1786
Dimensions: 25.0 x 20.5 (22.0 x 19.0) cm
Phys. Desc.: Printed, Uncolored
Language(s): English

Entry (No.): 742
Subject: Russia (18th, 19th, 20th c.)
Author: No Author
Title: Provincia Ecclesiastica di Mohilow Nell' Imperio Russo.
Year: 1870
Dimensions: 67.5 x 52.5 (59.0 x 46.5) cm
Phys. Desc.: Engraved, Colored
Language(s): Italian

Entry (No.): 743
Subject: Russia (18th, 19th, 20th c.)
Author: No Author
Title: Atlas kartogramm i diagramm k svodu statisticheskikh svedenii po sel'skomu khoziastvu Rossii k kontsu XIX veka. Vypusk III. S. Peterburg... 1905.
Year: 1905
Dimensions: 36.7 x 23.3 cm
Phys. Desc.: Book/Atlas
Language(s): Russian

Entry (No.): 744
Subject: Russia (18th, 19th, 20th c.)
Author: No Author
Title: Karta Evropeiskoi Rossii: Politicheskaia karta.
Year: ca. 1910
Dimensions: 24.7 x 42.8 cm
Phys. Desc.: Lithographed, Colored
Language(s): Russian

Entry (No.): 745
Subject: Russia (18th, 19th, 20th c.)
Author: No Author
Title: Fizicheskaia karta Evropeiskoi Rossii.
Year: 1912
Dimensions: 22.2 x 24.8 cm
Phys. Desc.: Lithographed, Colored
Language(s): Russian

Entry (No.): 746
Subject: Russia (18th, 19th, 20th c.)
Author: No Author
Title: Map of Russia
Year: 1918
Dimensions: 34.0 x 50.0 cm
Phys. Desc.: Lithographed, Colored
Language(s): English

Entry (No.): 747
Subject: Russia (18th, 19th, 20th c.)
Author: No Author
Title: Union of Soviet Socialist Republics in Europe.
Year: ca. 1920s–30s
Dimensions: 33.8 x 50.0 cm
Phys. Desc.: Lithographed, Colored
Language(s): English
Figure 82

Entry (No.): 748
Subject: Russia (18th, 19th, 20th c.)
Author: No Author
Title: Russia in Europe and its Border States.
Year: 1942
Dimensions: 51.0 x 60.5 cm
Phys. Desc.: Colored
Language(s): English
Figure 85

Entry (No.): 749
Subject: Russia (18th, 19th, 20th c.)
Author: No Author
Title: Union of Soviet Socialist Republics: International

boundaries according to Russian treaties and claims as of October 1, 1944. Boundaries of January 1, 1938 are shown in red. Compiled and Drawn in the Cartographic Section of the National Geographic Society for the National Geographic Magazine. Gilbert Grosvenor, Editor. Washington, December 1944.
Year: 1944
Dimensions: 101.4 x 63.3 cm
Phys. Desc.: Printed, Colored
Language(s): English

Entry (No.): 750
Subject: Russia (18th, 19th, 20th c.)
Author: No Author
Title: Karta narodov SSSR.
Year: 1953
Dimensions: 93.0 x 60.0 cm
Phys. Desc.: Colored
Language(s): Russian

Entry (No.): 751
Subject: Russia (18th, 19th, 20th c.)
Author: No Author
Title: Evropeiskaia chasti SSSR.
Year: 1954
Dimensions: 77.5 x 100.5 cm
Phys. Desc.: Colored
Language(s): Russian

Entry (No.): 752
Subject: Russia (18th, 19th, 20th c.)
Author: No Author
Title: Map of Location of the Slave and Concentration Camps in the Soviet 'Union.'
Year: 1959
Dimensions: 26.0 x 17.3 cm
Phys. Desc.: Uncolored
Language(s): English
Figure 93

Entry (No.): 753
Subject: Russia (18th, 19th, 20th c.)
Author: No Author
Title: Soiuz Sovestskikh Sotsialisticheskikh Respublik: Politiko-administrativnaia karta.
Year: ca. 1960
Dimensions: 92.5 x 59.0 cm
Phys. Desc.: Colored
Language(s): Russian

Entry (No.): 754
Subject: Russia (18th, 19th, 20th c.)
Author: No Author
Title: Amurskaia oblast´.
Year: 1966
Dimensions: 82.0 x 82.5 cm
Phys. Desc.: Colored
Language(s): Russian

Entry (No.): 755
Subject: Sevastopol
Author: Beggrov (Baggrov), A.
Title: Plan Sevastopolia s upravleniiami ot reki Bel´bek do Balaklavy, i s oznacheniem vsekh osadnykh raspolozhenii. Izdanie A. Beggrov s S. Peterburgi. 1854.
Year: 1854
Dimensions: 49.5 x 41.0 (41.8 x 35.5) cm
Phys. Desc.: Engraved, Colored
Language(s): Russian

Entry (No.): 756
Subject: Sevastopol
Author: Bibby, H.
Title: Sebastopol.
Year: ca. 1850s
Dimensions: 27.0 x 18.0 cm
Phys. Desc.: Printed
Language(s): English

Entry (No.): 757
Subject: Sevastopol
Author: Bodenehr, Gabriel (1664–1758)
Title: Mongats oder Munkatsch.
Year: 1720
Dimensions: 32.5 x 19.5 (21.7 x 13.7) cm
Phys. Desc.: Engraved, Uncolored
Language(s): German

Entry (No.): 758
Subject: Sevastopol
Author: Fer, Nicolas de (1646–1720)
Title: Plan de Narva Assiegé par [?] Moscovites, qui furent Attaqués et deffaits dans leurs lignes par Charles XII. Roy de Suede. Comandant son Armée emperfonne a l'age de dix huit an le 30 Novembre 1705. A Paris. Chez le St. De Fer, dan l'Isle du Palais a la Sphere Royale. Avec Priv. du Roy. 170[2?] Gravé par van Loon.
Year: ca. 1710s
Dimensions: 39.2 x 26.0 (34.0 x 22.7) cm
Phys. Desc.: Engraved
Language(s): French

Entry (No.): 759
Subject: Sevastopol
Author: Delafield, Major
Title: Sebastopol and its Environs showing the Russian Defences and the Approches and Other Works of Attack of the Allied Armies during the Siege of 1854–55 from the Report of Gen. Niel Chief of the French Engrs. Prepared to accompany the Report of Major Delafield on the art of War in Europe in 1854, 55, & 56.
Year: 1856
Dimensions: 105.6 x 74.0 (97.0 x 63.2) cm
Phys. Desc.: Lithographed
Language(s): English
Figure 67

Entry (No.): 760
Subject: Sevastopol
Author: Delafield, Major

Title: Theatre of War in the Crimea in 1854–55. From General Niel's journal of the Siege of Sebastopol. Prepared to accompany the report of Major Delafield to the Hon. Jefsn. Davis Secy. Of War on the Art of War in Europe in 1854 & 1855.
Year: 1854–55
Dimensions: 93.0 x 73.0 (88.0 x 63.5) cm
Phys. Desc.: Lithographed, Uncolored
Language(s): English

Entry (No.): 761
Subject: Sevastopol
Author: Delafield, Major
Title: Map // Shewing the // Military Roads // & Countries // between // Odessa & Perekop.
Year: 1857
Dimensions: 33.3 x 17.7 cm
Phys. Desc.: Engraved, Uncolored
Language(s): English

Entry (No.): 762
Subject: Sevastopol
Author: Dere, E.
Author: Bibby, H.
Title: Balaklava.
Year: ca. 1850s
Dimensions: 27.0 x 18.0 cm
Phys. Desc.: Printed, Uncolored
Language(s): English

Entry (No.): 763
Subject: Sevastopol
Author: Hind, R.
Author: Greatbach, R.
Title: Battle of Kars, Repulse of the Russians, Septr. 29th 1865.
Year: 1865
Dimensions: 27.0 x 18.0 (18.5 x 13.0) cm
Phys. Desc.: Lithographed, Uncolored
Language(s): English

Entry (No.): 764
Subject: Sevastopol
Author: Kostiuk, V.
Title: Panorama Oborona Sevastopolia
Year: 1964
Dimensions: 22.0 x 14.4 cm
Phys. Desc.: Book
Language(s): Russian

Entry (No.): 765
Subject: Sevastopol
Author: Merian, Matthäus (Mattheiu, Matheus, Matthew), the elder (1593–1650)
Author: Merian, Matthäus (Mattheiu, Matheus, Matthew), the younger (1621–87)
Title: Eigentliche abbildüng der Belägerte Statt Stawischze / welche dürch dess Königs in Polen, Generalen h. Stephan von Czarnez und [?ictoczie] Czarnezki / woiwoden zŭ Kiov, nach einer viertel Fährigen belägerüng zŭ der über-gab gezwŭngen worden, so geschehen den 8. Octobris. Anno 1664.
Year: ca. 1664–67
Dimensions: 39.0 x 33.5 (37.7 x 30.3) cm
Phys. Desc.: Engraved
Language(s): German

Entry (No.): 766
Subject: Sevastopol
Author: Prokhorow, Lt.
Title: Plan Sevastopolia i ego okrestnostei do Balaklavy s pokazaniem nepriatel´skikh osadnykh raspolozhenii 1855.
Year: 1855
Dimensions: 64.5 x 54.7 (59.0 x 49.0) cm
Phys. Desc.: Engraved, Colored
Language(s): Russian

Entry (No.): 767
Subject: Sevastopol
Author: Rapkin, John
Author: Winkles (Winckles), Henry
Title: The Crimea.
Year: ca. 1850s
Dimensions: 35.0 x 27.0 (33.0 x 24.0) cm
Phys. Desc.: Lithographed, Colored
Language(s): English

Entry (No.): 768
Subject: Sevastopol
Author: Rapkin, John
Title: Siege of Sebastopol
Year: 1854
Dimensions: 34.5 x 27.0 (33.0 x 24.0) cm
Phys. Desc.: Lithographed, Colored
Language(s): English

Entry (No.): 769
Subject: Sevastopol
Author: Rolffsen, I. I.
Title: Nouvelle Carte de la Crimée // Selon le Defsein des Ingenieurs Russien // dans la quelle sont marqués les marches // des Troupes Russiennes // dans l'an 1736 sous le Marechall Munnich // et dans l'an 1737 sous le Marechall Lascÿ
Year: 1730
Dimensions: 54.0 x 36.0 (46.2 x 29.0) cm
Phys. Desc.: Engraved, Uncolored
Language(s): French

Entry (No.): 770
Subject: Sevastopol
Author: Rolffsen, I. I.
Title: Plan de la Ligne de Perekop.
Year: ca. 1760–70
Dimensions: 48.5 x 32.0 (44.4 x 25.0) cm
Phys. Desc.: Engraved, Uncolored
Language(s): French

Entry (No.): 771

Subject: Sevastopol
Author: Rusell, W. H.
Title: Plan of the Heights and Bay of Alma.
Year: ca. 1850
Dimensions: 48.5 x 31.6 (45.2 x 29.0) cm
Phys. Desc.: Engraved, Uncolored
Language(s): English

Entry (No.): 772
Subject: Sevastopol
Author: Rusell, W. H.
Title: Plan of Balaklava Shewing the Cavalry Action of Oct. 25th 1854.
Year: 1854
Dimensions: 47.5 x 26.5 cm
Phys. Desc.: Engraved, Uncolored
Language(s): English

Entry (No.): 773
Subject: Sevastopol
Author: Rusell, W. H.
Title: Plan of the Battle of Inkerman, Novr. 5th 1854.
Year: 1854
Dimensions: 47.4 x 27.3 cm
Phys. Desc.: Engraved, Uncolored
Language(s): English

Entry (No.): 774
Subject: Sevastopol
Author: Rusell, W. H.
Title: Plan of the Fortifications of Sebastopol to Septr. 1st 1854.
Year: 1854
Dimensions: 44.5 x 36.0 (43.3 x 31.8) cm
Phys. Desc.: Engraved, Uncolored
Language(s): English

Entry (No.): 775
Subject: Sevastopol
Author: Rusell, W. H.
Title: Plan of the Battle of the Tchernaya. August 16th 1855. Reduced from the Q.M. General's Plan.
Year: 1855
Dimensions: 47.5 x 35.8 cm
Phys. Desc.: Engraved, Uncolored
Language(s): English

Entry (No.): 776
Subject: Sevastopol
Author: Rusell, W. H.
Title: Plan of the British Camp before Sebastopol. 1855.
Year: 1855
Dimensions: 35.4 x 25.0 (33.2 x 22.7) cm
Phys. Desc.: Engraved, Uncolored
Language(s): English

Entry (No.): 777
Subject: Sevastopol
Author: Rusell, W. H.
Title: Plan of the Defences of Sebastopol with the Lines of Attack of the Allied Armies Previous to the Final Assault on the 8th Septr. 1855.
Year: 1855
Dimensions: 65.5 x 46.8 (62.3 x 43.0) cm
Phys. Desc.: Lithographed, Uncolored
Language(s): English

Entry (No.): 778
Subject: Sevastopol
Author: Rusell, W. H.
Title: The Country around Sebastopol shewing the Positions of the Contending Armies after the Fall of the South Side and before the Peace of 1856.
Year: 1856
Dimensions: 36.0 x 38.8 (38.2 x 42.3) cm
Phys. Desc.: Engraved, Uncolored
Language(s): English

Entry (No.): 779
Subject: Sevastopol
Author: Tardieu, Pierre François (1711–71)
Title: Bender. Assiégé & pris par les Russes en 1770.
Year: 1770
Dimensions: 26.5 x 17.0 (50.5 x 31.8) cm
Phys. Desc.: Engraved, Uncolored
Language(s): French

Entry (No.): 780
Subject: Sevastopol
Author: Tymm, W. (Timm, V.)
Title: Vidy severnoi storony Sevastopolia. Rosov. S natury V. Timm, fevral 1855 goda. S vysochaishago soizvoleniia.
Year: 1855
Dimensions: 35.1 x 26.0 (31.0 x 18.5) cm
Phys. Desc.: Engraved, Colored
Language(s): Russian

Entry (No.): 781
Subject: Sevastopol
Author: Williams, Lt. Col.
Author: Nicholls, H. W.
Author: Greatbach, G.
Title: Bombardement of Odessa // by the English and French. // April 22, 1854 // The Expulsion of the Imperial Mole. From a Sketch by an Officer during the Action.
Year: 1854
Dimensions: 27.0 x 18.0 (19.9 x 11.8) cm
Phys. Desc.: Lithographed, Uncoloed
Language(s): English

Entry (No.): 782
Subject: Sevastopol
Author: Williams, Lt. Col.
Title: Retreat of the Russians from the South Side of Sebastopol, Sept. 8, 1855.
Year: 1855
Dimensions: 27.0 x 18.0 cm
Phys. Desc.: Lithographed, Uncolored
Language(s): English

Entry (No.): 783
Subject: Sevastopol
Author: Winkles (Winckles), Henry
Author: Rapkin, John
Title: Black Sea
Year: ca. 1850s
Dimensions: 34.5 x 27.0 (32.0 x 24.0) cm
Phys. Desc.: Lithographed, Colored
Language(s): English

Entry (No.): 784
Subject: Sevastopol
Author: Wyld, James, Jr. (1812–87)
Title: Bombardement de Sébastopol.
Year: 1854
Dimensions: 63.5 x 47.0 (60.0 x 41.0) cm
Phys. Desc.: Lithographed, Uncolored
Language(s): French

Entry (No.): 785
Subject: Sevastopol
Author: Wyld, James, Jr. (1812–87)
Title: Siege of Sebastopol. The // Environs // of // Sebastopol // with the // Batteries & Approaches // by // James Wyld, Geographer to the Queen & H.R.H. Prince Albert. London.
Year: 1855
Dimensions: 49.5 x 68.5 (47.0 x 66.5) cm
Phys. Desc.: Lithographed, Colored Outline
Language(s): English

Entry (No.): 786
Subject: Sevastopol
Author: Zündt (Cynthius, Zindt, Zündten, Zunthio, Zynder, Zynndt, Zynnthen), Matthias (ca. 1498–1572)
Title: Zygeth. Ware Conterfeit der be[u?]estigung Segt,/ sampt aller zugehörig darinder Graff. V. Serm // Difs 1566. Von dem Türkischen kaiser mit grosser macht belegert würt, Nurmberg. Mathis Zündt.
Year: 1566
Dimensions: 32.5 x 22.0 cm
Phys. Desc.: Engraved, Uncolored
Language(s): German

Entry (No.): 787
Subject: Sevastopol
Author: No Author
Title: Crimea.
Year: ca. 1850s
Dimensions: 24.0 x 20.0 cm
Phys. Desc.: Lithographed, Uncolored
Language(s): English

Entry (No.): 788
Subject: Sevastopol
Author: No Author
Title: Vid i plan Sevastopolia (s okrestnostiami). Map and Plan of Sebastopol.
Year: 1855
Dimensions: 67.2 x 55.5 cm
Phys. Desc.: Engraved, Uncolored
Language(s): Russian, English
Figure 69

Entry (No.): 789
Subject: Siberia – Old Maps Copied
Author: Beiton, Athanasius (Afanasy) von
Title: Yakutsk, Irkutsk
Year: 1710–11
Dimensions: 36.7 x 32.9 cm
Phys. Desc.: Reprint in Book
Language(s): Russian

Entry (No.): 790
Subject: Siberia – Old Maps Copied
Author: Blaeu (family publishing firm)
Title: Regiones Svb Polo Arctico.
Year: 1662
Dimensions: 37.4 x 33.3 cm
Phys. Desc.: Reprint in Book
Language(s): Latin

Entry (No.): 791
Subject: Siberia – Old Maps Copied
Author: Chirikov, Lieutenant Aleksei Ilyich
Title: Siberia
Year: 1746
Dimensions: 25.1 x 22.2 cm
Phys. Desc.: Reprint in Book
Language(s): Russian

Entry (No.): 792
Subject: Siberia – Old Maps Copied
Author: Evreinov (Yevreinov), Ivan Mikhailovich (geodesist; d. 1724)
Title: [Russia].
Year: 1722
Dimensions: 22.2 x 14.6 cm
Phys. Desc.: Reprint in Book
Language(s): Russian

Entry (No.): 793
Subject: Siberia – Old Maps Copied
Author: Homann, Johann Baptist (Baptista) (1664–1724)
Title: [Siberia & Pacific Coast]
Year: 1740
Dimensions: 22.2 x 14.6 cm
Phys. Desc.: Reprint in Book
Language(s): Latin

Entry (No.): 794
Subject: Siberia – Old Maps Copied
Author: Hondius, Jodocus (Josse de Hondt) (1563–1612)
Title: Tartaria.
Year: 1659
Dimensions: 32.4 x 24.8 cm
Phys. Desc.: Reprint in Book
Language(s): Latin

Entry (No.): 795

Subject: Siberia – Old Maps Copied
Author: Ides, Evert (Everard, Everardus) Isbrand (Ysbrants) (ca. 1660–1705)
Author: Witsen, Nicolas (1641–1717)
Title: Siberia
Year: 1948/1699
Dimensions: 22.2 x 14.6 cm
Phys. Desc.: Reprint in Book
Language(s): Latin

Entry (No.): 796
Subject: Siberia – Old Maps Copied
Author: Ides, Evert (Everard, Everardus) Isbrand (Ysbrants) (ca. 1660–1705)
Title: Siberia
Year: 1948/1704
Dimensions: 22.2 x 14.6 cm
Phys. Desc.: Reprint in Book
Language(s): Latin

Entry (No.): 797
Subject: Siberia – Old Maps Copied
Author: Ivanov, [Vasilii?]
Title: [Kamchatka area]
Year: 1742
Dimensions: 40.2 x 24.3 cm
Phys. Desc.: Reprint in Book
Language(s): Russian

Entry (No.): 798
Subject: Siberia – Old Maps Copied
Author: L′vov, Ivan
Title: Yakutsk
Year: 1736
Dimensions: 29.7 x 26.2 cm
Phys. Desc.: Reprint in Book
Language(s): Russian

Entry (No.): 799
Subject: Siberia – Old Maps Copied
Author: Remezov, Semen Ulianovich (ca. 1642–after 1720)
Title: [Map of Siberia]
Year: 1667
Dimensions: 22.2 x 14.6 cm
Phys. Desc.: Reprint in Book
Language(s): Russian

Entry (No.): 800
Subject: Siberia – Old Maps Copied
Author: Remezov, Semen Ulianovich (ca. 1642–after 1720)
Title: [Siberian towns and lands]
Year: 1700
Dimensions: 22.2 x 14.6 cm
Phys. Desc.: Reprint in Book
Language(s): Russian

Entry (No.): 801
Subject: Siberia – Old Maps Copied
Author: Remezov, Semen Ulianovich (ca. 1642–after 1720)
Title: Siberia
Year: 1705–6
Dimensions: 38.7 x 33.0 cm
Phys. Desc.: Reprint in Book
Language(s): Russian

Entry (No.): 802
Subject: Siberia – Old Maps Copied
Author: Remezov, Semen Ulianovich (ca. 1642–after 1720)
Title: Kamchatka
Year: 1712
Dimensions: 39.4 x 30.2 cm
Phys. Desc.: Reprint in Book
Language(s): Russian

Entry (No.): 803
Subject: Siberia – Old Maps Copied
Author: Remezov, Semen Ulianovich (ca. 1642–after 1720)
Title: Siberia
Year: 1713–14
Dimensions: 33.7 x 24.1 cm
Phys. Desc.: Reprint in Book
Language(s): Russian

Entry (No.): 804
Subject: Siberia – Old Maps Copied
Author: Remezov, Semen Ulianovich (ca. 1642–after 1720)
Title: [Europe with parts of Africa, Asia]
Year: 1737
Dimensions: 22.2 x 14.6 cm
Phys. Desc.: Reprint in Book
Language(s): Russian

Entry (No.): 805
Subject: Siberia – Old Maps Copied
Author: No Author
Title: [Pacific Coast]
Year: ca. 1710–30
Dimensions: 22.2 x 14.6 cm
Phys. Desc.: Reprint in Book
Language(s): Russian

Entry (No.): 806
Subject: Siberia – Old Maps Copied
Author: No Author
Title: [Siberia]
Year: ca. 1740
Dimensions: 22.2 x 14.6 cm
Phys. Desc.: Reprint in Book
Language(s): Russian

Entry (No.): 807
Subject: Siberia – Old Maps Copied
Author: No Author
Title: [Pacific Coast]
Year: 1764
Dimensions: 22.2 x 14.6 cm
Phys. Desc.: Reprint in Book
Language(s): Russian

Entry (No.): 808

Subject: Turkey/Ottoman Empire (18th, 19th c.)
Author: Guthrie, William (1708–1800)
Author: Dilly, C.
Title: Turkey, // in Europe, and // Hungary; from the best Authorities. Published as the Act directs June 15th 1785 by C. Dilly + G. Robinson London.
Year: 1785
Dimensions: 45.0 x 37.0 (37.1 x 33.5) cm
Phys. Desc.: Engraved, Colored Outline
Language(s): English

Entry (No.): 809
Subject: Turkey/Ottoman Empire (18th, 19th c.)
Author: Walker, John & Charles
Title: The // Turkish Empire // in Europe and Asia // with the Kingdom of Greece.
Year: ca. 1829–44
Dimensions: 42.7 x 35.5 (39.5 x 31.5) cm
Phys. Desc.: Engraved, Uncolored
Language(s): English

Entry (No.): 810
Subject: Turkey/Ottoman Empire (18th, 19th c.)
Author: Young, James H.
Title: Caucasian Countries // & Turkey in Asia.
Year: 1839
Dimensions: 27.6 x 22.5 (24.5 x 20.2) cm
Phys. Desc.: Engraved, Colored
Language(s): English

Entry (No.): 811
Subject: Turkey/Ottoman Empire (18th, 19th c.)
Author: No Author
Title: Map of Turkey in Europe Illustrating Berlin Congress Treaty, July, 1878.
Year: after 1878
Dimensions: 38.5 x 57.7 (36.0 x 49.5) cm
Phys. Desc.: Printed, Colored
Language(s): English

Entry (No.): 812
Subject: Ukraine (20th c.)
Author: Babicz, M.
Author: Sycz, A. (Sich, A.)
Author: Powolocky (Povolots´kyi), A. J.
Title: Nove na Karti Ukraïny. Al´bom Kartoskhem Ekonomichnykh Administratyvnykh Raioniv. URSR, 1961.
Year: 1961
Dimensions: 60.8 x 41.0 cm
Phys. Desc.: Book/Atlas
Language(s): Ukrainian

Entry (No.): 813
Subject: Ukraine (20th c.)
Author: Diachyshyn (Diaczyszyn), M.
Title: Zahal´na // Karta // Ukraïny. // Zladyv M. Diachyshyn // zakhodom i nakladom chas. "Svoboda," organu Ukraïns´koho Nar. Soiuza v Amerytsï. Rand McNally & Co., N.Y.
Year: 1950
Dimensions: 66.2 x 53.4 (79.0 x 45.4) cm
Phys. Desc.: Printed, Colored
Language(s): Ukrainian
Figure 89

Entry (No.): 814
Subject: Ukraine (20th c.)
Author: Dibrov, Ol. T.
Author: Lebedev, I. P.
Title: Karty do knyhy Ol. T. Dibrova Heohrafiia Ukraïns´koï RSR. Fizichna karta URSR. Heohrafichna karta URSR. Karta hruntiv URSR. Karta roslinnosti URSR. Politychno-administratyvna karta URSR. Ekonomichna karta Ukraïns´koï RSR i Moldavs´koï RSR. "Radians´ka Shkola", 1958.
Year: 1958
Dimensions: 37.9 x 26.3 cm
Phys. Desc.: Printed, Colored
Language(s): Ukrainian

Entry (No.): 815
Subject: Ukraine (20th c.)
Author: Drazniowsky, Roman
Author: Prokop, Liuda
Title: [Map of Ukraine]
Year: 1972
DimensionsL 34.5 x 24.6 cm
Phys. Desc.: Book with Maps
Language(s): Ukrainian

Entry (No.): 816
Subject: Ukraine (20th c.)
Author: Freytag, Gustav (1852–1938) & Wilhelm Berndt (1851–1920)
Title: Ukraïna.
Year: 1919
Dimensions: 99.5 x 70.0 (97.5 x 66.5) cm
Phys. Desc.: Printed, Colored
Language(s): Ukrainian

Entry (No.): 817
Subject: Ukraine (20th c.)
Author: Freytag, Gustav (1852–1938) & Wilhelm Berndt (1851–1920)
Title: Ukraïna.
Year: 1919
Dimensions: 94.5 x 82.0 cm
Phys. Desc.: Printed, Colored
Language(s): Ukrainian

Entry (No.): 818
Subject: Ukraine (20th c.)
Author: Gasenko, Georg. (Gassenko, Georges de)
Title: Svitova Mapa // z rozmishenniam Ukraïntsiv po svitu. // Z dodatkovymy tablytsiamy. Viden´ 1920.
Year: 1920
Dimensions: 61.2 x 47.5 cm
Phys. Desc.: Printed, Colored
Language(s): Ukrainian, German, French

Figure 80

Entry (No.): 819
Subject: Ukraine (20th c.)
Author: Gasenko, Georg. (Gassenko, Georges de)
Author: Rudnyts´kyi, S.
Title: Ukraïna v chyslakh zladyv G. Gasenko. Po nainoviishym ofitsiial´nym ukraïns´kym, rosiis´kym, frantsuz´kym i ynshym dzherelam, iak tezh po pratsiam E. Rekliu, Bonmariiazh-a, Sen´obos-a, Hrushevs´koho, Rudnyts´koho, Volkova, Penka, Frekha, Shefera i yn. — II. vydannia pry spivuchasty Prof. Dra S. Rudnyts´koho.
Year: 1920
Dimensions: 61.2 x 47.5 cm
Phys. Desc.: Printed
Language(s): Ukrainian
Figure 81

Entry (No.): 820
Subject: Ukraine (20th c.)
Author: Gerynovycz (Gerynovych), Volodymyr
Title: Heohrafichna karta zemel´, // de zhivut´ // Ukraïntsi.
Year: ca. 1950
Dimensions: 43.7 x 26.5 cm
Phys. Desc.: Printed, Colored
Language(s): Ukrainian

Entry (No.): 821
Subject: Ukraine (20th c.)
Author: Gerynovycz (Gerynovych), Volodymyr
Title: Heohrafichna karta zemel´, // de zhivut´ // Ukraïntsi.
Year: ca. 1950
Dimensions: 43.7 x 26.5 cm
Phys. Desc.: Printed, Colored
Language(s): Ukrainian

Entry (No.): 822
Subject: Ukraine (20th c.)
Author: Holubař, Jos.
Title: Post-Kurs-Karte // von // Galizien // (Blatt I.) Bezirk der k.k. Post und Telegraphen-Direktion Lemberg vom Post-Kurs-Bureau des k.k. Handelministeriums.
Year: before 1914
Dimensions: 72.0 x 52.0 cm
Phys. Desc.: Printed
Language(s): German

Entry (No.): 823
Subject: Ukraine (20th c.)
Author: Iakhinsov, J. B. (Jakhinsow, I. V., Jachinsow, J. B.)
Title: Ukraïns´ka Radians´ka Sotsialistychna Respublika. Fizychna karta.
Year: 1940
Dimensions: 46.5 x 31.5 cm
Phys. Desc.: Printed, Colored
Language(s): Ukrainian

Entry (No.): 824
Subject: Ukraine (20th c.)
Author: Karpa, William
Title: The Map of // Ukraine // Sponsored by the League of Americans of Ukrainian Descent, State Organization of Ukrainian Congress Committee of America. Chicago, Ill. John Duzansky, President. Compiled and drawn by William Karpa.
Year: 1955
Dimensions: 106.5 x 73.5 (103.0 x 69.5) cm
Phys. Desc.: Printed, Colored
Language(s): English
Figure 90

Entry (No.): 825
Subject: Ukraine (20th c.)
Author: Kolankowski, Dr. L.
Title: Mapa Kościelna Galicyi Wschodniej.
Year: early 20th c.
Dimensions: 66.5 x 56.0 cm
Phys. Desc.: Printed, Colored Outline
Language(s): Polish

Entry (No.): 826
Subject: Ukraine (20th c.)
Author: Kozak, Julius
Title: Ukraina.
Year: 19th c.
Dimensions: 26.0. x 17.5 cm
Phys. Desc.: Printed, Uncolored
Language(s): Polish

Entry (No.): 827
Subject: Ukraine (20th c.)
Author: Kubiiovych (Kubijovyč), Volodymyr
Title: Ethnographic Map of the Southwestern Ukraine (Halyčyna - Galicia).
Year: mid 20th c.
Dimensions: 141.3 x 139.4 cm
Phys. Desc.: Printed, Colored
Language(s): English, Ukrainian

Entry (No.): 828
Subject: Ukraine (20th c.)
Author: Kubiiovych (Kubijovyč), Volodymyr
Author: Kulyts´kyi, M. (Kulyckyj, M.)
Title: Physical Map of Ukraine. Fizychno karta Ukraïny.
Year: mid-20th c.
Dimensions: 44.8 x 25.0 (42.4 x 22.1) cm
Phys. Desc.: Printed, Colored
Language(s): English, Ukrainian

Entry (No.): 829
Subject: Ukraine (20th c.)
Author: Kubiiovych (Kubijovyč), Volodymyr
Author: Kulyts´kyi, M. (Kulyckyj, M.)
Title: Ethnographical Map of Ukraine 1930.
Year: 1968
Dimensions: 42.2 x 22.0 cm
Phys. Desc.: Printed, Colored
Language(s): English
Figure 88

Entry (No.): 830
Subject: Ukraine (20th c.)
Author: Kubiiovych (Kubijovyč), Volodymyr
Title: Ethnographic Map of the Southwestern Ukraine (Halyčyna-Galicia).
Year: 1952
Dimensions: 138.5 x 127.5 cm
Phys. Desc.: Printed, Colored
Language(s): English
Figure 87

Entry (No.): 831
Subject: Ukraine (20th c.)
Author: Kubiiovych (Kubijovyč), Volodymyr
Title: Ukraïna: Malyi Heohrafichnyi Atlias.
Year: 1971
Phys. Desc.: Atlas/Text
Language(s): Ukrainian

Entry (No.): 832
Subject: Ukraine (20th c.)
Author: Kubiiovych (Kubijovyč), Volodymyr
Author: Kulyts´kyi, M. (Kulyckyj, M.)
Title: Etnohrafichna Karta Ukraïny. Ethnographical Map of Ukraine.
Year: ca. 1950
Dimensions: 44.8 x 25.0 (42.4 x 22.1) cm
Phys. Desc.: Printed, Colored
Language(s): Ukrainian, English
Figure 86

Entry (No.): 833
Subject: Ukraine (20th c.)
Author: Kulyts´kyi, M. (Kulyckyj, M.)
Title: Pivdenno-Skhidna Evropa.
Year: 1946
Dimensions: 70.5 x 55.0 cm
Phys. Desc.: Printed, Colored
Language(s): Ukrainian

Entry (No.): 834
Subject: Ukraine (20th c.)
Author: Kushnir (Kuschnir), Dr. Volodymyr
Title: Ukrainische Rundschau.
Year: 1915
Dimensions: 21.9 x 15.0 cm
Phys. Desc.: Map and Text
Language(s): German
Figure 78

Entry (No.): 835
Subject: Ukraine (20th c.)
Author: Lura, Jurij (Lyra, Yurii)
Author: Bukowski, Lew (Bukovs´kyi, Lev)
Title: Chornomors´kyi prostir: Atlias.
Year: 1941
Phys. Desc.: Atlas
Language(s): Ukrainian

Entry (No.): 836
Subject: Ukraine (20th c.)
Author: Lura, Jurij (Lyra, Yurii)
Author: Bukowski, Lew (Bukovs´kyi, Lev)
Title: Chornomors´kyi prostir: Atlias. Schwarzmeer Raum.
Year: 1941
Phys. Desc.: Atlas
Language(s): Ukrainian

Entry (No.): 837
Subject: Ukraine (20th c.)
Author: Mankiw, H.
Title: Map of Soils of Ukraine.
Year: 1968
Dimensions: 42.2 x 22.0 cm
Phys. Desc.: Printed, Colored
Language(s): English

Entry (No.): 838
Subject: Ukraine (20th c.)
Author: Mankiw, H.
Title: Map of Soils of Ukraine. Karta Hruntiv Ukraïny.
Year: 1968
Dimensions: 42.3 x 22.0 cm
Phys. Desc.: Printed, Colored
Language(s): English, Ukrainian

Entry (No.): 839
Subject: Ukraine (20th c.)
Author: Marunchak, M. H. (Marunczak, M. H.)
Title: Map of Nazi Concentration Camps in Ukraine.
Year: 1963
Dimensions: 22.7 x 14.2 cm
Phys. Desc.: Printed, Uncolored
Language(s): English
Figure 95

Entry (No.): 840
Subject: Ukraine (20th c.)
Author: Niderle, L.
Title: Etnografichna karta Slavianshchyny.
Year: ca. 1910
Dimensions: 24.0 x 19.8 cm
Phys. Desc.: Printed, Colored
Language(s): Ukrainian
Figure 75

Entry (No.): 841
Subject: Ukraine (20th c.)
Author: Olesiiuk, Tymish
Title: Mapa Ukraïns´koï Metropoliï // iz poiasnenniamy.
Year: 1944
Dimensions: 41.5 x 22.5 cm
Phys. Desc.: Printed, Uncolored
Language(s): Ukrainian

Entry (No.): 842
Subject: Ukraine (20th c.)
Author: Poliansky, G.
Title: Geological Map of Ukraine
Year: 1968

Dimensions: 42.2 x 22.0 cm
Phys. Desc.: Printed, Colored
Language(s): English

Entry (No.): 843
Subject: Ukraine (20th c.)
Author: Polians´kyi, Iu. (Poljanskyj, J.)
Title: Heolohichna karta Ukraïny. Geological Map of Ukraine.
Year: ca. 1968
Dimensions: 44.8 x 25.0 (42.4 x 22.1) cm
Phys. Desc.: Printed, Colored
Language(s): Ukrainian, English

Entry (No.): 844
Subject: Ukraine (20th c.)
Author: Sichyns´kyi, V. (Sichynsky, W.)
Title: Ukraïna na holiands´kyi mapi kintsia XVII st.
Year: ca. 1930
Dimensions: 22.0 x 18.0 cm
Phys. Desc.: Printed, Uncolored
Language(s): Ukrainian

Entry (No.): 845
Subject: Ukraine (20th c.)
Author: Viryn, O. (Weeryn, O.)
Title: Karta ekonomichnykh raioniv Ukraïny.
Year: 1957
Dimensions: 21.5 x 14.5 cm
Phys. Desc.: Printed, Uncolored
Language(s): Ukrainian
Figure 91

Entry (No.): 846
Subject: Ukraine (20th c.)
Author: No Author
Title: Karta ukraïns´kykh zemel´, 44 milïoniv naselenia.
Year: ca. 1918
Dimensions: 13.5 x 8.5 cm
Phys. Desc.: Printed, Colored
Language(s): Ukrainian

Entry (No.): 847
Subject: Ukraine (20th c.)
Author: No Author
Title: Carpatho-Ukraine.
Year: ca. 1930
Dimensions: 27.5 x 18.6 cm
Phys. Desc.: Copy, Printed, Uncolored
Language(s): English

Entry (No.): 848
Subject: Ukraine (20th c.)
Author: No Author
Title: Map of Dialects in Ukraine.
Year: ca. 1930
Dimensions: 19.0 x 10.6 cm
Phys. Desc.: Printed, Uncolored
Language(s): English
Figure 83

Entry (No.): 849
Subject: Ukraine (20th c.)
Author: No Author
Title: Map of Green Ukraine. Mapa Zelenoï Ukraïny. Published by the Ukrainian Publishing Co., Harbin, Manchuk[?], Asia.
Year: 1937
Dimensions: 40.0 x 59.2 (39.0 x 57.0) (reduced from orig. size)
Phys. Desc.: Copy and Negative, Uncolored
Language(s): English

Entry (No.): 850
Subject: Ukraine (20th c.)
Author: No Author
Title: Map of Western Ukraine. Plan of Lviv. Mapa zakhidnykh ukraïns´kykh zemel´.
Year: 1946
Dimensions: 50.0 x 70.0 cm
Phys. Desc.: Printed, Uncolored
Language(s): English, Ukrainian

Entry (No.): 851
Subject: Ukraine (20th c.)
Author: No Author
Title: Karta Volyni. Map of Volhynia.
Year: 1947
Dimensions: 65.0 x 45.5 cm
Phys. Desc.: Printed, Uncolored
Language(s): Ukrainian, English

Entry (No.): 852
Subject: Ukraine (20th c.)
Author: No Author
Title: Mapa Ukraïny. Map of Ukraine.
Year: 1947
Dimensions: 97.0 x 67.0 cm
Phys. Desc.: Printed, Uncolored
Language(s): Ukrainian, English

Entry (No.): 853
Subject: Ukraine (20th c.)
Author: No Author
Title: Karta shosseinykh i gruntovykh dorog tsentra evropeiskoi chasti SSSR
Year: ca. 1950
Dimensions: 24.0 x 29.0 cm
Phys. Desc.: Printed, Colored
Language(s): Russian

Entry (No.): 854
Subject: Ukraine (20th c.)
Author: No Author
Title: Karta Ukraïny. Map of Ukraine.
Year: ca. 1950
Dimensions: 75.0 x 53.0 cm
Phys. Desc.: Printed, Uncolored
Language(s): Ukrainian, English

Entry (No.): 855

Subject: Ukraine (20th c.)
Author: No Author
Title: Map of Ukraine showing dialect used.
Year: ca. 1950
Dimensions: 19.0 x 10.5 cm
Phys. Desc.: Printed, Uncolored
Language(s): English

Entry (No.): 856
Subject: Ukraine (20th c.)
Author: No Author
Title: Map of Ukraine. The Percentage of Ukrainians of the Whole Population.
Year: ca. 1950
Dimensions: 42.5 x 22.8 cm
Phys. Desc.: Printed, Colored
Language(s): English

Entry (No.): 857
Subject: Ukraine (20th c.)
Author: No Author
Title: Natsional´nosti. Nationalities. Atlas of Ukraine + Adjoining Countries. Grafichne Zavedennia A. Hebendis, L´viv. Naklad: V-vo "Ukraïns´kyi Vydavnychyi Instytut u L´vovi."
Year: ca. 1950
Dimensions: 42.0 x 22.5 cm
Phys. Desc.: Printed, Colored
Language(s): English, Ukrainian

Entry (No.): 858
Subject: Ukraine (20th c.)
Author: No Author
Title: Podkarpatska Rus´.
Year: ca. 1950
Dimensions: 25.2 x 20.5 cm
Phys. Desc.: Copy, Uncolored
Language(s): Ukrainian

Entry (No.): 859
Subject: Ukraine (20th c.)
Author: No Author
Title: Polityko-Administratyvna Karta URSR.
Year: ca. 1950
Dimensions: 60.5 x 42.0 cm
Phys. Desc.: Printed, Uncolored
Language(s): Ukrainian

Entry (No.): 860
Subject: Ukraine (20th c.)
Author: No Author
Title: Polityko-Administratyvna Karta Ukraïns´koï RSR 1921 r.
Year: 1958
Dimensions: 26.5 x 18.0 cm
Phys. Desc.: Printed, Uncolored
Language(s): Ukrainian

Entry (No.): 861
Subject: Ukraine (20th c.)
Author: No Author
Title: Polityko-Administratyvna Karta Ukraïns´koï RSR 1940 r.
Year: 1958
Dimensions: 26.5 x 18.0 cm
Phys. Desc.: Printed, Uncolored
Language(s): Ukrainian

Entry (No.): 862
Subject: Ukraine (20th c.)
Author: No Author
Title: Polityko-Administratyvna Karta Ukraïns´koï RSR i Moldavs´koï RSR.
Year: 1958
Dimensions: 43.0 x 29.5 cm
Phys. Desc.: Printed, Uncolored
Language(s): Ukrainian

Entry (No.): 863
Subject: Ukraine (20th c.)
Author: No Author
Title: General Physical Chart // of the Ukraine // and the Contiguous Countries.
Year: ca. 1960
Dimensions: 45.5 x 28.5 cm
Phys. Desc.: Printed, Colored
Language(s): English

Entry (No.): 864
Subject: Ukraine (20th c.)
Author: No Author
Title: Skhema Avtomarshrut po Krymu.
Year: ca. 1960
Dimensions: 29.2 x 22.2 cm
Phys. Desc.: Printed, Colored
Language(s): Russian

Entry (No.): 865
Subject: Ukraine (20th c.)
Author: No Author
Title: Skhema Avtomarshrutov Moskva-Simferopol´ Khar´kov-Rostov.
Year: ca. 1960
Dimensions: 29.4 x 14.1 cm
Phys. Desc.: Printed, Uncolored
Language(s): Russian

Entry (No.): 866
Subject: Ukraine (20th c.)
Author: No Author
Title: Skhema Avtomarshrutov po Kavkazy.
Year: ca. 1960
Dimensions: 29.2 x 32.2 cm
Phys. Desc.: Printed, Uncolored
Language(s): Russian

Entry (No.): 867
Subject: Ukraine (20th c.)
Author: No Author
Title: Skhidna Halychyna.

Year: ca. 1960
Dimensions: 45.7 x 37.0 cm
Phys. Desc.: Printed, Colored
Language(s): Ukrainian

Entry (No.): 868
Subject: Ukraine (20th c.)
Author: No Author
Title: There Can Be No Peace in Europe until Ukraine Is Reunited and Forty-Five Million Ukrainians Are Free Again.
Year: ca. 1960
Dimensions: 14.5 x 9.5 cm
Phys. Desc.: Printed, Colored
Language(s): English
Figure 94

Entry (No.): 869
Subject: Ukraine (20th c.)
Author: No Author
Title: Administrative Map of Ukraine 1968.
Year: 1968
Dimensions: 42.2 x 22.0 cm
Phys. Desc.: Printed, Colored
Language(s): English

Entry (No.): 870
Subject: Ukraine (20th c.)
Author: No Author
Title: Ethnological Map of the Ukrainian SSR 1959.
Year: 1968
Dimensions: 34.0 x 24.0 cm
Phys. Desc.: Printed, Colored
Language(s): English

Entry (No.): 871
Subject: Ukraine (20th c.)
Author: No Author
Title: Physical Map of Ukraine.
Year: 1968
Dimensions: 42.2 x 22.0 cm
Phys. Desc.: Printed, Colored
Language(s): English

Entry (No.): 872
Subject: Ukraine (20th c.)
Author: No Author
Title: Ukraïns´ka RSR i Moldavs´ka RSR.
Year: 1968
Dimensions: 91.0 x 65.0 cm
Phys. Desc.: Printed, Colored
Language(s): Ukrainian

Index of Maps by Place Name

This index is taken from the Krawciws' original notes; their spelling of place names has been preserved. Maps that appear in figures are given in italics; 301 and 339 are missing from this index due to the Krawciws' own omission.

Subject Index

References to maps are given in italics; variant spellings of names are given in parentheses.